Understanding Terrorism Innovation and Learning

This book examines the role of terrorist innovation and learning in theory and practice, and in the context of three specific EU case studies.

It is often said that terrorist groups are relatively conservative in character operating in a technological vacuum – relying almost exclusively on bombs and bullets. This observation masks increasing complexity, creativity and innovation within terrorist groups; one of the most distinguishing features of al-Qaeda's terrorist operations is its propensity for remarkable innovation. This book examines how and why terrorist groups innovate more generally and al-Qaeda-related terrorist plots in Europe more specifically. The starting point for this book was twofold: first, to examine the issue of innovation and learning more generically both in theory, within specific themes, and within the context of al-Qaeda's influence on this process; second, this book examines the evolution of specific al-Qaeda-related plots in three specific northern EU states – the United Kingdom, Denmark and Germany – where there has been a significant volume of planned, failed and executed terrorist plots. In particular, these case studies explore signs of innovation and learning.

This book will be of much interest to students of terrorism and counter-terrorism, political violence, security studies and IR in general.

Magnus Ranstorp is Research Director at the Center for Asymmetric Threat Studies (CATS) at the Swedish National Defence College, Stockholm, and author/editor of five previous books, including, most recently, *Understanding Violent Radicalisation* (Routledge, 2010).

Magnus Normark is a Senior Analyst at the Division of CBRN Defence and Security at the Swedish Defence Research Agency (FOI). He is the co-editor of *Unconventional Weapons and International Terrorism* (Routledge, 2009).

Political Violence
Series Editor: David Rapoport

This book series contains sober, thoughtful and authoritative academic accounts of terrorism and political violence. Its aim is to produce a useful taxonomy of terror and violence through comparative and historical analysis in both national and international spheres. Each book discusses origins, organisational dynamics and outcomes of particular forms and expressions of political violence.

Aviation Terrorism and Security
Edited by Paul Wilkinson and Brian M. Jenkins

Counter-Terrorist Law and Emergency Powers in the United Kingdom, 1922–2000
Laura K. Donohue

The Democratic Experience and Political Violence
Edited by David C. Rapoport and Leonard Weinberg

Inside Terrorist Organizations
Edited by David C. Rapoport

The Future of Terrorism
Edited Max Taylor and John Horgan

The IRA, 1968–2000
Analysis of a secret army
J. Bowyer Bell

Millennial Violence
Past, present and future
Edited by Jeffrey Kaplan

Right-Wing Extremism in the Twenty-First Century
Edited by Peter H. Merkl and Leonard Weinberg

Terrorism Today
Christopher C. Harmon

The Psychology of Terrorism
John Horgan

Research on Terrorism
Trends, achievements and failures
Edited by Andrew Silke

A War of Words
Political violence and public debate in Israel
Gerald Cromer

Root Causes of Suicide Terrorism
The globalization of martyrdom
Edited by Ami Pedahzur

Terrorism versus Democracy
The liberal state response, second edition
Paul Wilkinson

Countering Terrorism and WMD
Creating a global counter-terrorism network
Edited by Peter Katona, Michael D. Intriligator and John P. Sullivan

Mapping Terrorism Research
State of the art, gaps and future direction
Edited by Magnus Ranstorp

The Ideological War on Terror
Worldwide strategies for counter-terrorism
Edited by Anne Aldis and Graeme P. Herd

The IRA and Armed Struggle
Rogelio Alonso

Homeland Security in the UK
Future preparedness for terrorist attack since 9/11
Edited by Paul Wilkinson

Terrorism Today, Second Edition
Christopher C. Harmon

Understanding Terrorism and Political Violence
The life cycle of birth, growth, transformation, and demise
Dipak K. Gupta

Global Jihadism
Theory and practice
Jarret M. Brachman

Combating Terrorism in Northern Ireland
Edited by James Dingley

Leaving Terrorism Behind
Individual and collective disengagement
Edited by Tore Bjørgo and John Horgan

Unconventional Weapons and International Terrorism
Challenges and new approaches
Edited by Magnus Ranstorp and Magnus Normark

International Aviation and Terrorism
Evolving threats, evolving security
John Harrison

Walking Away from Terrorism
Accounts of disengagement from radical and extremist movements
John Horgan

Understanding Violent Radicalisation
Terrorist and jihadist movements in Europe
Edited by Magnus Ranstorp

Terrorist Groups and the New Tribalism
Terrorism's fifth wave
Jeffrey Kaplan

Negotiating with Terrorists
Strategy, tactics, and politics
Edited by Guy Olivier Faure and I. William Zartman

Explaining Terrorism
Causes, processes and consequences
Martha Crenshaw

The Psychology of Counter-Terrorism
Edited by Andrew Silke

Terrorism and the Olympics
Major event security and lessons for the future
Edited by Anthony Richards, Peter Fussey and Andrew Silke

Irish Republican Terrorism and Politics
A comparative study of the official and the provisional IRA
Kacper Rekawek

Fault Lines in Global Jihad
Organizational, strategic and ideological fissures
Edited by Assaf Moghadam and Brian Fishman

Militancy and Political Violence in Shiism
Trends and patterns
Edited by Assaf Moghadam

Islamist Radicalisation in Europe
An occupational change process
Daniela Pisoiu

An International History of Terrorism
Western and non-Western experiences
Edited by Jussi Hanhimäki and Bernhard Blumenau

Democracy and Terrorism
Friend or foe?
Leonard Weinberg

State Terrorism and Human Rights
International responses since the end of the Cold War
Edited by Gillian Duncan, Orla Lynch, Gilbert Ramsay and Alison Watson

Prisons, Terrorism and Extremism
Critical issues in management, radicalisation and reform
Edited by Andrew Silke

The Psychology of Terrorism, Second Edition
John Horgan

Victims of Terrorism
A comparative and interdisciplinary study
Edited by Orla Lynch and Javier Argomaniz

Lone-Actor Terrorists
A behavioural analysis
Paul Gill

Understanding Terrorism Innovation and Learning
Al-Qaeda and beyond
Edited by Magnus Ranstorp and Magnus Normark

Understanding Terrorism Innovation and Learning

Al-Qaeda and beyond

Edited by Magnus Ranstorp and Magnus Normark

Routledge
Taylor & Francis Group
LONDON AND NEW YORK

First published 2015
by Routledge
2 Park Square, Milton Park, Abingdon, Oxfordshire OX14 4RN

and by Routledge
711 Third Avenue, New York, NY 10017

First issued in paperback 2016

Routledge is an imprint of the Taylor & Francis Group, an informa business

British Library Cataloguing-in-Publication Data
A catalogue record for this book is available from the British Library

Library of Congress Cataloging-in-Publication Data
Understanding terrorism innovation and learning : al-Qaeda and beyond / edited by Magnus Ranstorp and Magnus Normark.
pages cm. – (Political violence)
Includes bibliographical references and index.
1. Terrorism. 2. Terrorism–Case studies. 3. Weapons.
4. Technology. 5. Terrorism and mass media.
I. Ranstorp, Magnus, editor. II. Normark, Magnus.
HV6431.U3475 2015
363.325–dc23 2014043879

ISBN 13: 978-0-415-78718-5 (pbk)

ISBN 13: 978-1-138-84743-9 (hbk)

Typeset in Baskerville
by Wearset Ltd, Boldon, Tyne and Wear

Contents

Contributors

Gary Ackerman is the Director of the Special Projects Division at the National Consortium for the Study of Terrorism and Responses to Terrorism (START). Prior to taking up his current position, he was Research Director at START and before that the Director of the Weapons of Mass Destruction Terrorism Research Program at the Center for Nonproliferation Studies in Monterey, California. His research encompasses various areas relating to terrorism and counter-terrorism, including terrorist threat assessment, terrorist technologies and innovations, terrorist motivations for using chemical, biological, radiological and nuclear (CBRN) weapons and the modeling and simulation of terrorist behavior. He is the co-editor of *Jihadists and Weapons of Mass Destruction* (CRC Press, 2009), author of several articles on CBRN terrorism and has testified on terrorist motivations for using nuclear weapons before the Senate Committee on Homeland Security.

John Allison (MSc MA MIExpE FRGS) served in the British Army for 24 years, primarily in the field of Explosive Ordnance Disposal, counter-terrorist search and intelligence. He has a wealth of experience, having served as a bomb disposal officer in Bosnia, special advisor on counter-terrorism in Gaza/West Bank, IED intelligence analyst in the Combined Explosives Exploitation Cell, Baghdad and latterly as a security advisor/ intelligence SME in Pakistan.

Paul Cruickshank is CNN's terrorism analyst and the editor of *Al-Qaeda*, a five-volume collection of key scholarly works on the terrorist network. He is the co-author of the 2014 book *Agent Storm: My Life Inside al Qaeda and the CIA* (Atlantic Monthly Press).

Adam Dolnik, PhD, is Professor of Terrorism Studies at the University of Wollongong in Australia, and a former Professor of Counterterrorism at the George C. Marshall Center for European Security Studies in Germany. In the past, Dolnik has also served as Chief Trainer at the International Centre for Political Violence and Terrorism Research (ICPVTR) in Singapore, and as a researcher at the Weapons of Mass

Destruction Terrorism Research Project at the Monterey Institute of International Studies in California and at the United Nations Terrorism Prevention Branch in Vienna.

Søren Hove (PhD in Middle East Studies) works as a Special Consultant at the Danish Defense. Prior to this he was a research fellow for seven years at the Danish Institute for International Studies (DIIS, Copenhagen) and the Centre for Contemporary Middle Studies (University of Southern Denmark). His academic work focuses on al-Qaeda's propaganda and the relationship between al-Qaeda's central leadership and their regional affiliates.

Yassin Musharbash is an investigative reporter and terrorism analyst for German newsweekly *Die Zeit*. He holds an MA in Arab Studies and Political Sciences from the University of Göttingen. In 2006, his book *Die neue al-Qaida. Innenansichten eines lernenden Terrornetzwerks* ('The New al-Qaida. Insights into a Learning Terror Network') was published by Kiepenheuer & Witsch. It was followed in 2011 by his first novel, *Radikal*.

Magnus Normark is a senior analyst and project manager at the Swedish Defence Research Agency (FOI). His area of expertise is CBRN-terrorism, Counter proliferation and CBRN-threat assessments. Normark has been a national delegate in international regimes and mechanisms such as Australia Group, Proliferation Security Initiative, Global Initiative to Combat Nuclear Terrorism, and has working experience from international UN missions, including UNSCOM in Iraq (1994–1995). He has published several studies on countries of concern in the Middle East, arms control and international counter- and non-proliferation risks.

Raffaello Pantucci is Director of International Security Studies at the Royal United Service Institute. Prior to coming to RUSI, Raffaello lived for over three years in Shanghai where he was a visiting scholar at the Shanghai Academy of Social Sciences (SASS). Before that he worked at the International Institute for Strategic Studies (IISS) in London, and the Center for Strategic and International Studies (CSIS) in Washington. He has also held positions at the European Council of Foreign Relations (ECFR) and the International Center for the Study of Radicalisation (ICSR) at King's College, London. He is the author of a history of jihadism in the UK, *We Love Death As You Love Life: Britain's Suburban Mujahedeen* (Columbia University Press, 2015).

Nico Prucha is a principal at Cognitive Architectures and fellow at the Department of Near Eastern Studies, University of Vienna. His research focuses on textual and audio-visual content analysis of jihadist activity online, specifically focusing on jihadist Shari'a law interpretation of hostage taking and executions. He has done extensive fieldwork in the Middle East and Europe and works on lingual and cultural-anthropological elements of the jihadist phenomena for Cognitive Architectures.

Magnus Ranstorp is Research Director at the Center for Asymmetric Threat Studies at the Swedish National Defence College. Over the last 25 years, he has researched Islamic extremism, terrorism and counter-terrorism worldwide. Previously he was Director of St Andrew's Centre for the Study of Terrorism and Political Violence. He is EU Radicalisation Awareness Network (RAN) Working Group Leader on Foreign Fighters. In 2013 Ranstorp completed a major Swedish Cabinet assignment on Prevention of Violent Extremism in Third Countries.

Joshua Sinai is a senior manager, national security programs, Resilient Corporation, in Alexandria, VA. He has more than 25 years' experience in analyzing terrorism and counter-terrorism-related subjects in government, the corporate sector and academia, including teaching at the graduate level. He also serves as the Book Reviews Editor of the journal *Perspectives on Terrorism* (www.terrorismanalysts.com), for which he writes the regularly published "Counterterrorism Bookshelf" review column. He obtained his PhD in Political Science, with a specialization in Comparative Politics and the Middle East, from Columbia University.

Morten Skjoldager is a journalist at Danish Daily *Politiken* where he covers terrorism and counter-terrorism issues. He is also author of the acclaimed book *Truslen indefra* published in 2009 by Lindhardt og Ringhof which details the Danish terrorism cases.

Guido Steinberg is a senior fellow at the German Institute for International and Security Affairs (Stiftung Wissenschaft und Politik, SWP) in Berlin, specializing in Middle East Politics and Islamist Terrorism. He studied Islamic Studies, Medieval and Modern History, and Political Science in Cologne, Bonn, Damascus, and at the Free University of Berlin. He has previously worked as a research coordinator at the Free University Berlin (2001) and as an advisor on international terrorism in the German Federal Chancellery (2002–2005).

1 Introduction

Understanding terrorism innovation and learning – Al-Qaeda and beyond

Magnus Ranstorp and Magnus Normark

It is often said that terrorist groups are relatively conservative in character operating in a technological vacuum – relying almost exclusively on bombs and bullets. This observation masks increasing complexity and creativity and innovation within terrorist groups. One of the most distinguishing features of al-Qaeda's terrorist operations is its propensity for remarkable innovation. The out-of-the-box idea of using multiple, synchronized hijacked aircraft as missile systems against U.S. iconic targets in the 9/11 attacks of 2001 has been steadily matched by innovative al-Qaeda terror plots involving shoe bombs, liquid bombs targeting seven transatlantic aircraft in 2006, the Madrid train bomb in 2004, the multiple London 7/7 attacks with the train and Underground train bombs, to the 2010 knife attack on a British MP by Rohonara Choudhry and the 2010 Mumbai-style plot against Danish paper *Jyllands-Posten*, where terrorists planned to behead editors. Some of these terrorist plots are interconnected but many are unconnected.

Ever since the evolution of modern international terrorism in 1968, there has been a contagion effect and learning between different groups or attack modes.[1] In fact one could argue that the 9/11 idea of multiple, synchronized hijackings originated with the 1970 Dawsons Field hijackings of four aircraft (one other failed) from multiple airports. It is further not inconceivable that the 9/11 attackers had originally calculated that one of the plane attacks on 11 September 2001 would be captured on film in a similar fashion to the blowing up of aircraft on Dawsons Field in 1970 before the international press. Similarly it is possible to argue that the 2006 Transatlantic plot with liquid explosives was in some way inspired by strands of the 1994 Operation Bojinka in which Ramzi Yousef (behind the 1993 bombing of the World Trade Center) smuggled on board separate components of a bomb, including a modified Casio watch as power source, and assembled these into a workable explosive device.[2] One other striking similarity is that the 1994 Operation Bojinka involved blowing up simultaneously 11 aircraft over the Pacific. Although there is no hard evidence for these connections it is obvious a contagion factor could be present. One case where there are proven links is the case of David Headley, who participated as the principal reconnaissance for the 2008 Mumbai attack, after which he was arrested for plotting a significant terror attack against the Danish *Jyllands-Posten* in 2009.[3] One and a half years later four Swedes, one of whom had spent time in a terrorist training camp in Waziristan, were arrested for plotting to target the same paper using exactly the same *modus operandi* as earlier.

The diffusion of knowledge about improvised explosive devices (IEDs) is more precise as experts can gauge specific bomb signatures regarding how they were constructed and who was involved. For example, knowledge of IED construction migrated between terrorist groups and conflict zones after 9/11 from Chechnya and Algeria to the 2003 Iraqi conflict then to the Afghan theater.[4] Financial incentives were also a key catalyst for driving innovation in design and delivery of IEDs within the Iraqi and Afghanistan conflict.[5] Among the latest innovative tactics in Afghanistan is the use of women and children suicide operatives as well as explosives hidden inside the traditional Afghan headdress – the *lunge* – for assassinations.[6] A range of complex factors affects technology adoption decisions in terrorist groups, such as comparative advantage, compatibility, complexity, risk and cost.[7] Factoring in technology is only one key issue affecting terrorist decisions. Available human and physical resources, absorptive capacity, internal group decision structures, pressures and security precautions, target hardening, timing and earlier experience are only some of the factors that influence terrorist targeting behavior. Some scholars argue that there are three pathways to acquire necessary knowledge – learning by watching, learning by formal study and learning by doing – but that actual operational capability is closely related to practical training experience gained in the field.[8]

Since the September 11, 2001 terror attacks, significant focus and research has been devoted towards understanding the polymorphous al-Qaeda phenomenon as it has morphed, migrated and changed due to globalization and according to internal and external pressures as well as changing political development. This book examines how and why terrorist groups innovate generally and al-Qaeda-related terrorist plots in Europe more specifically. The starting point for this book was twofold: first, to examine the issue of innovation and learning more generically both in theory, within specific themes and within the context of al-Qaeda's influence on this process. Second, this book examines the evolution of specific al-Qaeda-related plots in three northern European Union (EU) states – the United Kingdom, Denmark and Germany – where there has been a significant volume of planned, failed and executed terrorist plots. In particular, these case studies look at signs of innovation and learning. This book is the cumulative result of three international workshops on terrorist innovation and learning, held at the Swedish National Defence College in April and June 2012 and in June 2013, where we gathered international expertise from academia and government. These workshops were sponsored by the Swedish Civil Contingency Agency.

The term "innovation"

The study of innovation in terrorist behavior is relatively embryonic and as some scholars have correctly observed "is hardly systematic, comparative, or oriented toward theory building."[9] Martha Crenshaw has usefully differentiated innovation on three levels: strategic, tactical and organizational. While strategic innovation involves "game-changing" shifts and tactical innovation involves changes in methods and operations, organizational innovation focuses on the role of hierarchies and decentralized structures and their impact on the adaptation, innovation and learning of terrorist tactics.[10] Crenshaw also argues that innovation is often the result of "problem-solving" rather than of mere inspiration or by accident.[11]

Available resources and skilled members play a defining role, together with new emerging technologies, both weapon technologies and information technologies that transform remote reconnaissance capability and secure communication methods.[12] As such there is a constellation of areas which could involve terrorist innovation.[13]

As there is no definitive scholarly consensus, this book has applied a broad interpretation of the term "innovation" in our effort to identify and elaborate on indications of evolution within terrorists' tradecraft and behavior. The term "innovation" is multifaceted, with a variety of definitions applied within the research community. The following set of definitions, taken together, could be viewed as representative of the broad approach taken by this study:

- "The adoption of new patterns of behavior" (Crenshaw);
- "An idea, practice, or object that is perceived as new by an individual or another unit of adoption" (Rogers);
- "The expressing of a given purpose in a different set of components" (Arthur);
- "The integration of an invention into one or more cultural, organizational, economic or political contexts" (Ackerman).

Literature on terrorist innovation and learning

There is a remarkable lacuna in available research on how and why terrorist groups innovate and possibly learn from their own and others' mistakes. Even fewer studies exist that focus on the role of innovation and learning in al-Qaeda-related plots in Europe. There are many different reasons for the lack of prioritized focus on this issue. First, data collection and data precision is difficult due to a variety of contextual factors, the challenges faced in studying clandestine groups and the disruption of many terrorist plots at different degrees or levels of preparation and execution. Most focus is naturally on successful attacks rather than foiled or failed operations. Another difficulty pertains to gaining the inside story of clandestine internal processes within terrorist cells that could explain personal relationships, attitudes towards risk, and their individual and collective decision-making process at different levels in attack preparation and execution.[14] Breaking down the components that constitute the process of terrorist planning to actual execution involves many variables and requires high resolution detail than is often not available from open-source literature.[15] Court documents are often the solution to this problem as they are rich in detail about operational dimensions.

Second, analytical inferences can often only be made on generalized levels as al-Qaeda-related plots do not build on each other but are often one-off attacks. Having said that, the violent extremist milieus are often relatively small in size and overlapping with high degrees of interconnectivity between cell structures. There is collaborative learning both online and offline in how to avoid detection and even to learn from past operational mistakes. It is much easier to detect technology trajectories and instances of innovation and learning in terrorist groups that have definable, hierarchical command structures and engage in limited geographic areas and in sustained terrorist campaigns against set targets.

Third, most academic focus on terrorist innovation has largely centered on the specter of CBRN terrorism, where the social and natural scientists often have a stovepipe understanding of threat convergence issues. Less effort exists in trying to understand terrorist innovation and learning in cases involving conventional weapons. Lastly, very few studies exist mapping out al-Qaeda-related plots in Europe, especially trying to gauge operational complexity and any signs of innovation.

Petter Nesser's chronology of al-Qaeda-related planned, prepared and executed terrorist attacks occurring between 1994 and 2007 reveal that 56 European-based plots occurred from 2001 until the end of 2007. This study gave a descriptive helicopter perspective of different *modus operandi* and targeting factors. It revealed that Great Britain, Denmark and the Netherlands were frequent targets and that support for the Iraqi conflict as well as the Danish cartoon controversy was at the center of the rationale for specific targeting.[16] This figure was later updated with an additional 23 incidents during the period 2008–2010.[17] Additionally, Nesser argues that very few cases involving self-radicalized "solo-terrorists" are in his data set. The reason is, he argues, that the jihadist usually conceive themselves as part of *jama'a* (group), *tanzim* (organization) or *haraka* (movement); most go abroad to fight before being re-directed home; and most lack the practical and technical skill sets such as logistics, bomb-making techniques, counter-surveillance, etc.[18]

Building on this dataset is Javier Jordan's analysis of 85 al-Qaeda-related attacks in Europe between 2001 and 2010 which examined choice of targets, weapons chosen for terrorist attacks, the influence of foreign conflict zones and social network linkages to established jihadi groups.[19] This study revealed that two-thirds of the terrorist plots involved homemade explosives, most often involving triacetone triperoxide (TATP). It also showed that preferable targets were "soft" public transport targets and places. One area that shows a high degree of terrorist learning and innovation is the obsession with targeting civil aviation, as revealed by the multiple shoe bomb plots, liquid explosives and car bomb attacks on terminal buildings.

Adam Dolnik's *Understanding Terrorist Innovation* (2007)[20] is probably among the most comprehensive efforts to assess terrorist innovation and the trajectories of their violent attacks. While the study includes a broad range of terrorist groups, it concludes that the most innovative groups were those who were ideologically flexible rather than having rigid posture (as ideology defined who their enemy were and means to pursue them); facing emerging competition with other groups in same operational area; target hardening and strong counter-terrorism measures forces problem solving and innovation; and finally degree of resources and technical awareness of terrorist groups influence their target selection.[21] Innovation may also be simply surreptitious in nature – discovering by accident combined with available human operative skill sets.

Similarly James J. F. Forest's *Teaching Terror* (2006) provides a useful overview of strategic and tactical terrorist learning, emphasizing that successful groups: find and operationalize knowledge; commit to long-term training and education of its members; exchange knowledge and learn from other groups or settings; and study "counterterrorism strategies and adapt accordingly."[22] Both individual and organizational learning is enhanced by knowledge transfer in terrorist training camps and collaboratively online in the proliferation of jihadi websites.

In contrast, others argue that it is possible to provide a conceptual framework for understanding malevolent creativity in terrorist organizations, with multi-level drivers of creativity focusing on levels of organizational structure; leadership and technical expertise; creativity in individuals and in teams; and the impact of the larger environment, including effects of counter-terrorism measures.[23]

A wide variety of academic perspectives exist as to which key factors drive innovation and learning in terrorist operational behavior. Based on the earlier work by Brian Jackson, Dolnik identities 11 relevant factors for innovation: "the role of ideology and strategy; the dynamics of the struggle; countermeasures; targeting logic; attachment to weaponry; group dynamics; relationship with other organizations; resources; openness to new ideas; durability; and the nature of technology."[24]

Ideological influence

As shown by Charlie Drake's pioneering work, terrorist targeting is often greatly influenced by ideology which sets out the moral framework and political objectives.[25] The interaction between the strategic environment and the tactical choices influences terrorists' decisions to engage in terrorism. Kydd and Walter identify five strategic reasons: attrition; spoiling; intimidation; outbidding; and provocation.[26] The peculiarities of al-Qaeda ideology, with an overarching strategy of recreating an Islamic caliphate territory, have resulted in multilayered violent tactics, whether prioritizing and focusing on near or distant enemies.[27] As such, al-Qaeda "is an organization that promotes a sweeping agenda of revolution."[28] Towards these ends, al-Qaeda has justified its violent acts directed towards Muslims and non-Muslims alike as religiously noble and just, although its ongoing concept of jihad challenges what is scripturally allowed.[29] Some have even argued that al-Qaeda has a 20-year plan to achieve victory, from 9/11 to the destabilization of disloyal regimes to final confrontation and victory.[30] In many ways this al-Qaeda plan prophetically anticipated the unfolding Arab Spring revolution which led to civil war in Syria and chaos in many Arab states.

Al-Qaeda's Ayman al-Zawahiri has on numerous occasions articulated a grand strategy with the various strands of the organization needed to work on four interrelated fronts: (1) inflicting losses on the Crusader West by making it bleed; (2) expelling the so-called Crusader-Zionist occupying forces from the lands of Islam; (3) overthrow of corrupt apostate regimes that sold out to the West and Israel; (4) *dawah* (popular missionary activity).[31] As such, it is important to monitor the statements of Zawahiri and other key leaders of al-Qaeda affiliate organizations to look for clues of changes or shifts in focus. Often these al-Qaeda special tactical operations, known as *ghazwah* (the Arabic term for Islamic raids during the time of the Prophet Mohammed),[32] seem to be much more operationally sophisticated if they are planned, organized and directed from senior operational leaders within al-Qaeda or al-Qaeda in the Arabian Peninsula (AQAP). Yet to conceptualize all al-Qaeda sophisticated terror attacks as directed from the top down belies the complexity of the process.

Some scholars have argued that "al-Qaeda's philosophy of managing terrorist activity was a doctrine imposed from the top down" combined with improvisation and innovation from the bottom up.[33] Similarly, Hayden argues that "the most likely organizations to exhibit innovation and learning will be those with core-periphery structures, where ideas from outliers can be quickly accessed and assimilated and exogenous shocks can be distributed.[34] While more formalized training in terrorist training camps in Pakistan seems to be crucial for more complex attacks, it is evident that providing directives through key al-Qaeda strategists and theoreticians has influenced the evolution of attack modes. As argued by Ryan, a genre of jihadist studies exists which "is not written to recruit but to teach jihadists how to avoid old problems and enact strategies for success."[35] Some of the most influential al-Qaeda strategists have been Abu al-Qureshi, who focused on fourth-generation warfare; Abd al-Aziz al-Muqrin and his book *A Practical Course for Guerrilla War*,[36] Abu Bakr al-Naji's *The Management of Savagery*, which is a blueprint for conducting a regional insurgency; and Abu Musab al-Suri and his *The Call to Global Islamic Resistance*, over 1,600 pages long, which provides detailed guidance on small-cell terrorism and an ideal system of jihad.[37]

The influence of Abu Musab al-Suri's thinking seems to be significant, as evidenced by his work being found in terrorists' or terrorist plotters' possession in Europe over the last decade. His central idea about disjointed terrorist cells, being protected from security forces and each other, operating in so-called 'open fronts' following inspiration, propaganda and central directives has evolved into a forceful reality with the advance and growing interconnectivity of globalisation. Nowhere has this been more evident than the growing interplay between online instruction and propaganda and offline violent extremist behavior. In many ways, the growing influence of AQAP and their online magazine *Inspire* is testimony to the growing "individualization of jihad" and the dangers of self-taught and self-radicalized jihadists.

AQAP's dozen or more online issues of *Inspire* magazine has provided direct "inspiration" and instruction for leaderless terrorist cells worldwide. The article "Make a Bomb in Your Mom's Kitchen", detailing instructions of workable pressure-cooker bombs, seems to have been used by the 2012 Boston marathon bombers. In Europe, this explosives instruction was used in building a bomb in a 2011 terrorist plot at a Territorial Army base in the United Kingdom.[38] *Inspire* magazine provide many operational shortcut pathways for those with less direct operational skillsets. For example, Ikrimah al-Muhajir from the Explosives Department provided technical details of the printer device in AQAP's Operation Hemorrhage and ways to bypass aviation security detection methods.[39] Other *Inspire* issues have outlined: demolition of buildings using gas cylinders and AK-47s (Winter 2010); training with handguns and remote-control detonation using a car alarm (Autumn 2011); processes and shopping list for selecting targets, as well as how to ignite forest fires to inflict economic damage (May 2012); targets who have offended Islam, and causing road accidents (Winter 2013); how to make car bombs (Spring 2014); an "ultimate moving machine" using a pickup truck with welded knives to cut down targets (Autumn 2010). Under the section "Open Source Jihad," *Inspire* magazine provides operational tips, from bomb-making instructions to countersurveillance tips and how to evade signals intelligence. Many of these and other detailed operational instructions were later published in the *Inspire* magazine's 64-page e-booklet *Lone Mujahid Pocketbook* (2013).

There is limited evidence for how important these instructional guides have been for actual terrorist operations. These types of instructions were available long before the digital age, circulating in videotapes and written manuals on how to conduct guerilla and terrorist attacks alongside bombmaking instructions which were found in terrorist training camps in Afghanistan before and after the 9/11 attacks in 2001. One of its earliest training manuals was *Military Studies in the Jihad against the Tyrants*,[40] as well as the 11-volume *Encyclopedia of Afghan Jihad*. Similarly, al-Qaeda-related terrorists have long been well-versed with electronic dead drops and different forms of coded communication as means to avoid detection. This spectrum ranges from human couriers or codes to encryption and one-time Satellite phones.

As Anne Stenersen argues in her study on al-Qaeda's e-learning, courses for bomb-making are a relatively marginal phenomenon, but the quality of online instructions are continuing to steadily improve which should be a cause of security concern.[41] Distance-learning courses are also part of the jihadi repertoire, as exemplified by a 51-page handbook by Abu 'Amar al-Qa'idi entitled *A Course in the Art of Recruitment*,[42] or *al-Battar*, al-Qaeda's online military magazine whose goal is to "spread 'military culture among the youth of Islam' and provide lessons in physical exercise and jihad preparedness, proficiency in use of light weapons and 'guerrilla group actions in the cities and mountains,' and 'important points in security and intelligence'."[43] Understanding the collaborative nature of these virtual training camps which exchange and share ideas and techniques becomes crucial as illustrated by the jihadi social media landscape.[44]

What is common over time is that al-Qaeda in all its forms "stresses the importance of lesson learning to warfare and repeatedly emphasizes thinking about war and the enemy."[45] The richness of the so-called jihadi strategic studies literature is indicative of this fact. As argued by Jackson and Loidolt, al-Qaeda's strategists and their texts stress and operate on an innovation doctrine.[46] It is also the case that al-Qaeda is acutely aware that it is necessary to learn from experience, as the emergence of conflict theaters creates new opportunities and challenges to keep cohesion and overall direction as to where it is heading and what violence to prioritize. Some conflicts such as Syria generate inter-jihadi discord at times and create tension between those wanting to engage in insurgency and fighting the near enemy rather than prioritize efforts to strike at the West. While al-Qaeda have "successfully convinced many local insurgents the world over that all Islamic insurgents are fighting individual battles that are part of a larger worldwide war,"[47] there is a constant struggle about where to leverage the focus of violence – towards local targets or the West.

Group dynamics and technical expertise

Some have argued that innovative changes can be attributed to internal group dynamics, which play a major role in pressuring terrorist groups towards altering or adopting new attack modes to demonstrate strength and relevance. These group dynamics could be the resulting influence of technical skill sets of specific bomb designers and bomb-makers that affect operational innovative processes. Within Europe before al-Qaeda appeared, the Provisional Irish Republican Army (PIRA) was renowned for its continuous creativity in its often advanced attack modes, especially the design and deployment of improvised explosive devices (IEDs). These incremental innovation processes were driven by trial-and-error, the presence of skilled bomb designers and makers, as well as organizational and technical pressures on PIRA from British countermeasures.[48] After September 11, 2001, the presence of key al-Qaeda operational innovators has been evident in more complex and deadly operations; examples are the critical roles played by Pakistani-based Rashid Rauf in the 2006 Transatlantic terrorist plot, in which seven airliners departing from the United Kingdom were to be brought down by peroxide-based liquid explosives, and by al-Qaeda in the Arabian Peninsula and their chief bomb-maker Ibrahim al-Asiri; the latter were allegedly responsible for the so-called 2009 Underwear bomber, Abdulmuttalab, targeting Northwest Airlines Flight 213, as well as the 2010 cargo plane bomb plot.

Often master bomb-makers are the exception rather than the norm. Innovation can be unintentional or intentional, incremental as a form of problem-solving process or even transformational. This does raise the question of what constitutes improvisation of IEDs. Gill, Horgan and Lovelace argue that this can be analyzed "across seven dimensions: degree of sophistication; device components; explosive ingredients; initiation/detonation type (i.e., the device used to trigger the explosives); delivery systems (i.e., the means by which the IED is carried to the target); perpetrator type; and what is the device designed to do?"[49] The 2007 RAND study provides further categories of technology characteristics that affect the terrorist absorptive process.[50]

The role and influence of conflict zones is often critical in providing physical terrorist training and key connectivity to core al-Qaeda trainers who maintain active influence on their networks. This influence and the role of enhanced capabilities has proven critical in the success rate of terrorist attacks. A study by Thomas Hegghammer on the effect of returning foreign fighters show that while only one in nine return home after fighting abroad they are considerably more capable and dangerous than those that do not fight in conflict zones.[51] As Paul Cruikshank shows in his contribution, there were 43 serious terrorist plots against the West between 2004 and 2013, of which 47 percent had direct links to training camps in Pakistan. Much has been written about terrorist training more generally,[52] and specifically looking at country-specific terrorist case studies. Apart from terrorist weapons training, al-Qaeda operatives are taught extensive tradecraft in counter-surveillance and other ways to avoid being detected by Western intelligence services.[53]

Other academic approaches argue that terrorist group leaders play a key role in demanding, funding and forcing new innovation. Assaf Moghadam argues that al-Qaeda operate under the principle of *Auftragstaktik* (mission command) whereby the top-down al-Qaeda structure or key leadership issues a directive for targeting while the al-Qaeda subordinates operate under considerable autonomy in actually executing the attack when it comes to target type, attack method, logistics and even timing.[54] In this model the top-down drivers of innovation are al-Qaeda's goal and ideology, leadership skills and obsession with martyrdom and specific types of targets, and the bottom-up drivers are trial and error. According to Moghadam, there are also "integrative drivers for innovation such as receptivity to proposals for terrorist attacks and centralization of decision, decentralization of execution."[55]

Group or terrorist cell dynamics are crucial to understanding, especially as there are different roles or hierarchies and specialized functions performed within cell structures.[56] Some of al-Qaeda's operations can be understood as direct command operational directives when terrorists have been in training camps abroad, while much of it follows Abu Musab al-Suri's doctrine of *Nizam, la Tanzim* (System, not Organization) where "the logistical procurement process begins and ends with the individual *jihadis.*"[57]

Terrorist attack complexity

Measuring terrorist attack complexity is difficult given the inherently unique features of each case and the dynamics on the meso and micro levels. However, Teun van Dongen provides a Terrorist Attack Complexity Index composed of 16 indicators relating to 21 jihadi terrorist attacks in Europe between 2004 and 2011.[58] Apart from providing a useful way to deconstruct variables in the process of planning, preparing and executing a terrorist attack, this study show that the complexity of jihadi terrorist attacks was closely related to expertise received from terrorist training camps outside Europe or from jihad experience in conflict zones abroad. In fact in the ten most complex terrorist operations there was direct participation in terrorist training camps abroad with the exception of one case. The study also shows that complexity could further correlate with larger size of terrorist cell structures. Additionally, terrorist attack complexity was also due to direct core al-Qaeda instructions.

Research into why terrorist operations succeed or fail show that for a group to successfully execute terrorist action there are specific operational requirements such as resources, situational awareness, number and quality of people available, technical skills, operational security skills, planning skills, command and control or leadership skills, intelligence gathering and surveillance, recruitment of people with right skill sets or knowledge, technology acquisition, production, development and testing, operational training and mission rehearsal and weapons training.[59] For example, Michael Kenney argues that the failure of jihadist plots in Britain and Spain could be explained by poor tradecraft and limited practical experience combined with their propensity for suicide attacks and ideological zeal.[60]

Other scholarly efforts have sought to model what drives al-Qaeda's targeting preferences, and developed four models: coercion; damage; rally; and franchise hypotheses.[61] The RAND study shows that al-Qaeda's target selection is first and foremost designed to coerce and to damage economies and to a lesser extent to rally supporters and potential supporters.[62]

Structure of the book

This book is divided into two parts; the first seven relevant chapters lay the conceptual framework of driving forces of terrorism, its tradecraft and the nature of innovation per se; then the next four chapters in Part II of the book highlight indications of innovation and learning drawn from terrorism events in three European countries since 2001.

Gary Ackerman introduces a valuable perspective of the underlying forces surrounding terrorism innovation in the light of the broader innovation literature, and thus draws some interesting conclusions about differences and analogs between the world of terrorist organizations and other sectors in society striving for innovative solutions. Ackerman guides us through some of the basic conditions that create fertile ground for innovation and furthermore highlights drivers that may stimulate terrorists to move beyond their traditional *modus operandi*.

John Allison provides an in-depth analysis of innovative characteristics in the context of improvised explosive devices (IED). As one of terrorists' and insurgents' most preferred weapons of choice, the distinctive hallmarks of IED techniques and tactics is explored from a counter-terrorism practitioner's perspective. Allison furthermore elaborates on the potential trends and trajectories of the IED threat.

Adam Dolnik contributes with an elaborative chapter by identifying some distinct characteristics of highly innovative terrorist organizations. He provides us with a valuable insight into the complex process and specific variables at work, distinguishing innovative organizations from traditional ones. Specifically he determines four distinct reasons serving as potential triggers to innovation amongst terrorist organizations.

Søren Hove continues with a chapter exploring the conceptual signatures of terrorism through a discussion on how al-Qaeda and its affiliates choose their targets and the influence from propaganda and ideological writings on target designation against the West. He guides us through the changes in al-Qaeda strategic writings from the 1990s to the present and elaborates on the extent and nature of correlations between the propaganda and al-Qaeda attacks against Western targets. Hove concludes with an interesting perspective on the shifting trends of targets and methods al-Qaeda is promoting and furthermore sheds light on the ambiguous causal links between al-Qaeda propaganda and the terrorist acts conducted by the organization.

Terrorists' adaptation to new social technological development and services for their propaganda is in focus in Nico Prucha's excellent chapter. Prucha provides a detailed and highly relevant picture of how al-Qaeda and its affiliates have developed skillsets and structures in order to utilize the internet as the prime facilitator for propaganda and media outreach to an audience on a global scale. Through his research on the terrorist use of social media platforms such as Facebook, YouTube and Twitter, Prucha paints an insightful picture of the unprecedented quantity and quality of contemporary jihadi media activities.

Paul Cruickshank explores the threat from the overseas jihadi training of Western militants and conveys in his chapter the trends in jihadists' traveling, the character of the training activities and not least the effects in terms of capabilities. Cruickshank furthermore analyzes the trends in future training activities and highlights some regional areas of specific concern to the future threat against Europe and the West.

Joshua Sinai concludes the first part of this book by examining innovation in tradecraft by al-Qaeda and its affiliates in terms of pre-incident counter-surveillance. In this novel perspective on al-Qaeda's operational tradecraft Sinai analyses surveillance and counter-surveillance measures employed by various terror cells in selected cases and paints an interesting picture of the trends and developments in the capabilities of al-Qaeda affiliates to conduct their business undetected and with its intended effect.

The second part of this book is dedicated to in-depth analysis of al-Qaeda linked (jihadist) terror plots in Europe since 2001, specifically highlighting the level of innovative elements of their activities. In the first chapter Raffaello Pantucci provides a rich elaboration of the multitude of British terror plots from 2001 to 2013 and draws some very illustrative conclusions regarding the terror cells' abilities to adapt and develop their capabilities and methods in order to conduct successful attacks. Pantucci specifically looks at six defined phases or aspects of each plot to elaborate on trends and patterns of innovation and learning over time.

Morten Skjoldager continues with a chapter examining the methods and skills used in the Danish terrorist plots and elaborates on the innovative dimensions and visible trajectories. Skjoldager's contribution lays out the aspects of the terrorist plots in Denmark which display innovative change in strategies and methods but concludes that target designation has been profoundly conservative.

The final contribution on case studies in Europe is divided into two separate chapters, authored by Yassin Musharbash and Guido Steinberg, specifically penetrating two of the most spectacular jihadist terror plots in Germany during the last decade, namely the Sauerland case and the Europlot. These two chapters elaborate on the similarities and differences between these two plots in terms of recruiting process, target designation and weapons choice, particularly illustrating what could possibly be the inside views of al-Qaeda central's last efforts and schemes to renew its ability to perpetrate large-scale attacks against the West through the documentation found when apprehending the leading figures behind the Europlot in late 2011.

In the concluding chapter the editors elaborate on some of the aspects raised in previous chapters, aiming to capture some of the overarching trends and tendencies in terrorists' innovative behavior. The editors then reflect on the potential ramifications of European returnees from conflict zones such as the Syrian war in relation to possible innovative and learning patterns and what effect this potential phenomenon may have on future jihadi terrorist attempts.

Notes

1 Bruce Hoffman, *Inside Terrorism* (Columbia University Press, 2006).
2 Rohan Gunaratna, "Al Qaida's Lose and Learn Doctrine: The Trajectory from Oplan Bojinka to 9/11," *Teaching Terror: Strategic and Tactical Learning in the Terrorist World* (2006): pp. 171–188.
3 Kaare Sørensen, *Halshug* (People's Press, 2013).
4 For example, see Thomas H. Johnson, "Taliban adaptations and innovations," *Small Wars and Insurgencies*, Vol. 24, No. 1 (2013): p. 10.
5 Mohammed M. Hafez and Maria Rasmussen, *Terrorist Innovations in Weapons of Mass Effect, Phase II* (Naval Postgraduate School, January 2012).
6 Ibid.
7 Kim Cragin, Peter Chalk, Sara A. Daly and Brian Jackson, *Sharing the Dragon's Teeth: Terrorist Groups and the Exchange of New Technologies* (RAND Corporation, 2007).
8 Luke M. Gerdes, Kristine Ringler and Barbara Autin, "Assessing the Abu Sayyaf Group's Strategic and Learning Capacities," *Studies in Conflict and Terrorism*, Vol. 37 (2014): pp. 267–293.
9 Mohammed M. Hafez and Maria Rasmussen, *Terrorist Innovations in Weapons of Mass Effect, Phase II*, PASCC Report Report Number 2012 003 (Monterey, January 2012): p. 39.
10 Maria Rasmussen and Muhammed Hafez, *Terrorist Innovations in Weapons of Mass Effect* (Defence Threat Reduction Agency, August 2010).
11 Ibid.
12 Brian A. Jackson and David R. Frelinger, *Rifling through the Terrorist Arsenal: Exploring Groups' Weapon Choices and Technology Strategies* (RAND, October 2007).
13 This was already recognized in 1993 with "Special Issue – Technology and Terrorism," *Terrorism and Political Violence*, Vol. 5, Issue 2 (1993).
14 For example, see Brian Jackson and Bryce Loidolt, "Considering al-Qa'ida's Innovation Doctrine: From Strategic Texts to 'Innovation in Practice'," *Terrorism and Political Violence*, Vol. 25, No. 2 (2013): p. 285.
15 For a useful outline of this, see Brian Jackson, "Organizational Decisionmaking by Terrorist Groups," in Paul K. Davies and Kim Cragin, *Social Science for Counterterrorism: Putting the Pieces Together* (RAND, 2009).
16 Petter Nesser, "Chronology of *Jihadism* in Western Europe 1994–2007: Planned, Prepared, and Executed Terrorist Attacks," *Studies in Conflict and Terrorism*, Vol. 31 (2008): pp. 924–946.
17 Petter Nesser, "Chronology of Jihadism in Western Europe: Update 2008–2010." Available at: www.ffi.no/no/Prosjekter/Terra/Publikasjoner/Documents/Petter

%20Nesser%20-%20Chronology%20of%20Jihadism%20in%20Western%20 Europe%20Update%202008–2010%20for%20FFI%20web.pdf.
18 Petter Nesser, "Individual Jihadist Operations in Europe: Patterns and Challenges," *CTC Sentinel*, Vol. 5, Issue 1 (January 2012): pp. 15–18.
19 Javier Jordan, "Analysis of Jihadi Terrorism Incidents in Western Europe, 2001–2010," *Studies in Conflict and Terrorism*, Vol. 35, No. 5 (2012): pp. 382–404.
20 Adam Dolnik, *Understanding Terrorist Innovation: Technology, Tactics and Global Trends* (Routledge, 2007).
21 George Michael, "Review of Adam Dolnik, Understanding Terrorist Innovation: Technology, Tactics and Global Trends," *Terrorism and Political Violence*, Vol. 21, Issue 1 (2009): pp. 179–182.
22 James J. F. Forest, *Teaching Terror: Strategic and Tactical Learning in the Terrorist World* (Rowman & Littlefield, 2006).
23 Paul Gill, John Horgan, Samuel T. Hunter and Lily D. Cushenberry, "Malevolent Creativity in Terrorist Organizations," *The Journal of Creative Behavior*, Vol. 47, Issue 2 (2013): pp. 125–151.
24 Leonard Weinberg, "Two Neglected Areas of Terrorism Research: Careers after Terrorism and How Terrorists Innovate," *Perspectives on Terrorism*, Vol. 2, No. 9 (2008).
25 C. J. M. Drake, "The Role of Ideology in Terrorists' Target Selection," *Terrorism and Political Violence*, Vol. 10, No. 2 (1998): pp. 53–85.
26 Jacob N. Shapiro, "Terrorist Decision-Making: Insights from Economics and Political Science," *Perspectives on Terrorism*, Vol. 6, Nos 4–5 (2012).
27 Jeevan Deol and Zaheer Kazmi (eds), *Contextualising Jihadi Thought* (Hurst, 2012).
28 Dan Byman, *The Five Front War: The Better Way to Fight Global Jihad* (John Wiley & Sons, 2008): p. 9.
29 Shireen K. Burki, "Jihad or Qatal? Examining Al Qaeda's Modus Operandi," *Defence and Security Analysis*, Vol. 29, No. 3 (2013): pp. 234–252.
30 Radwan Mortada, "Al-Qaeda's 20-Year Plan: From 9/11 to Final Victory," *Al-Akhbar*, 29 January 2014.
31 Mark E. Stout, Jessica M. Huckabey and John R. Schindler with Jim Lacey, *The Terrorist Perspectives Project: Strategic and Operational Views of Al Qaida and Associated Movements* (Naval Institute Press, 2008): pp. 138–140.
32 Kevin A. O'Brien, "Assessing Hostile Reconnaissance and Terrorist Intelligence Activities: The Case for a Counter Strategy," *RUSI Journal*, Vol. 153, No. 5 (October 2008): pp. 34–39.
33 Assaf Moghadam, "How Al Qaeda Innovates," *Security Studies*, Vol. 22, No. 3 (2013): p. 491.
34 Nancy K. Hayden, *Innovation and Learning in Terrorist Organisations: Towards Adaptive Capacity and Resiliency*. Paper presented at 31st International Conference of the System Dynamics Society, July 2013. Available at www.systemdynamics.org/conferences/2013/proceed/papers/P1407.pdf.
35 Michael W. S. Ryan, *Decoding Al-Qaeda's Strategy* (Colombia University Press, 2013): p. 8.
36 Norman Cigar, *Al-Qa'ida's Doctrine For Insurgency: 'Abd Al-'Aziz Al-Muqrin's – A Practical Course for Guerrilla War* (Potomac Books, 2009).
37 Brynjar Lia, *Architect of Global Jihad: The Life of al-Qaeda Strategist Abu Mus'ab al-Suri* (Columbia University Press, 2008).
38 Anthony F. Lemieux, Jarret M. Brachman, Jason Levitt and Jay Wood, "Inspire Magazine: A Critical Analysis of its Significance and Potential Impact through the Lens of the Information, Motivation, and Behavioral Skills Model," *Terrorism and Political Violence*, Vol. 26, Issue 2 (2014): pp. 354–371.
39 *Inspire* magazine, November 2010.

40 For example, see Ben Venzke and Aimee Ibrahim, *The al-Qaeda Threat: An Analytical Guide to al-Qaeda's Tactics and Targets* (Tempest Publishing, 2003).
41 Anne Stenersen, "'Bomb-Making for Beginners': Inside Al-Qaeda E-Learning Course," *Perspectives on Terrorism,* Vol. 7, Issue 1 (February 2013): pp. 25–37.
42 Brian Fishman and Abdullah Warius, "A Jihadist's Course in the Art of Recruitment," *CTC – Sentinel,* Vol. 2, Issue 2 (February 2009).
43 Presentation by Michael Knapp, *"The War of the Ether": Al-Qaeda's PSYOPS Campaign against the Western and Muslim Worlds* (United States Army, no date).
44 For an excellent overview of these forums, see Ezzeldeen Khalil, "Jihad 2.0 – Internet Forums Remain Key to Militant Islamists," *Jane's Intelligence Review* (September 2013): pp. 14–19.
45 Stout *et al.*, op. cit.: p. 119.
46 Brian A. Jackson and Bryce Loidolt, "Considering al-Qa'ida's Innovation Doctrine: From Strategic Texts to "Innovation in Practice," *Terrorism and Political Violence,* Vol. 25, No. 2 (2013): pp. 284–310.
47 Dan Byman, *The Five Front War: The Better Way to Fight Global Jihad* (John Wiley & Sons, 2008): p. 30.
48 See Magnus Ranstorp and Hans Brun, *Terrorism Learning and Innovation: Lessons from PIRA in Northern Ireland – A Closed Workshop Summary* (Swedish National Defence College, April 2013).
49 Paul Gill, John Horgan and Jeffrey Lovelace, "Improvised Explosive Device: The Problem of Definition," *Studies in Conflict,* Vol. 34 (2011): pp. 732–748.
50 Kim Cragin *et al.*, op. cit. (2007): p. 17.
51 Thomas Hegghammer, "Should I Stay or Should I Go? Explaining Variation in Western Jihadists' Choice between Domestic and Foreign Fighting," *American Political Science Review* (February 2013): pp. 1–15.
52 James J. F. Forest (ed.), *The Making of a Terrorist: Recruitment, Training, and Root Causes* (Praeger Security International, 2006).
53 Blake W. Mobley, *Terrorism and Counter-Intelligence: How Terrorist Groups Elude Detection* (Columbia University Press, 2012).
54 Asaf Moghadam, op. cit.
55 Ibid.
56 For example, see Murat Ozer, "The Impact of Group Dynamics on Terrorist Decision Making," in S. Ozeren (ed.) *Understanding Terrorism: Analysis of Sociological and Psychological Aspects* (IOS Press, 2007).
57 Arabinda Acharaya and Sonal Marwah, "*Nizam, la Tanzim* (System, not Organisation): Do Organizations Matter in Terrorism Today? A Study of the November 2008 Mumbai Attacks," *Studies in Conflict and Terrorism,* Vol. 34, No. 1 (2011): p. 2.
58 Teun van Dongen, "The Lengths Terrorists Go To: Perpetrator Charcteristics and the Complexity of Jihadist Terrorist Attacks in Euope, 2004–2011," *Behavioral Sciences of Terrorism and Political Aggression,* Vol. 6, No. 1 (2014): pp. 58–80.
59 Brian A. Jackson and David R. Frelinger, *Understanding Why Terrorist Operations Succeed or Fail* (RAND, 2009).
60 Michael Kenney, "'Dumb' Yet Deadly: Local Knowledge and Poor Tradecraft among Islamist Militants in Britain and Spain," *Studies in Conflict and Terrorism,* Vol. 33 (2010): pp. 911–932.
61 Martin C. Libicki, Peter Chalk and Melanie Sisson, *Exploring Terrorist Targeting Preferences* (RAND, 2007).
62 Ibid.

Part I

Perspectives on terrorism innovation

2 The theoretical underpinnings of terrorist innovation decisions

Gary Ackerman

Introduction: assessing innovation in the context of terrorism[1]

One of the few features common to terrorism in all its incarnations, besides the obvious desire to have a psychological impact on wider audiences than the immediate victims, is that it is almost never a static phenomenon. Highly dynamic processes present themselves at every turn – from the strategic and tactical interactions with the state, to competitive and collaborative relations with other non-state actors, to intra-organizational evolution of membership, leadership and goals, and, of course, to the unrelenting changes in the social, political and technological environment in which the terrorists operate. These dynamics create almost Darwinian pressures for a terrorist organization – it must adapt and evolve if it wants to enjoy any type of longevity. This makes the topic of terrorist learning and innovation key to understanding terrorism, particularly if security forces would like to "get ahead" of terrorist adversaries and shape their evolution in ways that decrease the overall threat.

Yet, until recently, there has been relatively little scholarly study of how innovation and learning occur in terrorist organizations.[2] Even the most comprehensive studies of terrorist innovation thus far have left many unanswered questions. For example, Dolnik's excellent foray into this topic identified some important drivers of innovation, but his case studies were unable to provide much insight into several of these.[3] Moreover, although existing studies have described some of the factors underlying terrorist innovation, none of them have provided a detailed explanation of the *process* of terrorist decision making in this area.

However, outside the rather narrow realm of terrorism studies, there is an immense corpus of prior work on innovation in other domains, not only in the military context, but more generally in disciplines as diverse as business management, public policy and sociology. This vast literature incorporates a treasure trove of insights relating to innovation in a variety of human activities, but has remained largely untapped by terrorism researchers; to all appearances there has been only a single attempt – and a partial one at that – by terrorism scholars to systematically draw on the broader social and behavioral sciences.[4] This is somewhat odd, considering that scholars are less likely to derive spurious inferences from a limited number of case studies if these can be tested against a coherent set of deductively or inductively derived hypotheses and theories, both of which abound in the broader innovation literature. After all, since, at their most basic, terrorists are human beings and terrorist groups are social organizations, there is a *prima facie* case to be made that this wider body of literature is relevant to the terrorist context.

At the very least, the general literature on innovation, upon comparison with the empirical record of terrorism explored through case studies, might reveal that terrorists are *sui generis* with respect to innovation, which would precipitate further investigation into why this is so. This chapter will therefore highlight the prominent findings on terrorist innovation, while supplementing them with findings and theories drawn from organizational innovation and learning[5] in the general innovation literature, which can in turn be utilized to guide our interpretation of the case study evidence of terrorism.[6]

The nature of innovation and the innovation process

In the context of terrorism, Crenshaw has quite broadly defined innovation as "the adoption of new patterns of behavior,"[7] but by considering two alternative conceptions of innovation from other contexts and examining what scholars in those contexts hold to be the essence of innovation, we may be able to derive insights relevant to our own inquiry. Everett Rogers, a sociologist and communications scholar, is widely regarded as the father of the study of innovations and how these diffuse through populations. He defines innovation as "an idea, practice, or object that is perceived as new by an individual or another unit of adoption,"[8] thus making the conceptual focus one of *subjective determination.* By orienting innovation around the perceptions of the creator or user, he also draws attention to the inherent uncertainty in all innovation decisions about whether the ostensibly novel practice is really "new" in any significant sense or will prove superior to existing ways of doing things. This forms the impetus for the ubiquity of information-seeking behaviors about a prospective innovation in order to address this uncertainty.[9] Brian Arthur, an economist and scholar of technological development, focuses on innovation in situational terms, by equating innovation – or at least significant innovations beyond incremental improvements – with the concept of a *redomaining* of technology.[10] In this sense he describes innovations as "the expressing of a given purpose in a different set of components,"[11] i.e., a different domain of practice or material technology. This implies that innovations often involve achieving the same end with qualitatively different means, which in turn suggests that anticipating the new patterns of behavior we have associated with terrorist innovation may require looking beyond the strategic, tactical or material attributes traditionally associated with terrorist actors.

The above notions of innovation highlight an additional concept that pervades any discussions about terrorist innovation, namely the centrality of knowledge in the innovation process and the connection between innovation and invention, broadly construed. First, knowledge is inseparable from innovation in the sense that the generation or transfer of knowledge about a new idea or behavior is a prerequisite to implementing it, and often even to attempting to adopt it. Yet knowledge is a multifaceted phenomenon and scholars of innovation have drawn distinctions between different species of knowledge. Mokyr distinguishes *propositional* knowledge ("what" knowledge or beliefs about how nature works) from *prescriptive* knowledge ("how" knowledge or techniques for accomplishing something).[12] Further, several scholars categorize knowledge according to its transferability. They differentiate between *explicit* knowledge (that which can be codified in manuals, videos and the like) and the more pervasive *tacit* knowledge[13] (that which cannot be so codified and must be transferred from person to person).[14] More recently, Kenney has argued for an additional distinction between the concepts of *techne* (abstract technical knowledge) and *metis* (knowledge acquired from practical experience, a type of "cunning" that allows one to circumvent unforeseen obstacles).[15] For our purposes at this point, it suffices to note that these authors appear to argue strongly that successful innovation requires proficiency in or transfer of both elements of each aforementioned knowledge pair.

An additional comment worth making concerns the distinction between innovation and invention, and in particular whether this might be significant in the terrorist context. Invention is usually alluded to as "the process by which a new idea is discovered or created,"[16] in other words "an increment in the set of the total technological knowledge of a given society."[17] Innovation, on the other hand, is taken in the majority of the literature to constitute an intrinsically social process, involving the integration of an invention into one or more cultural, organizational, economic or political contexts.[18] One might argue that terrorists rarely, if ever, invent a tactic, organizational principle or weapon from scratch and hence that the factors undergirding invention as such are less important to explaining terrorist innovation. Yet, revolutionaries from Karl Heinzen and Carlos Marighella to Ramzi Yousef and Abu Hamza al-Muhajir have built upon existing weapons and tactics, modifying and adapting them to such an extent that most dispassionate observers would apply the label of novelty. Therefore, even those factors traditionally associated with the basic invention of new technologies or principles may inform our understanding of how and why terrorists innovate. At least for the purposes of the current enterprise, then, it makes sense to look at both concepts.

More simply, and without significant loss in conceptual clarity, we can treat invention and innovation together in the current review.

The last note on the general nature of innovation relates to one of the major observations in the literature that innovations only rarely emerge *de novo* from the epistemological ether. Much of the genesis of new technologies[19] and other types of innovation[20] are argued to lie in a combination or synthesis of previous ideas, technologies or practices, often using components borrowed from other domains of activity.[21] Apart from harnessing the relatively few truly new discoveries of or ways to control natural phenomena, innovations are thus *recombinant*, but also *recursive* in that they are themselves made up of other sub-innovations and so on.[22] In the specific context of technology, Arthur postulates that therefore a mechanism of *evolution by combination* operates in a quasi-Darwinian fashion,[23] leading invariably to greater complexity in new innovations and (usually) enhanced usefulness. Technology, from this perspective, thus becomes largely *autopoietic* (or self-creating).[24] I see no immediately apparent reason why Arthur's arguments should not be applied to the concept of innovation and learning more generally,[25] which would be in accordance, in the terrorism context, with the commonly observed behavior of terrorists in combining well-understood components (say, dynamite and garage door openers) into novel and more efficient means of destruction. Further, it has long been recognized that innovation in a particular area of activity is often inextricably connected to (and sometimes dependent on) parallel technological or behavioral developments in related or constituent domains.[26] Canonical examples include the fact that Leonardo da Vinci's prescient designs for tanks and aircraft (in the domain of military technology) were unproduceable without long-subsequent advances in the metallurgical domain, and the observation that tremendous leaps in computing power have enabled the practical, low-cost sequencing and manipulation of genomes, thus opening synthetic biology up to rapid commercialization.[27]

What the recombinant, recursive and parallel nature of innovation suggests in terms of understanding terrorist innovation is that terrorists, in order to innovate to even a large extent, need not possess within their ranks a creative genius on the order of a Leonardo or an Einstein, but only someone with sufficient mental alacrity to scan the horizon of existing methods and materials and think of new ways of combining them to suit organizational or personal goals. The bar for even potentially dramatic innovation is thus, theoretically at least, lower than the level at which it is often popularly portrayed.

The process of terrorist innovation can be modeled as a series of steps, encompassing: (1) both general preconditions and specific drivers for innovations; (2) awareness of a particular innovation; (3) the decision whether to innovate or not; (4) the wherewithal to actually implement the process of innovation; and, if successful, (5) the decision whether to maintain or discard the innovation after initial usage.[28] The process can be initiated at either step (1) or step (2), the latter occurring when terrorists are presented with an innovation or learning opportunity which they had not previously considered. As Figure 2.1 illustrates, this process can fail at various points, and current innovation decisions and capabilities are shaped in no small way by previous ones (with sample feedback loops depicted as dotted lines).

The remainder of this chapter will focus on the initial stages of the innovation process, namely the underlying conditions for and proximal drivers of innovation, the awareness of innovation opportunities and the decision whether to innovate or not.

Antecedent conditions

Human beings are remarkably creative and adaptive to environmental stresses. However, at the same time, the options available to terrorists and other actors are not infinite, but are constrained by their position in space-time, a concept that Stuart Kaufman has termed the "adjacent possible."[29] There are thus some basic conditions that, while not necessary in a strictly causal sense, create fertile ground for innovation. Conversely, in the vast majority of cases, the absence of these conditions makes it very difficult for terrorists to even consider innovating.

Practical feasibility

The most basic delimiter of an adoption decision is that the adopter must perceive the technology to be within their technical abilities and resources, or at least within their expected capacity to acquire these.[30] In addition, economists have argued that the innovation must, in most cases, also be

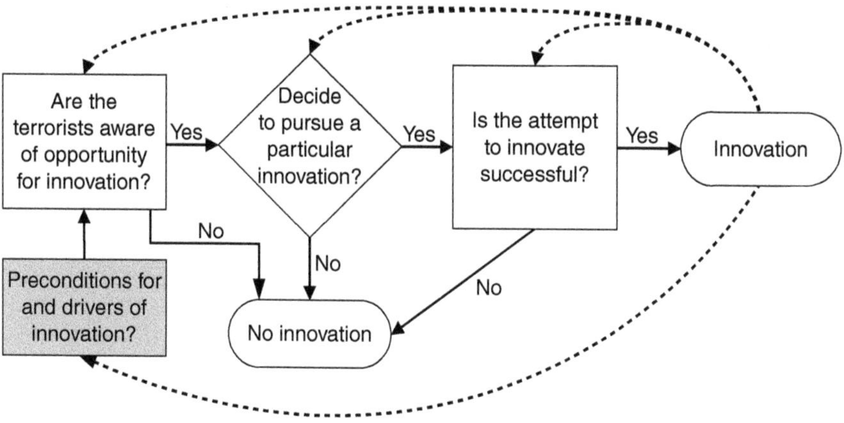

Figure 2.1 The terrorist innovation process.

economically feasible, that is, at least as efficient as existing ways of operating.[31] However, even *prima facie* technical, financial and economic feasibility may be insufficient because the potential adopter can shy away from secondary obstacles and costs beyond the innovation itself, such as when an adoption requires an expensive restructuring of an entire organization and its institutional practices.[32] On the other hand, ample resource reserves can facilitate innovation, allowing, for example, for numerous people with different expertise to work on the innovation (thus bolstering network effects) or for extensive experimentation with different methods and materials. Overall, in the case of terrorist innovation, initial empirical results seem to suggest that while a minimum level of resources (and presumably technical skills) are necessary, the possession of copious resources by no means guarantees that adoption will be pursued.[33]

Ideological and cultural compatibility

At the very least, for adoption to be pursued the use of the proposed novel weapon or practice must not be anathema to deeply held cultural values or the worldview and/or self-image of the terrorist group.[34] In certain cases, it may not be the actual innovation itself, but the ideas, practices and structural changes its adoption might engender that are resisted by the group's ideologues.[35] Similar arguments apply with respect to attitudes towards innovation in general. While an ideology or an organizational or national culture that embraces innovation can facilitate a positive adoption decision, ultra-conservative belief systems that decry innovation of any kind are far less likely to approve of adopting a new, untried weapons system or tactic.[36] However, this is not absolute – in certain cases ideological resistance to innovation can be overridden by other concerns[37] or even adapted to permit the adoption.[38]

Hospitable environment

Writers in the context of commercial enterprises have observed that so-called "non-market forces" (including intellectual property law, government regulation, industry standards and labor practices) can stifle innovation by reducing both the motivation and capability to innovate.[39] The policies of the Ming Dynasty in China (1368–1644) and the Tokugawa shogunate in Japan (1600–1867) are legendary examples of political and legal environments that discouraged innovation.[40] Analogs in the terrorism sphere that could stifle innovation might be an oppressive security regime or underdeveloped industrial and communications infrastructures. Conversely, in cases where terrorist organizations enjoy some degree of assistance or toleration from the country in which they reside, or operate in a highly developed technical environment, innovation can be expected to be easier and more prevalent, *ceteris paribus.*

Risk tolerance

Since any innovation requires deviating from the *status quo* to embrace an uncertain alternative, in order for innovation to succeed there needs to be a sufficient degree of tolerance for taking risks and accommodating deviance by the organization's decision makers.[41] Given that the natural posture of many organizations is a conservative one (in terrorist organizations, especially with respect to tactical deployments),[42] several scholars have asserted that the most likely terrorist organizations to innovate are those that have a high tolerance for risk, i.e., are willing to tolerate operational failure.[43] The amount of risk a group is willing to tolerate is dynamic and can alter according to the circumstances or environment.[44]

Drivers

Although it might very well be that "technological gimmicks have long fired terrorist imaginations,"[45] organizations rarely, if ever, innovate solely for the sake of innovating.[46] As alluded to previously, a fairly common observation among scholars of terrorism is that, however varied their goals and beliefs, most terrorists tend to be conservative in their choice of weapons, preferring to use the most proven, cost effective, and reliable means at their disposal that prove sufficient to achieve their operational objectives.[47] This organizational inertia thus stands as an impediment to innovation that must be overcome. Several straightforward explanations have been given for this dynamic, including: (a) a general predilection for minimizing the complexity of terrorist operations overall;[48] (b) uncertainties stemming from operating in a context of incomplete information[49] regarding either the organization's ability to successfully adopt a new, often unfamiliar, practice or the reliability of the innovation in achieving its desired effects;[50] (c) additional resources that must be devoted to the adoption process;[51] and (d) potential concerns regarding the safety of members of the terrorist organization during production or handling of an unfamiliar weapon, or the implementation of a new tactic.[52] The most obvious reason for tactical conservatism in the vast majority of cases, however, is the adequacy of the existing, widely available weapons, targets and tactics. For example, the tried and true explosive or firearm suffices to accomplish the terrorists' aims for violence,[53] even when the objectives entail substantial numbers of casualties or property damage.[54]

Therefore, even assuming that the basic preconditions for innovation and learning are in place, some stimulus is usually needed to overcome this innate organizational inertia and move an actor to consider abandoning the *status quo*. At a fundamental level, one of the primary pathways to a decision to innovate is initiated when an actor perceives that there is a problem to be solved that is not addressed, or not addressed as efficiently as it could be, by existing practices or technologies.[55] For example, in the case of Northern Ireland, the construction of high walls and fences around police stations in the early 1970s to prevent frontal attacks on them presented the Provisional Irish Republican Army (PIRA) with a tactical problem. This in turn prompted the PIRA to pursue a home-made mortar program, which would allow them to circumvent these countermeasures by lobbing explosives over the barriers.[56] The perceived discrepancy between an actor's expectations and its current performance is known as the performance gap and is said to be a major activator of innovation.[57]

The variety of "drivers" that occasionally coalesce to overcome the inertia against innovation can be broadly categorized into those that stem primarily from concerns and dynamics specific to the organization itself ("internal drivers"), and those emerging from changes that occur mainly in the organization's broader environment ("external drivers"). Although useful for conceptual clarification, it should be noted that internal and external drivers are rarely present in a purely dichotomous form and, in practice, most motivational factors represent to some degree a confluence of forces both internal and external to the organization. The following list describes several of the motivational drivers most likely to be relevant for terrorists, presented as they are likely to manifest in the terrorism context.

Internal drivers of innovation

Overcoming countermeasures

Terrorist efforts to coerce their enemies occur in a dynamic context, in which the adversary government and the populace both adapt to minimize and mitigate the damage and resulting political leverage from terrorist attacks. Therefore, terrorists may pursue new modes and methods (usually weapons or tactics) in order to thwart defensive countermeasures.[58] As Hoffman argues, terrorists' "fundamental organizational imperative to act" forces them to respond to this behavior by outside actors.[59] In the area of weapons innovation, this is a manifestation of the commonly described "offense–defense" co-evolutionary dynamic, which some have described as a primary driver in weapons development,[60] and which has occasionally led to "arms races" between terrorists and governments,[61] in what Hoffman elsewhere describes as a "technology treadmill."[62] At the same time, counter-terrorist pressures might induce a terrorist organization to innovate organizationally, for example, when PIRA transitioned from a military-type structure to the more diffuse Active Service units in an (ultimately only partially successful) attempt to minimize the damage wrought by informers, so-called "touts."

Audience desensitization

A particular concern for terrorists seeking to coerce is that populations and the media often become desensitized to violence. As asymmetric purveyors of a technique of violent manipulation, true terrorists[63] concern themselves with the publicity they receive from their attacks.[64] It is widely argued that an increasingly crowded and fragmentary global media environment, in which a given type of terrorist attack can become almost routine, creates pressure for terrorists to expand their operational objectives in what Schmid and de Graff refer to as a "built-in escalation imperative."[65] This is no doubt responsible for some of the observed choreography of terrorist violence in recent times, such as choosing a target where the attack is almost certain to be captured on video (such as the World Trade Center), or in some cases even filming and distributing footage of the operation itself (as has become common in the case of Hizb'allah and the Chechen rebels). Innovating by utilizing a novel weapon or tactic (especially a particularly sophisticated or frightening one) may be one component of such an escalation.

Inter- or intra-organizational competition

In circumstances where multiple political organizations (both violent and non-violent) acting against a common antagonist are vying for sympathy, support and recruits, terrorists might pursue new weapons or tactics in an effort to distinguish themselves from potential or actual competitor groups.[66] Since technological proficiency is celebrated to a large extent in modern globalized cultures, a terrorist organization may believe that adopting advanced weapons before others do will enhance, or at least preserve, its status relative to competitors.[67] Similarly, if a competitor has already adopted a novel, high-profile weapon or tactic, even a well-resourced terrorist organization with strong support might feel compelled to follow suit[68] for fear of losing its current standing in the dissident milieu.[69] At the same time, internal rivalries or factionalization within a terrorist group might induce either incumbent leaders or emerging challengers to push for adopting new weapons, tactics or structures as a means of bolstering their own positions within the organization by appearing to be forward-thinking and daring.[70]

Non-strategic affinities

Strategic concerns related to overcoming countermeasures or forestalling competition are not the only internal drivers of innovation. An important factor that can drive the adoption of new weapons or practices is the presence of a non-strategic affinity for particular technologies or tactics, or a generalized desire to be on the "cutting edge."[71] This orientation can be rooted in the organization's ideological doctrines,[72] but can also emerge from the idiosyncrasies of its leaders or key operational decision makers, whether these reflect personal delusions or ego-driven desires to be regarded as a terrorist "mastermind."[73]

External drivers of innovation

Lowered costs

There is a strong finding in the literature that reduced costs encourage technological adoption. Technological advances in the broader society can make the acquisition of a particular lethal capability easier or cheaper to acquire, thus encouraging the adoption of a new weapon or tactic.[74] For those technologies with economies of scale, greater levels of adoption by other actors may make it more affordable for terrorists to adopt it themselves. For example, the advent of chemical microreactors, which make chemical synthesis simpler, safer and less conspicuous, can decrease the risks and material costs of producing chemical weapons and thus attract a terrorist group towards pursuit of these weapons. In the broader literature, this situation has been described, in an inversion of the familiar aphorism, as "invention being the mother of necessity."[75] Yet, it has been suggested that the effects of emerging technology on innovation behavior are far from automatic – technology "opens doors; it does not force society to walk through them."[76]

Serendipity

In addition to actual technological change, serendipity can play a role – the opportunistic or accidental acquisition of specialized skills or weapons materials can rebalance the cost-benefit equation and prompt a terrorist group to seriously consider adopting a particular weapon or tactic for the first time.[77]

Security pressures and their absence

Similarly, broader environmental factors may influence the resource requirements specific to a given actor. For example, there are indications that when experienced terrorist organizations feel secure,[78] with a safe haven[79] and/or a surfeit of resources,[80] they can investigate and possibly adopt new technologies or practices as a strategic hedge against a future time when circumstances change, and their existing arsenal or operational system,[81] despite being currently effective, becomes unable to fulfill their operational objectives.[82] The latitude to undertake this kind of exploratory weapons development is akin to successful companies engaging in speculative research and development in order to diversify their portfolio in the long term. Somewhat counter-intuitively given the discussion above, increased costs from security force pressure may also encourage innovation.[83] Scholars assert that periods of increased action and success by security forces can stimulate innovation on the part of terrorist groups as they struggle desperately to survive in a pressure cooker of counterterrorist activity.[84] After conducting several case studies, Cragin, *et al.* found some empirical support for both of these viewpoints.[85]

Emulation of other actors

An important element of an actor's innovation calculus is the confidence it has in its ability to actually acquire the technology or practice it is considering adopting. The uncertainty associated with attempting to adopt a new weapon can be reduced considerably if a terrorist learns that the new practice or weapon technologies have been developed or adopted successfully and reliably used by another actor (even if it is a different type of actor).[86] This can in turn lead to a cascade of adoptions among terrorists and imitative diffusion thus forms an integral part of the story of innovation and learning. Indeed, in addition to their being cast as conservative, Hoffman and others have characterized terrorists as tending to be imitative.[87]

It has been argued that the more actors in a social system are alike (*homophily*) in terms of structural or cultural factors, e.g., a common language, sets of norms and so forth, the more likely adoption of an innovation becomes.[88] This pattern could result from increased trust in the initial user's experience or his adequacy as a model. However, this is far from a necessary condition since, in the case of weapons adoption, emulation has occurred across quite disparate terrorist organizations, and even across ideological milieus.[89]

Limited evidence suggests that in the political context, both observing the *practice* of an innovation by another and gaining knowledge of the *outcomes* of a prior adoption provide important signaling mechanisms about the feasibility of adoption.[90]

In almost all social domains, outside actors often attempt to influence, if not indeed drive, the adoption process, and there is no intrinsic reason to suppose that terrorism is any different. *Opinion leaders* are members of the system itself who are able – generally by a combination of reputation and demonstration – to influence other members towards or away from adopting an innovation. They are often situated at the hub of a social network. There is some indication that larger, more successful and better resourced organizations are more likely to serve as models and thus have greater success as opinion leaders, should they so desire.[91] At lower levels within organizations, "invisible colleges" can form among technical experts across different groups,[92] which in the terrorism context might serve as conduits for raising awareness and encouraging adoption of innovation between terrorist "colleagues" in different organizations.

Change agents are actors from outside a defined social domain who seek to encourage adoption of innovations. At the level of states, civilian producers of weapons have been known to fairly frequently "push" new weapons systems onto military organizations.[93] Analogs in the terrorism domain might include arms dealers, other transnational criminal organizations, or state-sponsors that encourage the adoption of certain weapons or practices for their own pecuniary or political purposes. It should be noted that change agents and opinion leaders can encourage adoption not only by demonstrating or declaring a weapon or practice's usefulness, but also the technology or tactic's legitimacy – ideologically or morally – in the armed struggle.

General theorists of technology have described how diffusion often experiences increasing returns to scale,[94] which can encourage feedback loops of more and more adoption (whether self-initiated or through the efforts of change agents and opinion leaders). Even when this is not the case, if one or more actors adopt an innovation for purely idiosyncratic reasons, other potential adopters might misperceive this adoption decision and believe that adoption by others signifies that the innovation is advantageous. If they in turn adopt the innovation, this might spur more bandwagon-style adoptions, possibly leading to a non-linear adoption cascade reminiscent of the proverbial onlookers of the Emperor who has no clothes. In this way, even a weapon or other practice that is sub-optimal with respect to almost all other features can serendipitously become "locked" into widespread usage by terrorists.

Gaining awareness of innovations

Adoption of an innovation can only proceed if a terrorist organization becomes aware of an alternative to its existing weapons or practices. Recognition that another option is available is an inherently subjective determination, implying that what might be regarded as a potential innovation by one terrorist actor might not be by another.[95] Moreover, it also means that this stage of the adoption process can be affected by various biases, including confirmatory bias, where actors might "filter out" ideas that are not concordant with their values, heuristics or existing routines.[96]

The first mechanism by which terrorists might become aware of candidate innovations for adoption is through a self-initiated search after a realization by a terrorist organization's decision makers, as detailed in the previous section, that there is an operational or strategic need that is unmet by existing practices or materiel. For example, there are many cases of terrorists proactively seeking out new weapons, from the PIRA deciding that they required their own mortars, to Prabhakaran's desire to equip the Liberation Tigers of Tamil Eelam (LTTE) with seaborne attack capability through his Sea Tigers fleet of fast attack boats. Although the extant terrorism studies literature does not explicitly detail this process, it can be expected that, just as has occurred in other areas of innovation adoption (such as in military procurement), this internally sparked search for a new weapon or tactic might originate either from lower organizational levels, as field or cell commanders recognize an operational deficit, or as a result of a top-down, semi-institutionalized process whereby leaders explicitly set up an infrastructure to develop new weapons or tactical capabilities.[97] In any event, the prospects for a self-initiated search for new tactics and weapons will undoubtedly be colored by existing organizational or ideological predispositions towards or against innovation in general.[98]

The second avenue, by which knowledge of an innovation might reach a terrorist organization or individual, is through demonstration of its usefulness or potential usefulness in the external environment. This can obviously occur when other terrorist groups employ a new weapon or tactic successfully, but might also occur when an innovation is used by non-terrorist actors such as states, or conceivably when a relevant technology appears in commercial markets.[99]

The third major mode by which a terrorist organization might become aware of a new weapon or tactic is through the conscious efforts of actors external to the organization (either opinion leaders or change agents) seeking to encourage the adoption of a particular weapon or process. Social bias can also come into play here, where a previous adopter of an innovation, having already expended resources on adopting it, will be more likely to extol its virtues of the innovation than to highlight its problems.[100]

The channels through which terrorist groups gain awareness and knowledge of innovations depend to some extent on the mechanism by which the awareness is gained, but can include a variety of means, from direct observation of a collaborator's use of a weapon or tactic, through written descriptions of new processes or technologies in electronic or printed manuals, and even to broadcast media, such as when new weapons or weapons-relevant technologies are discussed in the press or on television, or other terrorist groups produce YouTube videos of their newest techniques. The general innovation literature, based largely on observations of commercial enterprises, suggests that mass-media communications channels are more effective at creating awareness of an innovation, while interpersonal ties are better avenues for actually persuading an entity to adopt an innovation.[101] Although the internet and other modern communications technologies enable even isolated terrorist groups to have access to copious amounts of information,[102] the ubiquity of communications media does not guarantee awareness or interest in any particular weapon.[103] At the same time, there has been little research even in the general innovations diffusion literature on certain key aspects of information flows in this context, such as whether it is more important for a potential adopter to observe first-hand an innovation in operation, versus hearing about it second-hand through a variety of media.

Several factors can act to facilitate or retard the possibility of a terrorist organization becoming aware of a novel weapon or tactic. In the general literature on the spread of innovations, it has been found that networks of actors facilitate the diffusion of knowledge about an innovation to others within their network,[104] and there is no reason not to conclude that a similar dynamic operates within networks of terrorist groups, especially when one considers such examples as the relative alacrity with which new techniques for constructing and utilizing IEDs spread from jihadist insurgents in Iraq to those in Afghanistan in recent years.[105] This is related to the idea that knowledge tends to spread more easily across organizations sharing similar cultures and objectives, i.e., where homophily exists.[106] Conversely, divergent cultures and aims can in general impede the transfer of knowledge,[107] but so can infrastructural or political barriers[108] that hamper information flows in the broader society within which an organization finds itself. Lastly, knowledge of an innovation can be expected to be easier to come by for a terrorist group that functions as a "learning organization," i.e., that appears to possess the "ability to identify knowledge useful to its long-term success and incorporate that knowledge into the operations and future plans of the organization."[109]

With respect to weapons innovation in particular, while there is little direct discussion in the literature, the general history of weapons development and adoption, as a major type of innovation, suggests that features inherent to the weapon itself might also facilitate awareness thereof. The most obvious examples would be weapons whose effects are unique and whose advantages over existing methods of combat are readily observable and immediate, together with weapons associated with the latest scientific and technological advances, in other words are seen as being on the "cutting edge."

Information about a new weapon, tactic or organizational process is regarded as being of little value without the following step, that of interpretation of this information.[110] Interpretation assigns meaning to the information about the potential innovation, and is described by Jackson *et al.* as consisting of making judgments about the value of the group's current activities, its possible future activities and the continued relevance of older knowledge and practices.[111] Important factors affecting a group's ability to effectively interpret new information include an existing knowledge base through which to make sense of the information, sufficient time and opportunity to properly consider the information, as well as the breadth of distribution of information within the group, the extent to which members share common frames of reference and the depth and feedback speed of the media through which the information is transmitted to and within the group.[112]

Numerous factors influence the ability of terrorists to develop the requisite knowledge to become aware of a new technology, much less recognize it as an opportunity for innovation. The channels by which information about technologies spread and the ubiquity of the technology within the terrorist's network and larger environment are the main drivers of a terrorist's exposure to the new technology. However, it is characteristics internal to the group that translate this exposure to awareness and recognition as an opportunity for acquisition. It should be noted that factors facilitating awareness do not imply either a positive decision to adopt or success in subsequent adoption of an innovation; for example, an innovation might permeate a terrorist group's network – and thus garner its attention – despite being wholly unsuited for the group.

The decision process

Even with the presence of the necessary preconditions and positive drivers towards innovation, terrorists are still faced with determining whether attempting to adopt the innovation is worthwhile in light of all the other ideological, instrumental and environmental demands on the group's attention and resources. Most importantly, this implies that the adoption of new tactics or activities is the result of a conscious decision that devoting the time and effort to incorporate a new technology or practice that may or may not bear fruit is preferable to expending those same resources on maintaining or expanding existing capabilities.[113] Further, the choice may not simply be a binary one of adoption versus rejection – as Cragin *et al.* point out, the choice often includes the option of adopting now or postponing adoption (i.e., deferring) to a later date.[114]

As such, at least some consideration must be put into the nature of the decision-making process itself. At least at an operational level, terrorists are widely regarded as employing some sort of decision calculus, albeit on occasion one appearing to bear little resemblance to formal decision-theoretic analysis.[115] Rational actor models, which are common in all of the literatures under review, emphasize the instrumental aspects of the potential innovation, e.g., its strategic implications or its financial costs. Some organizations may be less influenced by these instrumental concerns for various reasons, as discussed below, but none can afford to ignore them indefinitely unless survival is not a priority.

Once again focusing on the example of weapons innovations – because the terrorism literature contains relatively more discussion of this area than other types of innovation – we see that, in line with this pragmatic perspective, the literature identifies three primary elements with respect to terrorist weapons selection: (1) desired attack outcomes (i.e., the terrorists' operational goals and preferences, which are determined by a multitude of variables, among which ideology features prominently);[116] (2) the nature of the weapon or weapon technology itself;[117] and (3) the characteristics of the intended target.[118] While the relationship between these three elements is by no means unidirectional, it should be theoretically possible, at least, for a terrorist group to select the weapon that best matches its operational goals with respect to a particular target or set of targets, including desired levels of casualties, physical damage, publicity or psychological impact.[119] If their current arsenal is perceived to be inadequate to achieve their attack goals against a given target, or is at least judged to be inefficient in doing so, terrorists may initiate a search for new weapons or decide to pursue a weapon of which they are already aware.[120]

If other terrorist innovation decisions exhibit similar considerations, such decisions appear to reflect the same pattern of military problem solving observed with respect to states and other political collectives,[121] as well as the notion of performance gaps as a major driver of technological innovation more broadly.[122] Since most terrorist groups will seek to redress their perceived performance gap as efficiently as possible (in other words, maximize the "bang for their buck"), the decision to adopt new technologies or practices can be expected to take on many of the characteristics of the familiar rational choice paradigm, including the weighing of costs and benefits.[123]

While the general decision process is likely to remain more or less the same, there will obviously be some variation in the influences on adoption decisions depending on the type of innovation being considered. Looking at innovations in terrorist targeting, for example, a terrorist group's ideology generally limits the space of possible targets by determining what is a legitimate focus of attack.[124] Yet, within this set of permissible targets, there is usually plenty of room for variation and innovation. Many of the same drivers that encourage innovation in other operational areas influence the selection of new types of targets, including defensive countermeasures, intergroup competition and media or public desensitization to attacks on existing target types. For instance, where innovating in weapons or attack methods is not an option in the face of target hardening measures, terrorists might shift to a completely different type of target, sometimes referred to as target displacement. However, especially when it comes to targets, one should also remain cognizant of non-rational factors, such as antipathy directed towards a particular target by a terrorist leader who believes it to be associated with or represent those responsible for some perceived personal slight or group degradation. At the same time, awareness of a potential candidate for a new target is often correlated with the degree of prominence and publicity that the target receives.[125] For example, extensive media coverage of a new facility can both alert terrorists to the target's vulnerability to attack and increase its symbolic stature and thus its attractiveness to terrorists seeking to maximize the public profile of their attacks.

One aspect requiring emphasis is that the impetus to adopt new weapons is ultimately subjective and can emerge wherever a need is perceived,[126] thus highlighting the role played by cognitive and affect-based biases and other framing effects.[127] In particular, some violent non-state actors may be unconcerned with political considerations, such as maintaining the support of terrestrial constituencies, and instead be oriented purely towards murder and mayhem as ends in themselves. While not strictly speaking terrorists, the decisions of such (relatively rare) nihilists might still be rational within their frame of reference;[128] weapons and tactics adoption decisions might be as simple as the most destructive means that the group can acquire within its resource constraints and perceived window for action. Yet, the behavior of both these actors and terrorists proper can also be affected by a variety of cognitive, emotional, cultural and other "expressive" factors[129] that operate either consciously or unconsciously, and are often responsible for different tactical decisions of terrorist organizations within the same ideological milieu.[130] This implies that the spread of an innovation, even under favorable structural conditions in a network, is hardly deterministic.

For example, the terrorists' own perceptions of their prior attempts at innovation are important in their future decisions, but not necessarily in a linear fashion – some terrorists might become more leery of attempting to innovate in the wake of a prior failure to adopt an innovation, or if the innovation, albeit successfully acquired, failed to provide the expected benefits. At the same time, other terrorist entities, perhaps exhibiting a distorted sense of their own abilities or an extreme case of cognitive dissonance, might be undaunted by multiple failures, and enthusiastically pursue future innovative projects.

Furthermore, in the general innovation literature, there is the hypothesis that the early stages in the life cycle of an innovation (i.e., with few prior adopters) predominantly reflect rational cost-benefit adoption decisions, while such decisions in later stages are determined more by the faddish imitative dynamics described above, even when this represents adoption of inefficient innovations.[131] In situations where adoption decisions are driven by their own momentum, DiMaggio and Powell believe that the primary element of the decision becomes which actor to emulate rather than which innovation to adopt.[132]

It is necessary to note that it is possible in certain circumstances for innovation to occur without any discernible motivating factors. One might find this occurring in overly bureaucratic organizations where innovation has become so institutionalized or routine, that it occurs without a conscious decision. While probably rare in practice, it is not unheard of in the terrorist context. For example, in some cases the IRA altered the construction of explosive devices with no clear reason or obvious improvement in performance.[133]

One important issue to explore is whether the decision with respect to adopting a potential innovation varies according to the decision-making structures within the terrorist group itself. How might the decision calculus differ, say, between a cohesive, hierarchical organization like Hizb'allah or the PIRA of the early 1970s and a more diffuse network like the broader al-Qaeda movement? While structure is believed to play an important role,[134] there is no simple answer to this question – each type of organizational structure affords both advantages and disadvantages to a would-be innovator and much depends on the specific attributes of the terrorist network or hierarchy.[135]

One advantage of having a centralized leadership within a strict hierarchy is that, providing the leadership is aware of an innovation and amenable to it, naysayers in the remaining parts of the organization can be overruled; in many cases a leader can act with single purpose to drive the organizational adoption of a new weapon or tactic.[136] Terrorist leaders from Shoko Asahara of Aum Shinrikyo to James Ellison of the Covenant, Sword and Arm of the Lord have taken their organizations along in adopting non-traditional weapons and tactics. Perhaps this is why Dolnik opines that "radical innovation is virtually always a product of a top-down approach."[137] Moreover, terrorist outfits modeled hierarchically after military organizations often tend to specialize functionally, in other words "bureaucratize," making it more likely that a specific organ will emerge that focuses on research and development, thus generating a constituency for innovation.

These very dynamics, however, can present a double-edged sword for innovation. For if a leader is risk averse or wedded to the status quo, there is little scope for innovation to develop organically in a rigid organizational structure. In addition, the very existence of bureaucratic constituencies within a hierarchical organization runs the risk of these entities mobilizing to secure their position and purposefully disregarding a new innovation that does not suit their interests. After all, the varied strata in hierarchical organizations, especially for those groups somewhat isolated from the broader society, can provide multiple opportunities for agenda-setters at all levels to put up barriers to the diffusion of knowledge about innovations to key decision makers. And even the most centralized leadership cannot decide to adopt an innovation that it does not know about, thus allowing bureaucratic organs to stifle innovation.[138] In hierarchical terrorist organizations, therefore, the decision to innovate is very much dependent on the attitudes towards innovation espoused by leaders or other key gatekeepers of the hierarchy, either in general or with respect to a specific weapon or tactic.

Decentralized terrorist networks, while often smaller and less well-resourced than their hierarchical cousins, do possess some strength when it comes to recognizing and pursuing innovations. First, these networks may be more likely to rapidly adapt to environmental changes and respond to technological and tactical opportunities.[139] In essence, individual cells in a decentralized network may be more attuned to local conditions and thus more likely to recognize an innovation opportunity. Second, decentralized networks are generally more likely to have the freedom to experiment – for every cell with a leader who disdains innovation, there may be another with a leader that embraces it. Those innovations that prove successful can then percolate to other elements of the network, especially given the internet and other modern communications tools that allow disparate groups to maintain contact and share tactics. This allows for a more organic approach to innovation with little risk to the resources or the prospects of the overall organization.

Terrorists in more amorphous groups do face several important constraints on their ability to embrace new weapons and tactics. Such organizations, being generally smaller than the well-defined centralized hierarchies, may simply have fewer individuals available overall to devote time and energy to becoming aware of and considering potential innovations. Size by itself, however, is not necessarily a limiting factor; rather, it is the breadth of its talent pool – those with sufficient acumen to recognize new innovations and navigate any obstacles to raising the possibility of adoption – that is key, so even a relatively small, decentralized organization might possess the type of members who could facilitate innovation. A more important weakness of decentralized organizations is that disparate members who interact sporadically and remotely may find it difficult "to engage in a deliberative concerted decision-making process,"[140] and thus may not be collectively cognizant of the costs or benefits of an innovation for the organization as a whole. While modern communications technologies might make decentralized awareness and decision making feasible, it does not necessarily make it as seamless as operations under spatial proximity, which ensure direct interaction and influence between actors.[141]

Thus, at least in terms of awareness of innovation and the adoption decision, the effects of organizational structure are not clear-cut, but instead highly contextual. Some terrorist groups have borrowed a leaf from the commercial high-technology sector's attempts to reconcile this duality by creating "skunkworks," units that are tasked specifically with innovation and insulated to some degree from the broader organizational structure. Although relatively rare in the annals of terrorism, groups such as the PIRA, Hizb'allah and the LTTE have adopted this approach.

Overall, it appears that most terrorists are likely to engage in something like a strategic decision whether or not to adopt an innovation, and will behave rationally, at least within their frame of reference with respect to perceptions and goals. Even with this basic rationality in adoption decisions, however, there is still much scope for decisions, whether in favor of or against adopting an innovation, to depart substantially from the "objectively" optimal choice for the group.

Innovation decision modifiers

In many cases actors will reject innovations in spite of clear strategic, tactical, political, ideological or psychological benefits and low apparent costs. This implies that other factors condition the decision to adopt innovations, factors that scholars have identified as being rooted in the inherently social and psychological nature of the adoption process.[142] Although not acting as drivers of adoption themselves, there may thus be several features of terrorist organizations that can retard, facilitate or otherwise shape the decision to adopt an innovation, on occasion even overriding the core rational cost-benefit calculus of weapons adoption described above. Most of the antecedent conditions discussed above can function in this capacity, in that the absence of these conditions can retard the prospects for innovation and thus dissuade less risk tolerant actors from pursuing that innovation. More active barriers to and facilitators of innovation that are likely to operate during the decision process itself are introduced below.

Opposition by guardians of the status quo

The conservative tendencies and resulting organizational inertia discussed earlier combine to produce a reticence to adopt new tactics or weapons, perhaps especially among self-appointed guardians of the *status quo* whose status within the organization might diminish if extant practices are displaced by novel means of combat.[143] In large, hierarchical organizations, this is typified by bureaucratic coalitions that emerge to protect the status quo. A clear example in the terrorism context is the opposition expressed by some of the "old guard" officers commanding (OC's) in the Provisional IRA during the 1970s to the use of new electronic timers and detonators in place of previous chemical safety fuses.[144]

Internal rivalries

Another way in which elements of an organization can influence the adoption of an innovation is when internal rivalries between organizational factions or individuals within organizations push for or against the adoption of innovations irrespective of their actual relative advantage to the adopting organization. Possible alternative reasons for opposing or embracing an innovation might include such objectives as strengthening the faction's power base, mounting a challenge to another faction, gaining favor with the leadership, or any of a variety of other internal power plays.

Near-term demands

Even though an organization might regard future innovation as essential, it still has mouths to feed today,[145] whether these are current shareholders in a company, or, in the case of terrorism, constituencies that require constant demonstration of attack prowess. In such cases, current exigencies may not allow for the diversion of resources towards development or adoption of an innovation that might bestow benefits only at some future date.

Nature of the weapon, tactic or other practice

The physical, logistical and other characteristics of the innovation under consideration can also be expected to influence the probability of adoption by terrorists, based on several findings from the broader innovation diffusion literature. It has been argued that, all else being equal, innovations perceived to have greater compatibility, trialability and observability, and less complexity than others are more likely to be adopted.[146] In the specific case of weapons innovation, Palfy describes several other aspects as important in this regard, including life-span, stability, environmental persistence and methods of dispersal.[147] However, he does not state how these factors might facilitate or retard adoption.

Just as in the case of preconditions, none of the obstacles mentioned here is necessarily insurmountable – after all, the rapid growth in production, technology and social connections over the past two centuries bear witness to the successful generation and diffusion of countless innovations, including among terrorists. The strategies adopted by organizations and individuals to overcome innovation barriers include: purposive organizational changes to facilitate innovation; the creation of separate, independent "skunkworks" organizational entities to avoid bureaucratic inertia and vested interests;[148] and the use of specialized brokers to bridge various domains and seek out and nurture new technologies and practices.[149]

Conclusion

This chapter has examined the underlying and motivating forces surrounding terrorist innovation through the lens of the broader innovation literature in the expectation that these more general insights would also apply to some degree in the context of modern terrorism. Based on initial indications, admittedly anecdotal, many of the basic dynamics surrounding innovation in business, the public sector and state militaries appear to find analogs in the world of terrorist organizations. Primarily, initial indications seem to confirm that the central logic surrounding terrorist innovation is the same as in other contexts, namely that innovations in weapons, tactics and organizational processes are pursued when they exhibit relative advantage over existing methods, provided their acquisition would be minimally feasible and compatible with the organization's central ethos. Further, this central logic must incorporate the social, ideological, organizational and decision-making biases of the organization, which might cause the ultimate adoption decision to deviate from a purely objective utility optimalization procedure. Beyond these essential results, none of the motivational factors discussed above have been unequivocally shown to be either necessary or sufficient for the pursuit of innovation and thus none can be construed as strictly causal in nature.

Yet, terrorists are in many ways very different from such actors as firms or state militaries. After all, terrorists are often far more diffuse and far less bureaucratic than these entities. Also, their goals are generally more complex than maximizing profit or securing victory on the battlefield and they generally experience far less direct competition than occurs in markets or among states. Most importantly, they are clandestine actors under extreme environmental pressures, not least under constant threats to their physical survival. We must therefore be cautious in directly applying the results from other contexts to the domain of terrorism. What is thus required is thorough testing of the hypotheses offered here, to determine whether they can stand up empirically, beyond mere anecdote, against the record of terrorist innovation.

It must be noted that this discussion has focused only on the initial stages of the innovation process; it has not broached the question of what leads to a successful innovation adoption, or how the ultimate utility for the adopter is to be determined. While many of the same factors that have been discussed here can influence these latter stages, additional factors must be taken into account, such as the requirement to transfer tacit knowledge and *metis*,[150] or the impact of networks[151] on adoption. Moreover, the same factors might not have similar effects at all stages of the adoption process; some factors that act to facilitate the decision to adopt an innovation might act to retard the successful adoption of that innovation.

As long as these aspects are borne in mind, the present discussion can serve as a theoretical basis against which the empirical experience of terrorist innovation can be tested and compared. It can therefore be used to guide our thinking and deepen our understanding of the decision by terrorists when faced with the opportunity to embrace new weapons, tactics or organizational processes.

Notes

1 Portions of the first two sections of this chapter are drawn from an earlier paper, presented at the Naval Postgraduate School in August 2010. See Gary Ackerman, "Understanding Terrorist Innovation through the Broader Innovation Context," in Maria J. Rasmussen and Mohamed Hafez (eds), *Terrorist Innovations in Weapons of Mass Effect: Preconditions, Causes, and Predictive Indicators*, The Defense Threat Reduction Agency, Advanced Systems and Concepts Office, Report Number ASCO 2010–019 (October 2010).

2 The most in-depth studies that have been undertaken thus far are Rasmussen and Hafez, op. cit.; Adam Dolnik, *Understanding Terrorism Innovation: Technology, Tactics and Global Trends* (New York: Routledge, 2007) and Brian A. Jackson, John C. Baker, Kim Cragin, John Parachini, Horacio R. Trujillo and Peter Chalk, *Aptitude for Destruction: Vol. 1. Organizational Learning in Terrorist Groups and Its Implications for Combating Terrorism; Vol. 2. Case Studies of Organizational Learning in Five Terrorist Groups* (Santa Monica: RAND, 2007). Crenshaw, as usual, was way ahead in this regard with a conference paper in 2000,

but this remained unpublished until inclusion as Martha Crenshaw, "Innovation: Decision Points in the Trajectory of Terrorism," in Rasmaussen and Hafez, op. cit.

3 See Dolnik, op. cit., pp. 173–175. Among his more equivocal results were the impact of intergroup competition (pp. 162–163) and the precise relationship between resources and innovation (p. 164).

4 The only attempt to explicitly relate broader insights of innovation to the terrorism realm of which I am aware has been that of Brian Jackson and even then his discussion was limited to the domains of technology adoption and private sector behavior. Brian Jackson, "Technology Acquisition by Terrorist Groups: Threat Assessment Informed by Lessons from Private Sector Technology Adoption," *Studies in Conflict and Terrorism*, 24 (2001).

5 Although most of the literature focuses on organizations rather than individuals, many of the theories and findings might also apply to so-called "lone wolves."

6 This chapter will focus on the terrorist organization as the unit of analysis, rather than the particular innovation itself. While lending itself more easily to empirical testing in the context of terrorism (where there is usually a limited universe of organizational actors), this approach does not engage the diffusion of an innovation throughout a population of potential consumers, a topic that constitutes a large proportion of the theoretical literature on innovation in business and public administration. Although touching on some of these dynamics, this paper will therefore de-emphasize the "epidemiological" approaches and theories that have received much attention from scholars. Chief among these is the ubiquity of the so-called "S-curve" model of innovation diffusion and the numerous mathematical functions developed to describe this. For more information on these topics, see Vijay Mahajan and Robert Peterson, *Models for Innovation Diffusion* (Beverly Hills: Sage Publications, 2004).

7 Martha Crenshaw, "Innovation: Decision Points in the Trajectory of Terrorism," paper presented at the conference *Trajectories of Terrorist Violence in Europe*, Harvard University (March 2001), p. 3. This is very similar to the definition favored by Ronald Kostoff, "Systematic Acceleration of Radical Discovery and Innovation in Science and Technology," *Technological Forecasting and Social Change*, 73, no. 8, p. 924, where innovation "reflects the metamorphosis from present practice to some new, hopefully 'better' practice."

8 Everett M. Rogers, *Diffusion of Innovations*, 5th edn (New York: Free Press, 2003), p. xx. He reiterates this by stating explicitly that "If an idea seems new to the individual, it is an innovation" (ibid., p. 12).

9 Ibid., p. xx.

10 Brian W. Arthur, *The Nature of Technology: What it is and How it Evolves* (New York: Free Press, 2009), pp. 73–74.

11 Ibid., p. 73.

12 Joel Mokyr, *The Gifts of Athena: Historical Origins of the Knowledge Economy* (Princeton: Princeton University Press, 2002), p. 4. Mokyr denotes propositional knowledge by Ω and refers to it as *episteme*, while prescriptive knowledge is denoted by λ and referred to as *techne*. One of Mokyr's primary contentions is that increases in useful, prescriptive knowledge of how to do things (λ) is predicated in the long run on an expansion of Ω-knowledge, which includes scientific discovery (ibid., p. 5). Also see Martin Heidegger for a more obtuse expression of a similar idea (Martin Heidegger [William Lovitt, trans.], *The Question Concerning Technology and Other Essays* (New York: Harper & Row, 1977), p. 13).

13 Donald MacKenzie, *Knowing Machines: Essays on Technical Change* (Cambridge: MIT Press, 1998), p. 11. MacKenzie characterizes tacit knowledge as "informal 'know-how' ... unverbalized and perhaps unverbalizable" (ibid.).

14 Jackson, op. cit., pp. 187–188.
15 Michael Kenney, *From Pablo to Osama: Trafficking and Terrorist Networks, Government Bureaucracies, and Competitive Adaptation* (University Park, PA: Penn State University Press, 2008), p. 4; Michael Kenney, "'Dumb' Yet Deadly: Local Knowledge and Poor Tradecraft among Islamist Militants in Britain and Spain," *Studies in Conflict and Terrorism*, 33, no. 10 (2010), p. 912.
16 Rogers, op. cit., p. 181.
17 Joel Mokyr, *The Lever of Riches: Technological Creativity and Economic Progress* (Oxford: Oxford University Press, 1990), p. 10.
18 Chesbrough refers to this in the business context as "invention implanted and taken to market" (Henry Chesbrough, *Open Innovation: The New Imperative for Creating and Profiting from Technology* (Boston: Harvard Business School Press, 2003), p. ix). See also Mokyr (1990), op. cit., pp. 10–11. At the same time, the majority of scholars who have looked into the matter with some scrutiny acknowledge that this distinction is far from absolute. See Rogers, op. cit., p. 181; Mokyr (1990), op. cit., p. 190.
19 Arthur, op. cit., pp. 203–204.
20 Andrew Hargadon, *How Breakthroughs Happen: The Surprising Truth about How Companies Innovate* (Boston: Harvard Business School Publishing Corporation, 2003), p. viii. Hargadon cites the well-known historian of technology, Abbot Payton Usher, who wrote in 1929 that "Invention finds its distinctive feature in the constructive assimilation of pre-existing elements into new syntheses, new patterns, or new configurations of behavior" (cited in ibid., p. 24).
21 Ibid., p. viii and Chesbrough, op. cit., p. 60. Indeed, a National Academies of Science panel on forecasting technologies refers to innovations stemming from "crossover advances" to hold the greatest potential for yielding surprising consequences. (Committee on Forecasting Future Disruptive Technologies, *Persistent Forecasting of Disruptive Technologies* (Washington, D.C.: National Research Council, 2010), p. 45).
22 Arthur, op. cit., p. 3.
23 Ibid., pp. 21–23.
24 Francisco J. Varela, Humberto R Maturana and R. Uribe, "Autopoiesis: The Organization of Living Systems, Its Characterization and a Model," *Biosystems*, 5 (1974), pp. 187–196.
25 Mokyr (1990, op. cit., pp. 163–165), however, cautions against viewing technology and innovation as purely path dependent.
26 David Landes, *The Unbound Prometheus* (Cambridge: Cambridge University Press, 2003), p. 2.; Mokyr (2002), op. cit., p. xi and Arthur, op. cit., p. 134.
27 Landes provides another historical example, in that the steam engine only became possible when superior methods of metalworking enabled the production of more accurate cylinders (Landes, op. cit., p. 2).
28 This is similar to, but differs in some respects from the "innovation-decision process" model put forward by Rogers (op. cit., p. 37), which posits an agenda-setting phase, whereby decision makers identify and prioritize their needs and initiate search efforts for means by which to address these needs (p. 422), followed by knowledge, persuasion, decision, implementation and confirmation phases, with each with each stage representing a potential point at which the innovation might be rejected (p. 177).
29 Stuart Kaufman cited in Steven Johnson, *Where Good Ideas Come From: The Natural History of Innovation* (New York: Riverhead Books, 2010), p. xx.
30 Clayton M. Christensen, Erik A. Roth and Scott D. Anthony, *Seeing What's Next* (Boston: Harvard Business School Press, 2004), p. 21. FARC was thus able to innovate when it came to mortars and rockets, since it could easily afford to pay the IRA for instructional assistance (Kim Cragin, Peter Chalk, Sara A. Daly

and Brian A. Jackson, *Sharing the Dragon's Teeth: Terrorist Groups and the Exchange of New Technologies* (Santa Monica, CA: RAND Corporation, 2007), p. 89; Roman Ortiz, "Innovation and Strategy of the Revolutionary Armed Forces of Colombia," in James J. F. Forest, ed., *Teaching Terror: Strategic and Tactical Learning in the Terrorist World* (Lanham, MD: Rowman & Littlefield Publishing Group, 2006), p. 215).

31 Mokyr (1990), op. cit., pp. 291–292.

32 Arthur, op. cit., p. 139.

33 Dolnik, based on his case studies, concludes that the relationship between resources and terrorist innovation is relevant but is not necessarily unconditional (op. cit., p. 164).

34 Bruce Hoffman, *The Modern Terrorist Mindset: Tactics, Targets, and Technologies*, Center for the Study of Terrorism and Political Violence (Scotland: St. Andrews University, 1997), pp. 171–172; Jackson, op. cit., p. 193; Mokyr (2002), op. cit., p. 249. In certain cases, it can therefore even be advantageous for a terrorist group to have an ill-defined or amorphous ideology. Ortiz, for example, identifies FARC's capacity for operational innovation with the organization's ideological flexibility (in Ortiz, op. cit., p. 209).

35 Mokyr (2002), op. cit., p. 241.

36 Jeremy Black describes this dynamic in the interstate context ("Determinisms and Other Issues," *The Journal of Military History*, 68, no. 4 (2004), p. 1223). One of the most famous examples is that of Islam, which in its initial phases embraced a host of new ideas from Greek and Roman civilization and proceeded to expand upon them. Though a sentiment present to some extent since the beginning of Islamic history, especially after the theologian al-Ghazali (1058–1111) determined that science and technology posed a threat to the preservation of the faith, the concept of *bidaa* (innovation) took on heretical connotations, and resulted in a subsequent resistance to and tight control over all new technologies, with arguably deleterious consequences for the military and political power of the Islamic world. For a detailed discussion, see Bernard Lewis, *The Muslim Discovery of Europe* (New York: W. W. Norton, 2001), pp. 224–226; Landes, op. cit., pp. 28–29; Ernest Volkman, *Science Goes to War: The Search for the Ultimate Weapon, from Greek Fire to Star Wars* (New York: John Wiley & Sons, 2002), p. 60.

37 Cragin *et al.*, op. cit., pp. xv, 44.

38 The decision by the traditionally patriarchal Islamist terrorists to allow female suicide bombers is one example of how practical exigencies can lead to a rethinking of long-held beliefs. Jenkins depicts a contemporary "democratization of extremism" where it is possible to "in a sense shop for belief systems," which suggests that even an ideological proscription against a particular weapon might not be absolute if certain adherents desire the weapon badly enough (Brian Michael Jenkins, "The New Age of Terrorism," in Russell D. Howard and James J. F. Forest, eds, *Weapons of Mass Destruction and Terrorism* (Dubuque, IA: Dushkin/McGraw-Hill, 2007), p. 29.

39 Christensen *et al.*, op. cit., pp. 20–21; 290; Landes, op. cit., p. 19; Mokyr (1990), op. cit., 180–181.

40 Geoffrey Parker, *The Military Revolution* (New York: Cambridge University Press, 2007), p. 137; John Keegan, *A History of Warfare* (New York: Vintage Books, 1993), pp. 44–45.

41 Arthur, op. cit., p. 108; Mokyr (2002), op. cit., p. 157. Researchers have also asserted that, in general, younger adults tend to be more risk tolerant than older adults (Committee on Forecasting Future Disruptive Technologies, op. cit., pp. 44–45).

42 See the discussion below in the section on drivers of innovation.

43 Jackson, op. cit., p. 203; Trujillo and Jackson, "Organizational Learning and Terrorist Groups," in Forest, op .cit., p. 61; Dolnik, op. cit., p. 167.

44 Ibid., p. 168; Indeed, Trujillo and Jackson maintain that "Risk tolerance is a function of the group's ideology and the external environment, as well as other variables" (op. cit., p. 61) only some of which are likely to be relatively stable over time.

45 Stephen Maurer, "Introduction: Worrying about WMD Terrorism," in Stephen Maurer, ed., *WMD Terrorism: Science and Policy Choices* (Cambridge: MIT Press, 2009), p. 7.

46 Jackson, op. cit., p. 189.

47 See, *inter alia*, Bruce Hoffman, *Terrorist Targeting: Tactics, Trends, and Potentialities* (Santa Monica, CA: RAND, 1992), p. 15; G. Woo, *Understanding Terrorism Risk* (IDS-Emergency Management, n.d.); R. Clutterbuck, "Trends in Terrorist Weaponry," in P. Wilkinson (ed.), *Technology and Terrorism* (London: Frank Cass, 1993), pp. 130–139; Brian Jenkins, "Defense against Terrorism," *Political Science Quarterly 101, Reflections on Providing for "The Common Good"* (1986), pp. 777–778; and Baruch Fischhoff, Roxana M. Gonzalez, Deborah A. Small and Jennifer S. Lerner, "Judged Terror Risk and Proximity to the World Trade Center," *Journal of Risk and Uncertainty*, 26, nos 2/3 (2003), p. 138. Among the few researchers to have attempted a concerted empirical investigation of this issue, Dolnik finds that small arms and explosives have dominated terrorist tactical repertoires (op. cit., pp. 26, 36) and asserts that, tactically speaking, "when one surveys the last 50 years of terrorist operations case by case, very few incidents strike the observer as creative *in any way*" (p. 56, emphasis in original).

48 Palfy describes a "relationship between simplicity and success ... because terrorist organizations, similar to military units in combat, become vulnerable to factors outside their sphere of control as soon as the mission enters its executions phase." (Arpad Palfy, "Weapon System Selection and Mass-Casualty Terrorism," *Terrorism and Political Violence*, 15 (2003), p. 87).

49 Kenney (2008), op. cit., p. 104.

50 Cragin *et al.*, op. cit., p. 13; Charles Ferguson, "Influence diagram analysis of nuclear and radiological terrorism," in Magnus Ranstorp and Magnus Normark (eds), *Unconventional Weapons and International Terrorism Challenges and New Approaches* (New York: Routledge, 2009), pp. 128–129; Stephen M. Maurer and George Rutherford, "The New Bioweapons: Infectious and Engineered Diseases," in Maurer (ed.), op. cit., p. 136. Extensive programs of testing a new weapon can reduce these uncertainties, although not completely eliminate all of them. Yet the testing process can consume resources and is itself often risky (Stephen Maurer, "Technologies of Evil: Chemical, Biological, Radiological, and Nuclear Weapons," in Maurer (ed.), op. cit., pp. 75–76).

51 Maurer, "Technologies of Evil," op. cit., pp. 75–76. The business literature also describes cases where despite desperately wanting to innovate, organizations are hamstrung by current demands on their resources that cannot be diverted from current constituencies in favor of potential future benefits (Eric von Hippel, Stefan Thomke and Mary Sonnack, "Creating Breakthroughs at 3M," in *Harvard Business Review on Innovation* (Boston: Harvard Business School Publishing Corporation, 2009), p. 33). In the terrorism context, the "mouths that need feeding today" might take the form of sympathizers among a local populace who demand consistent exhibitions of operational capabilities in order to maintain their financial support of the terrorists.

52 Mullins, *Sourcebook on Domestic and International Terrorism: An Analysis of Issues, Organizations, Tactics and Responses*, p. 307. Cited in Jackson, op. cit., who

maintains that up to 30 percent of all would-be bomb-makers blow themselves up. See also Maurer ("Technologies of Evil," op. cit., p. 62), who cites Fay, Morrissey and Smyth's data (1998), which suggests that 47 percent of all Provisional IRA members killed in Northern Ireland between 1970 and 1998 died as a result of bomb or firearm accidents.

53 Brian A. Jackson and David R. Frelinger, "Rifling through the Terrorists' Arsenal: Exploring Groups' Weapon Choices and Technology Strategies," *Studies in Conflict and Terrorism*, 31 (2008), pp. 15–16; Jenkins (1986), op. cit., pp. 777–778; Dolnik, op. cit., pp. 26, 36, 56; Hoffman (1992), op. cit., p. 15; and Fischhoff *et al.*, op. cit., p. 138; Maurer, "Technologies of Evil," op. cit., p. 47. An example of this conservatism is the encouragement in online jihadist forums to scale down grandiose attack plans to ones that will have more limited effects, but greater chances of success. One such posting noted that a hand grenade that explodes in one of New York's streets "is better than a nuclear bomb capable of exploding half of New York that does not explode!" (Anne Stenersen, "Al-Qaeda's thinking on CBRN," in Ranstorp and Normark, op. cit., p. 59).

54 There are many cases, of course, where terrorists intentionally circumscribe their violence and operate below their maximal attack capacity. Proffered reasons include self-imposed moral constraints on indiscriminate or large-scale killing, a concern about the sensibilities of less violent constituencies and instances where "primitive" methods are used specifically for the visceral shock value they might have on "civilized" audiences (Jenkins (2007), p. 24; also Dolnik, op. cit., pp. 23–24).

55 Arthur, op. cit., pp. 108, 154.

56 See, for example, A. R. Oppenheimer, *IRA: The Bombs and the Bullets, A History of Deadly Ingenuity* (Dublin: Irish Academic Press, 2008), p. xx.

57 Rogers, op. cit., p. 422; Eric Abrahamson, "Managerial Fads and Fashions: The Diffusion and Rejection of Innovations," *Academy of Management Review*, 16, no. 3 (July 1991), p. 592.

58 Hoffman (1997), op. cit., p. 15; Jenkins (1986), op. cit., pp. 777–778; Dolnik, op. cit., pp. 152 and 174; Kenney (2008), op. cit., p. 162.

59 Hoffman (1997), op. cit., p. 15.

60 The indefatigable Harry Turney-High goes so far as to argue that "the offense thinks up new weapons or improves the old ones, so that the defense's genius must think up new defense or be crushed out of existence. There is nothing new nor old in this. The entire history and prehistory of weapons is summarized in this cycle" (Harry Turney-High, *Primitive War: Its Practice and Concepts* (Columbia: University of South Carolina, 1949), p. 7).

61 Examples abound, including the technical "arms race" in explosive triggers between the PIRA and British government forces (Jackson *et al.*, op. cit., p. 99), the similar dynamic with respect to IEDs in the Iraqi insurgency (James Forest, in Forest (ed.), op. cit., pp. 21–22); and perhaps even the adoption by al Qaeda of box cutters on 9/11 as a means of circumventing the introduction, since 1973, of metal detectors into American airports.

62 Bruce Hoffman, *Inside Terrorism* (New York: Columbia University Press, 2006), p. 252.

63 As distinguished from other violent non-state actors, such as guerillas or idiosyncratic mass murderers.

64 Daniel Gressang IV, "Audience and Message: Assessing Terrorist WMD Potential," *Terrorism and Political Violence*, 13, no. 3 (Fall 2001), pp. 88–89.

65 Alex P. Schmid and Janny de Graaf, *Violence as Communication: Insurgent Terrorism and the Western News Media* (Beverly Hills, CA: Sage Publications, 1982), p. 172; see also Hoffmann (1997), op. cit., p. 4; Jackson, op. cit., p. 185.

66 From a set of four case studies, Dolnik was unable to confirm competition as a source for innovation, but found this factor causally suggestive (op. cit., pp. 162–163). Bloom identifies competition amongst terrorist groups as a major driver for adopting new tactics (in her study, suicide bombing) in a process she terms "outbidding" (Mia Bloom, *Dying to Kill: The Allure of Suicide Terror* (New York: Columbia University Press, 2005), pp. 162–163). Sawyer expanded the concept of competition that drives this outbidding and found that intra-constituent competitive forces were a major factor determining the level of violence Irish Republican paramilitaries produced in Northern Ireland (*Competition in the Market for Political Violence*, Unpublished Doctoral Dissertation, Georgetown University, 2011).

67 Cragin *et al.*, op. cit., p. 45. In the commercial sphere, Hagel *et al.* describe how "shaping companies" embark on bold acts in order to signal their position in a market (John Hagel, John Brown and Lang Davison, "Shaping Strategy in a World of Constant Disruption," *Harvard Innovation and Business Review*, 86 (October 2008), p. 88). Indeed, innovation has been shown to thrive in highly competitive environments in a number of contexts, including firms in commercial markets (Paul Osterman, "How Common is Workplace Transformation and Who Adopts it?" *Industrial and Labor Relations Review*, 47, no. 2 (1994), pp. 173–188), educational institutions (Erich Studer-Ellis, "Organizational Responses to Adversity: Evidence from Higher Educational Organizations," Presented at Annual Meeting of the American Sociological Association (Toronto, 1997) and states in the international system (Mokyr (1990), op. cit., pp. 206–207; Landes, op. cit., p. 31).

68 In the international system, neo-realists like Kenneth Waltz contend that "the possibility that conflict will be conducted by force leads to competition in the arts and instruments of force" among states and further that all participants in this competition eventually tend towards pursuit of the same types of weapons (Kenneth N. Waltz, *Theory of International Politics* (New York: McGraw-Hill, 1979), pp. 127–128).

69 Jeffrey Bale hypothesizes that "top-rung terrorist groups (such as al-Qaeda) have to keep pushing the envelope in terms of attacks to stay relevant and 'on top'", which might not apply to less influential or more localized terrorists ("Conceptual Background and Literature Extracts," in Gary Ackerman *et al.*, *Assessing Terrorist Motivations for Attacking Critical Infrastructure* (Monterey, CA: Center for Nonproliferation Studies, 2007), p. 54).

70 It should be noted that the perceived need for attack escalation does not necessarily drive a terrorist group to adapt through innovation (See Dolnik, op. cit., p. 149 and Palfy, op. cit., p. 91); they may simply do what they are already doing more efficiently. However, even incremental improvements of existing practice will usually necessitate some degree of organizational learning.

71 Dolnik found this factor to be relevant in each of the four case studies of innovation he conducted, declaring it to be "the strongest and most universal pre-indicator of the propensity of a terrorist group to innovate" (ibid., pp. 156, 158, 175).

72 This is based on the notion that "Ideology provides a motive – and possibly a formula – for action" (C. J. M. Drake, *Terrorists' Target Selection* (New York: St. Martin's Press, 1998), p. 16).

73 There are several prominent examples of this dynamic. One is Shoko Asahara's – of Aum Shinrikyo infamy – fascination with exotic weapons, including not only odes to sarin but also sending operatives to explore the building Nikola Tesla's earthquake generating machine (David Kaplan, "Aum Shinrikyo," in Jonathan Tucker ed., *Toxic Terror* (Cambridge, MA: MIT Press, 2000),

p. 212). Another example is the almost metaphysical regard displayed by late nineteenth-century radicals towards dynamite as a "scientific" force for change (David F. Ronfeldt and William Sater, *The Mindset of High-Technology Terrorists: Future Impplications from an Historical Analog* (Santa Monica, CA: RAND Corporation, 1981), pp. 2, 11). Furthermore, Ahmad Jibril, the founder of the PFLP-GC (Popular Front for the Liberation of Palestine – General Command), regarded himself as something of an inventor and wanted his organization to be the technical vanguard among terrorists (Dolnik, op. cit., p. 94) while, two decades later, Ramzi Yousef similarly sought personal prestige in his creativity with explosives.

74 The Irish Fenians, for example, apparently regarded dynamite as a "gift of science" (Ronfeldt and Sater, op. cit., p. 14 fn. 23), and embraced it enthusiastically. This has occurred many times in the general history of weapons development, with the canonical example perhaps being the Manhattan project, which resulted to a large extent from political leaders becoming aware of the implications of the dramatic advances in nuclear physics in the 1920s and 1930s.

75 Mokyr ((1990), op. cit., p. 151); MacKenzie (op. cit., p. 109); and Arthur (op. cit., p. 110) express this notion in similar terms.

76 Mokyr (2002), op. cit., p. 162.

77 Dolnik, op. cit., p. 175; Nancy Hayden, "Terrifying Landscapes: Understanding Motivations of Non-State Actors to Acquire and/or Use Weapons of Mass Destruction," in Ranstorp and Normark, op. cit., p. 167.

78 Jackson and Frelinger, op. cit., p. 25.

79 Jackson *et al.*, op. cit., p. 75.

80 The availability of copious financial resources has been identified as a major determinant in the FARC's ability to experiment with the development of complex weapons systems such as long-range mortars and anti-aircraft rockets. Forest, op. cit., pp. 15–18.

81 In this regard, Brian Jackson avers that "the centrality of technology to all terrorist and counterterrorist operations represents an important incentive for individual groups to seek out and master new techniques and weapons" (op. cit., p. 184.).

82 Jackson and Frelinger (op. cit., pp. 27–28) discuss technology strategies, with groups being faced with the decision to invest in a variety of technologies in order to take advantage of attack opportunities that present themselves, versus maintaining a narrow arsenal. They posit that groups that do not pursue a variety of weapons will seek a smaller number of versatile weapons that are "good enough" to exploit most opportunities. Organizations are also faced with the decision of whether or not to invest resources in customizing weapons to better address a particular attack opportunity.

83 Here we refer to a general increase in security force pressure rather than a specific countermeasure, thus distinguishing this driver from (1) above.

84 Jackson *et al.*, op. cit., p. 15. See also Cragin *et al.* (op. cit., pp. 59 and 67) for a discussion of the example of the Palestinian terrorist organizations.

85 Ibid., p. 98.

86 This is related to the concepts of fads and fashions that have been articulated in the business literature (see Abrahamson, op. cit.).

87 Hoffman (1992), op. cit., p. 15 and Jenkins (1986), op. cit., pp. 777–778.

88 David Strang and John Meyer, "Institutional Conditions for Diffusion," *Theory and Society*, 22 (1993), p. 490.

89 An interesting case is the discovery that some of the (often imperfect) recipes for extracting ricin found in al-Qaeda manuals appear to be word–for-word translations of instructions published in underground right-wing

literature in the United States (Anne Stenerson, "Al-Qaeda's Thinking on CBRN: A Case Study," in Magnus Ranstorp and Magnus Normak (eds), *Unconventional Weapons and International Terrorism* (New York: Routledge, 2009), p. 53).

90 David Strang and Sarah Soule, "Diffusion in Organizations and Social Movements: From Hybrid Corn to Poison Pills," *Annual Review of Sociology*, 24 (1998), p. 269. Research conducted on the diffusion of French coal-mining strikes and hijacking attempts suggests that both factors can play an important role (C. Connell and S. Cohn, "Learning from Other People's Actions: Environmental Variation and Diffusion in French Coal Mine Strikes, 1890–1935," *The American Journal of Sociology*, 101, no. 2 (1995), p. 366; Robert T. Holden, "The Contagiousness of Aircraft Hijacking," *The American Journal of Sociology*, 91, no. 4 (January 1986), p. 874).

91 See Hagel *et al.*, op. cit., p. 88. This is a similar concept to that of bandwagoning by weaker or allied states with respect to weapons adoption between militaries (Emily Goldman and Andrew Ross, "Conclusion: The Diffusion of Military Technology and Ideas – Theory and Practice," in Emily Goldman and Leslie Eliason (eds), *The Diffusion of Military Technology and Ideas* (Stanford: Stanford University Press, 2003), pp. 373–374). Assuming that this observation remains relevant in the terrorist context, it is little surprise then that the attack methods and weapons of groups like Hizballah and al-Qaeda have been adopted by a variety of other terrorist groups.

92 Derek J. De Solla Price and Donald Deb Beaver, "Collaboration in an Invisible College," *American Psychologist*, 21, no. 11 (1966), pp. 1011–1018.

93 Examples of such change agents include artillery producers in the late nineteenth century or arguably defense contractors today (Goldman and Ross, op. cit., p. 374; James Dunnigan, *Digital Soldiers: The Evolution of High-Tech Weaponry and Tomorrow's Brave New Battlefield* (New York: St. Martin's Griffin, 1998), p. xv).

94 Arthur, op. cit., p. 2. For example, in a commercial context, the more people buy (i.e., adopt) a given consumer product, the cheaper it becomes to produce and the greater the incentive for firms to make the product more marketable, which in turn encourages more adoption, and so on.

95 Everett Rogers, stated explicitly that "if an idea seems new to the individual, it is an innovation" (op. cit., p. 12).

96 Ibid., p. 171.

97 There is also the possibility that tinkerers within a terrorist organization might stumble upon a new weapon without a conscious intention to do so.

98 The anarchists and early socialists clearly believed that it was necessary to stay abreast of the latest scientific and technical developments in the hope of acquiring new weapons technologies, as seen in the declaration by a speaker at the Fifth Conference of Social Revolutionaries in 1909 that "Terror will be terror in the true sense of the word only if it represents the revolutionary implementation of the most advanced technical sciences at any given moment" (Ronfeldt and Sater, op. cit., 1).

99 This mechanism of awareness need not be completely distinct from the first mechanism mentioned above, as organizations can make the decision to scan the environment for new weapons prospects, even if they are not actively seeking a weapon at the time. In the general innovation context, Rogers therefore maintains that "Most organizations engage in an opportunistic surveillance by scanning the environment for new ideas that might benefit the organization ... most organizations continuously scan for innovations and match a promising innovation with one of their relevant problems" (Rogers, op. cit., pp. 422–423).

100 David Strang and Michael W. Macy, "In Search of Excellence: Fads, Success Stories, and Adaptive Emulation," *American Journal of Sociology*, 107 (2001), p. 155.
101 Strang and Soule, op. cit., pp. 270–272; Rogers, op. cit., p. 18; Mark S. Granovetter, "The Strength of Weak Ties," *American Journal of Sociology*, 78 (1973), p. 1360; Hargadon, op. cit., p. 59.
102 Dolnik, op. cit., p. 167.
103 For example, Stenerson (op. cit., p. 51) notes that CBRN (chemical, biological, radiological and nuclear) weapons receive very little attention among al-Qaeda's online followers, and even those knowledge objects (including manuals and discussion threads) that are available indicate a lack of knowledge of these weapons among jihadists.
104 Hargadon, op. cit., p. 60; Chesbrough, op. cit., p. x; and Arthur, op. cit., p. 108.
105 Recent empirical research has indicated that highly embedded and dense networks have a significant, substantial effect on terrorist organizations' decisions whether or not to pursue CBRN weapons, at least partly because of the informational advantages they bestow. Victor Asal, Gary Ackerman and R. Karl Rethemeyer, "Connections Can Be Toxic: Terrorist Organizational Factors and the Pursuit of CBRN Weapons," *Studies in Conflict and Terrorism*, 35, no. 3 (2012), p. 243.
106 Georges Romme and Ron Dillen, "Mapping the Landscaping of Organizational Learning," *European Management Journal*, 15, no. 1 (1997), pp. 68–78, cited in Jackson *et al.*, op. cit., p. 11.
107 K. K. Jones, "Competing to Learn in Japan," *McKinsey Quarterly*, 1 (1991): pp. 45–57, cited in Jackson *et al.*, op. cit., p. 11. However, it should be recognized that just because a potential adopter shares different cultural values, or indeed lacks scientific literacy, this does not mean that they cannot recognize the advantages of a new weapon. An excellent illustration is the Native Americans of New England in the seventeenth century who, despite having no previous experience with firearms, rapidly realized that a musket was far superior to a bow (and even, unlike the majority of European colonists, that a flintlock gun was better than a matchlock) and proceeded to adopt firearms with zeal (Patrick M. Malone, *The Skulking Way of War: Technology and Tactics among the New England Indians* (Toronto: Madison Books, 2000), p. 35.)
108 Committee on Forecasting Future Disruptive Technologies, op. cit., p. 43; Mokyr (2002), op. cit., p. 231.
109 Forest, op. cit., p. 18.
110 Trujillo and Jackson, op. cit., p. 57.
111 Jackson, *et al.*, op. cit., pp. 12–13
112 Jackson, op. cit., pp. 12–13; Trujillo and Jackson, op. cit., p. 57.
113 As an example, see Waller and George's discussion in Forest and Howard regarding uncertainties and opportunity costs with respect to familiar versus exotic weaponry (Forest Waller, Jr. and Michael George, "Emerging WMD Technologies," in Howard and Forest, eds, op. cit.).
114 Cragin *et al.* op. cit., p. 14).
115 Martha Crenshaw, "The Logic of Terrorism: Terrorist Behavior as a Product of Strategic Choice," in Walter Reich (ed.), *Origins of Terrorism: Psychologies, Ideologies, Theologies, States of Mind* (Washington, D.C.: Woodrow Wilson Center Press, 1998), pp. 7–24; Gordon H. McCormick, "Terrorist Decision Making," *Annual Reviews in Political Science*, 6 (2003), p. 481.
116 Gressang, op. cit., pp. 89–90; Jackson and Frelinger, op. cit., p. 23.
117 Jackson and Frelinger, who proceed to list various attributes that differentiate weapons, namely stand-off distance, capacity for use in parallel and the option

of trading human for technical resources (ibid., p. 27). To this list can be added several more attributes, including level of concentration of lethal effect, capacity for covert delivery, amount of training required for proficiency, to name just a few.

118 Ibid., p. 15; Palfy, op. cit., pp. 83–84.

119 See ibid., pp. 86–87, for a similar notion. Jackson and Frelinger (op. cit., p. 25) take this idea further by classifying alternate terrorist weapons *strategies* as either "versatility" strategies, where a limited repertoire of more versatile weapons like firearms and bombs is maintained, versus "variety" strategies where a multiplicity of weapons are acquired in order to address a range of possible attack configurations.

120 There are other options, of course, including scaling back desired outcomes or shifting targets.

121 Historical examples abound, from the development by ancient armies of siege engines in order to overrun walled towns (for a superb illustration of this dynamic, see Thucydides, *A History of the Peloponnesian Wars*, sections 2.75, 2.76 and 4.100), through efforts to equip vehicles with tracks and armor as a means of overcoming World War I entrenchments, to attempts to equip munitions with guidance systems as a means to counteract unaffordably high levels of inaccuracy in the aerial bombing campaigns of the first half of the twentieth century.

122 The performance gap can be defined as the perceived divergence between a given actor's expectations and its current level of performance. Rogers, op. cit., p. 422; Abrahamson, op. cit., p. 592.

123 The pure interpretation of this approach in the general diffusion literature maintains that the innovation that is perceived to supply the greatest benefit for the least cost will be adopted (Strang and Meyer, op. cit., p. 489). In the general context of technical innovation, MacKenzie stresses that the inherent uncertainties lead actors to rely on heuristics and "satisficing" in choosing whether and how to innovate, which can introduce a range of non-rationalistic (from the point of view of formal decision theory) influences on the adoption decision, such as culture, personality, social relations and national circumstances (op. cit., pp. 51–53; also see Chesbrough, op. cit., p. 70 and Rogers, op. cit., p. 116). We discuss several of these "non-rational" factors below.

124 Drake, op. cit., pp. 23–24, 34, 175.

125 Ibid., p. 39, 42.

126 This can occur even when such a need is self-created, such as when it is based on a delusional goal. Conversely, a genuine need that goes unrecognized by decision-makers may not prompt a consideration of weapons adoption, at least not for some time.

127 Marissa Reddy Pynchon and Randy Borum, "Assessing Threats of Targeted Group Violence: Contributions from Social Psychology," *Behavioral Sciences and the Law*, 17 (1999), p. 345.

128 Gressang, op. cit., p. 96.

129 One example of an expressive motivational factor is personal glorification, as described by Borowitz (Albert Borowitz, *Terrorism for Self-Glorification: The Herostratos Syndrome* (Kent, OH: Kent State University, 2005)). For others, including historical prejudice, limbic conditioning processes and in-group maladaptations, see Forest, op. cit., p. 250.

130 Bale, op. cit.

131 Abrahamson, op. cit., pp. 605–606.

132 Paul J. DiMaggio and Walter W. Powell, "The Iron Cage Revisited: Institutional Isomorphism and Collective Rationality in Organizational Fields," *American Sociological Review*, 48 (1983), pp. 147–160.

133 Jackson *et al.*, op. cit., p. 105. Oppenheimer noted that the bombers would often get ahead of themselves in innovating new designs before effective countermeasures could be developed, which sometimes was actually counter-productive (Oppenheimer, op. cit., p. 200).

134 Trujillo and Jackson, op. cit., p. 60.

135 It should be noted that the effects of organizational structure on the prospects for successful adoption of an innovation, rather than the awareness or decision-making phases, are far more clear cut, with multiple authors indicating that centralized hierarchies are more efficient at implementing an adoption than decentralized networks. (For example, Jackson, op. cit., pp. 200, 202; Rosen, op. cit., p. 39; C. Marlene Fiol and Marjorie A. Lyles, "Organizational Learning," *The Academy of Management Review*, 10, no. 4 (October 1985), pp. 803–813, cited in Jackson *et al.*, op. cit., p. 9.)

136 Historical examples of top-down military innovation abound, from the Assyrians at Nineveh and Dionysus at Syracuse, to the rulers of the medieval Italian city-states and the nineteenth-century British Ordnance Boards (see Volkman, op. cit., pp. 20–21).

137 Dolnik, op. cit., p. 178. Of course, leadership in any organization is never absolute, and there are inevitably opportunities for functionaries to influence or frustrate leadership decisions.

138 For example, with respect to FARC, Ortiz suggests that:

> The group's strict discipline tended to stagnate its behavior in both political and military terms, limiting possibilities to adapt to changes in the strategic environment: At the same time, an excessive hierarchy established barriers to the circulation of information and ideas between the base and its commanders. Consequently, there was a reduction in the capacity of leaders to perceive the tactical needs of the field units and to receive suggestions on the introduction of potential innovations regarding their manner of operating.
>
> (Roman Ortiz, "Renew to Last: Innovation and Strategy of the Revolutionary Armed Forces of Colombia (FARC)," in Forest (ed.), op. cit., pp. 218–219, cited by Sawyer in Gary A. Ackerman and John P. Sawyer, *Embracing Technologies of Lethality* (College Park, MD: START, 2012), p. 13 fn. 60)

139 See Kenney (2008), op. cit. and Trujillo and Jackson, op. cit., p. 60. Eilstrug-Sangiovanni and Jones disagree with this assessment, however (M. Eilstrug-Sangiovanni and C. Jones, "Assessing the Dangers of Illicit Networks," *International Security*, 33, no. 2 (Fall 2008), pp. 7–44).

140 Sawyer, op. cit., p. 13.

141 Strang and Soule, op. cit., p. 275; L. A. Brown, *Innovation Diffusion: A New Perspective* (New York: Methuen, 1981); Rogers, op. cit., p. 88; J. Petras and M. Zeitlin, "Miners and Agrarian Radicalism," *American Sociological Review*, 32 (1967), p. 578.

142 In the political science literature, these effects are well-understood from analyses of foreign policy decision making, where process, status, personal considerations and psychological framing can all influence decisions. For an example, see Allison and Zelikow's classic study of the Cuban Missile Crisis (Graham T. Allison and Philip Zelikow, *Essence of Decision: Explaining the Cuban Missile Crisis* (New York: Longman, 1999)).

143 While the terrorism literature does not specifically describe instances of this internal opposition to new technologies occurring within terrorist groups, the phenomenon is extensively detailed in the general technology adoption literature and is described as being especially acute in the context of the historical

adoption of weapons by militaristic entities, due in part to military commanders' traditional emphasis on discipline, martial values and risk aversion where survival, both personal and institutional is on the line. For a more in-depth discussion, see Emily O. Goldman, "Receptivity to Revolution: Carrier Air Power in Peace and War," in Goldman and Eliason (eds), op. cit., p. 269; Keegan, op. cit., p. xvi; Steven Peter Rosen, *Winning the Next War: Innovation and the Modern Military* (Ithaca: Cornell University Press, 1991), pp. 19, 25; and Dunnigan, op. cit., p. 6.

144 Author interview with former PIRA bomb-maker, Athlone, Republic of Ireland, June 20, 2012

145 Von Hippel, *et al.*, op. cit., p. 33.

146 *Compatibility* can be regarded as the extent to which the innovation is perceived to be compatible with existing values, goals and past experiences of the potential adopters. *Trialability* represents the degree to which the innovation can be experimented with on a trial basis. *Observability* is the ease with which the results of the innovation and its prior adoption are observable to others (Rogers, op. cit., pp. 15–17).

147 Palfy, op. cit., p. 88.

148 Rogers, op. cit., p. 149.

149 Hargadon, op. cit., p. 26.

150 Kenney (2008), op. cit., p. 4, Arthur, op. cit., pp. 159–160; MacKenzie, op. cit., pp. 216, 235.

151 Hargadon, op. cit., pp. ix; 60, Chesbrough, op. cit., p. x; Arthur, op. cit., p. 108; In the terrorism context, networks have been shown to have a significant and large impact on terrorist organizations' decision as to whether to pursue CBRN weapons and whether or not to engage in mass-casualty attacks; see Asal *et al.*, op. cit, p. 243 and Victor Asal and R. Karl Rethemeyer, "The Nature of the Beast: Organizational Structures and the Lethality of Terrorist Attacks," *Journal of Politics,* 70, no. 2 (April 2008), p. 443.

3 Innovation and the improvised explosive device (IED)

John Allison

Introduction

A review of the literature relating to terrorism and innovation suggests that there is much work to be done. Undoubtedly prompted by the events of 9/11, the academic analysis of terrorism and innovation appears to have been initiated over a decade ago[1] and yet there has been limited progress to date. There has been some reinvention of the wheel and an acknowledgement[2] that there is no common understanding amongst academics, of what innovation, in this context, actually is, where it takes place or who drives the innovation process. That said, the work of the Rand Corporation[3] and Dolnik[4] provides us with valuable open-source insight.

Much of the current research is based on assumptions or practices from the commercial/business world. Where the literature examines innovation within a campaign of violence, those case studies are often very dated. With numerous emerging threat streams in countries like Pakistan, Yemen, Somalia, Nigeria and Mali there is a pressing requirement to examine the process of innovation in these areas, if we are to produce findings that will have immediate, real-time application. The security situation is evolving rapidly; in Somalia for example, Person Borne IED (PBIED) tactics have become more sophisticated in a bid to defeat security force Tactics, Techniques and Procedures (TTPs).[5] The innovation *triggers* in such locations should be examined, identified and captured. This task is not beyond the reach of the academic community if Open-Source Intelligence (OSINT) collection methods are utilised or engagement with the security agencies is increased.

There has been a preoccupation with innovation in the context of Weapons of Mass Destruction (WMD) or Weapons of Mass Effect (WME). Some argue that it would be more appropriate to examine the ability of terrorist groups to revise conventional methods of attack, either by adjusting the technology or TTPs, on the grounds that this is the more 'imminent' threat.[6] With regards to WME, it takes limited *tweaking* on the part of a bomb designer or terrorist tactician to turn a small hand-carried device into a weapon of mass, strategic effect. In 1999 David Copeland emplaced an IED fabricated from basic components in the Admiral Duncan pub in London.[7] The attack killed three people, 26 suffered serious burns, four required amputations and another 53 were injured. Had that type of device been modified and redeployed elsewhere, the results could have been even more catastrophic. Critical infrastructure can also be effectively targeted with small devices if the vulnerable points (VPs) are known.

Two studies conducted in 2010/2011 looked at innovation in relation to WME.[8] One scholar suggested that the Colombian terrorist organisation, FARC, was not 'particularly innovative'.[9] Throughout its campaign of violence, the FARC repeatedly demonstrated that, in terms of IED fabrication, delivery and emplacement, it was a highly innovative group. They standardised the large-scale production of high quality improvised munitions and developed the technology to deliver and initiate a car bomb using radio-control. They also prepared a number of unusual payloads for improvised mortars,[10] including a glue and sulphuric acid mix designed to incapacitate security force personal prior to launching an assault on a base.[11] They were experts in improvised mine warfare too, which accounted for a high number of Colombian Armed Forces casualties.

Another chapter in the same publication looks at Loyalist paramilitary groupings in Northern Ireland and their propensity for innovation in relation to WME. The author argues that the Loyalists 'seemed uninterested in innovation'.[12] This analysis is troubling. First, the Loyalist groupings in Northern Ireland never had a requirement to attack armoured vehicles or to breach heavily fortified police stations and SF bases, unlike PIRA. Therefore one of the main drivers of innovation within PIRA was absent in Loyalist circles. In actual fact prior to the 1994 ceasefire there was an acknowledgement that a small group of Loyalist bomb-makers were becoming increasingly capable.[13] Elements of the grouping were also experimenting with what could be described as a radical innovation, the use of Passive Infrared (PIR) initiated munitions.[14] In the same section, lip service is paid to the fact that the Royal Ulster Constabulary (RUC) eventually discovered a Loyalist improvised small arms manufacturing facility on an unprecedented scale.[15] More than 800 sub-machine guns were in the final stages of completion and it is believed that the gunsmith, Samuel McCoubry, had been fabricating weapons for the Loyalists for over 20 years. The Loyalist groupings had a long history of producing high quality 'improvised' firearms in small workshops throughout Northern Ireland. This was in keeping with their strategy of carrying out Close Quarter Attacks (CQA) with small arms. A high level of Operational Security (OPSEC) and engineering skill would be required to fabricate such large quantities of weapons, for an extended period. Had it been their intent, these weapons could indeed have resulted in a mass effect.[16]

Within academia there has also been some debate[17] about the importance of safe-havens and external support in relation to IED innovation.

Suffice to say, Southern Ireland was a critical safe-haven for PIRA bomb designers just as the tribal areas in Pakistan are to the Taliban. Safe havens give the extremist the time and space to experiment with and develop and test new improvised munition systems. Regarding external support, it is surprising that no mention is made of the strategic impact of Iranian Explosively Formed Projectile (EFP) technology being passed on to insurgents in Iraq.[18] In this instance, the external support proved, quite literally, catastrophic for UK/US forces operating in the region.[19] The manner in which the technology was delivered also proved to be highly innovative and lead to a rapidly enhanced insurgent capability.

Dolnik, in his pioneering study of terrorist innovation,[20] concludes with the surprising observation that 'when one surveys the last 50 years of terrorist operations case by case, very few incidents strike the observer as creative in any way'. In IED fabrication, delivery and emplacement terms this statement is misleading. Throughout the PIRA campaign of violence, for example, the grouping consistently demonstrated a high level of creativity. For one, the sophistication of its munitions caching (hides) ability within the 'IED system' was so sophisticated/creative, that it forced British defence scientists to develop a new range of counter-terrorist search equipment and it forced the British Army to revise its search procedures. The grouping also developed highly creative *come-on* scenarios, designed to lure the security forces into a particular area. Some of the incidents were almost stage-managed. On one occasion, using a small explosive charge, PIRA blew up a car in the middle of a country road and then quickly booby-trapped the rubble. This was intended to make the bomb disposal officer think that the device had either *partialled* or detonated prematurely. Arguably the situation in Afghanistan and Iraq is all too similar and the security forces have to defeat a cunning and highly creative enemy on a daily basis. All too often, innovation within other areas of the IED system, such as weapons hides, are overlooked, something which will be remedied in this chapter.

The previous paragraphs were not intended to be a hostile critique of the important work being carried out in academic circles. As one retired Army officer explained,[21] academia has a valuable role in bringing a 'non-military and non-state perspective' to the table. Rather it is designed to illustrate that there is a need to examine the issue from a fresh perspective. This chapter will clearly outline the nature of the IED threat and the various points at which innovation can and does occur within the IED system. The chapter will examine innovation in relation to the Improvised Explosive Device (IED) through the lens of a counter-terrorism practitioner. The IED is only one of a repertoire of attack methods available to the insurgent or terrorist, however it is now widely regarded as a weapon of choice.[22] There is also a growing, if misguided, perception amongst extremist groupings worldwide that this field expedient weapon system has been instrumental in defeating the military might of a number of countries – the United States and United Kingdom in Iraq and Afghanistan and Israel in Southern Lebanon following the withdrawal in 2000. This will undoubtedly continue to inspire the next generation of extremists to adopt this method of attack.

It is necessary first to clarify the definition of an IED and discuss the term 'improvised'. Thereafter the chapter will explain why IEDs are so prevalent, how they are used and who uses them. In a bid to further understanding, the chapter will then examine some of the generic factors that influence IED design, fabrication and emplacement. Innovation theory will be discussed and applied to the IED system. A new, all-encompassing definition of IED innovation will be proposed. The elements that make up an IED system will subsequently be examined in detail and it is relevant to discuss current C-IED (Counter-IED) doctrine. Real world examples of extremist innovation will underpin this analysis throughout. In conclusion a number of recommendations for future areas of research will be presented.

The Improvised Explosive Device and improvisation

In military parlance the IED is defined as: 'A device[23] placed or fabricated in an improvised manner incorporating destructive, lethal, noxious, pyrotechnic or incendiary chemicals and designed to destroy, incapacitate, harass or distract. It may incorporate military stores, but is normally devised from non-military components.'[24] Typically when the term IED is used, people tend to think of an explosion, blast and fragmentation. In Northern Ireland and on the UK mainland PIRA made extensive use of 'cassette incendiary' devices to target commercial premises. These were small items that could be easily concealed in a jacket pocket, so the bomb layer was able to gain access to the target with minimal risk of detection. The cassette[25] was then carefully secreted in and amongst combustable items, such a clothes rack in a department store and the bomb layer would make good their escape. An electronic timer would then initiate the incendiary material in the cassette, often after closing time to maximise damage. This simple device which cost no more then a few pounds to manufacture caused millions of pounds of damage to high-street stores throughout the UK. It was well designed and fit for purpose. In June 2007 a jeep containing gas canisters and petrol was rammed into the departure hall at Glasgow airport with the intention of causing a fireball and mass casualties, although the incendiary bomb failed to function as intended.[26]

A review of the academic literature reveals that the term 'improvised' has not been widely discussed. If one is to fully understand the IED system and the decision-making process behind it, a working knowledge is vital. The phrase 'improvise, adapt and overcome' is frequently bandied around in a military environment particularly when troops are deployed overseas with limited resources and are required to come up with a solution quickly and efficiently with whatever happens to be available at the time.[27] In the Western world we are gradually loosing our ability to effectively improvise, adapt and overcome because most of us now live in a throwaway society. If something breaks in the home, all too often that item is simply discarded, without a second thought. The recycling centres in the UK provide ample evidence of this. In the developing world or in a country where resources are limited, the situation is radically different. People are often forced to 'make do and mend'. As a result they become experts at foraging, salvaging and modifying materials, usually in order to find a solution to a specific problem. Years of sanctions in Iraq and high levels of deprivation in Afghanistan have created generations of highly resourceful individuals. The addition of combat experience and terrorist training makes for a potent mix.

In deprived regions, necessity is the mother of invention. To a lesser extent, people living in remote or rural locations in the Western world, farmers for example, also demonstrate an ability to improvise. Many of the heavy mortar systems and large VBIEDs (Vehicle-borne IED) fielded by PIRA were fabricated in rural locations in places like South Armagh in Northern Ireland. Not only did this location provide a degree of OPSEC, it was also home to a large farming community, who were used to making running repairs to agricultural machinery with welding and other light fabrication equipment. Equally, most of us probably know someone who might be described as a DIY enthusiast or someone who enjoys spending much of their spare time *tinkering* in a garden shed or garage. In April 1998, Edgar Pearce, the self-styled 'Mardi Gra' bomber was arrested by police in London. In an extortion bid he single-handedly waged a three-and-a-half-year terror campaign in and around the city by planting 36 IEDs outside branches of Barclays bank and Sainsburys. He was a DIY enthusiast and 'regularly tinkered with old mantlepiece clocks'.[28] Most of his bomb components came from car boot sales and the assembled devices were innovative in their simplicity.[29] Pearce had the *capability* and the *intent* to manufacture and emplace IEDs. For an IED attack to be initiated, both capability[30] and intent must be present. A number of would-be jihadists in the UK have the intent but not the capability, equally someone can have the capability but no real intent.

In addition to gradually losing our ability to improvise, very few of us now have to physically hunt for food to survive, so we have lost tracking skills, our basic instincts are not so finely honed and we have a general lack of situational awareness. Observation will show that most commuters heading to work on a typical day in London do not even notice the person sitting next to them. In a bid to combat the IED threat in Afghanistan, British soldiers are now being taught 'Ground Sign Awareness' (GSA) by expert trackers.[31] This is testament to the skill and tactical cunning with which Victim Operated IEDs (VOIED) are emplaced by the Taliban.

The recent conflict in Libya provides us with a valuable insight into the art of battlefield improvisation. The rebels were outgunned by well equipped government forces and, with little or no hope of receiving large quantities of conventional weapons from an external source, there was only one solution. Thinking logically and rationally, they decided to improvise, adapt and overcome, using locally available resources in order to become an effective fighting force. Engineering students suddenly found themselves designing improvised munitions, alongside civilian structural engineers and mechanics.[32] The results were impressive, helicopter rocket pods were mounted on pick-up trucks, Rocket Propelled Grenades (RPG) were converted in light engineering shops and the high-explosive fill was extracted from anti-tank mines to make IEDs. The anti-government forces in Syria are now engaged in similar activities.[33] There is also ample evidence to suggest that improvised directional charges are being fabricated to defeat Syrian government armoured vehicles.[34] Putting one's emotions to one side for a minute, the improvised munitions engineering and tactics utilised in Libya and Syria, be it improvised rocket launchers or directional charges, are not dissimilar to those used by insurgents in Iraq and Afghanistan.[35] Whilst some technology transfer has undoubtedly taken place, when faced with a well equipped security force and lack of resources, improvisation is a natural and sensible reaction. The ideology and target set may be different but the principles behind the decision to improvise often remain constant. Stepping back once again from the highly emotive and tragic consequences of a suicide bombing, there is good reason to call this method of attack a 'poor man's cruise missile'.[36]

Having looked at improvisation in a generic sense it is necessary to drill down and examine why an extremist would resort specifically to producing an improvised munition, including IEDs. First, it may not be physically possible to procure conventional weapons, it may be financially prohibitive and there may be significant security risks in attempting to procure commercial weapons.[37] Attempting to attribute an improvised device is potentially more difficult for the security forces,[38] it may be easier to conceal the improvised munition and procure the components without compromise. Conventional weapon systems require conventional spares, ammunition and maintenance, all of which may be difficult to obtain, or carry out. Most importantly a specific target may require a specific method of attack with a bespoke weapon system. When the security forces in Northern Ireland upgraded the fortifications of its bases, PIRA attempted to defeat the technical mitigation measures by fielding the MK13 spigot mortar, essentially a 45 gallon oil drum filled with homemade explosives (HME). It was literally lobbed over the walls of the base using a black powder propellant charge. In Iraq, insurgents produced the Improvised Rocket Assisted Mortar (IRAM) to breach the perimeter security of US bases in a similar fashion.[39]

At this stage a note of caution needs to be sounded. It is important to remember that whilst someone may produce what appears to be a well designed improvised munition, its effectiveness will depend, not only upon how it functions on target but also the ease with which it can be fabricated, replicated, tactically delivered, emplaced and initiated. Once again an increasingly important design factor in a non-permissive environment is the ease with which the components of the device can be procured without arousing suspicion. PIRA produced a horizontal mortar system known as the MK16, designed to defeat armoured vehicles in Northern Ireland. It was primarily fabricated from sheet steel and mild steel tubing used for industrial plumbing. The individual components were cheap and readily available, purchasing and transporting the items would not arouse suspicion, the design could be replicated quickly and with limited engineering knowhow/equipment. This is demonstrative of the fact that innovation can occur at various stages within the IED system.

Improvised munitions, including IEDs, are not the sole preserve of the enemy. First, military personnel are actively encouraged to improvise on a day-to-day basis as this contributes to ensuring the success of the commander's mission, particularly when faced with a problem and limited time or resources.[40] In World War II there was a requirement to develop an underwater mine to destroy shipping in Germany. Scientists struggled to find an appropriate time delay mechanism for the munition. Eventually it was discovered that a sweet called an aniseed ball would slowly dissolve in water providing a sufficient delay element,[41] a remarkable feat of improvisation. Today combat engineers are trained to fabricate improvised demolition charges from everyday items and during the Vietnam War American special forces ran an extensive improvised munitions programme for indigenous friendly forces. It was seen to be a cost-effective force multiplier. Extremist elements continue to refer to technical manuals written for this unconventional warfare (UW) programme today.[42] Improvised weapon systems may also be used where there is a requirement to carry out a 'false flag', deniable operation. In this instance the remnants of a conventional weapon system may help to identify the perpetrator or sponsor.

In short, an effective IED attack is usually the result of careful, cost-benefit analysis. It is a 'planned operation', not a random act of violence.[43] On today's asymmetric battlefield, in Afghanistan for example, IEDs are increasingly being fielded by the Taliban at the strategic level, to diminish the appetite of the international community for the campaign. At the tactical level the weapon is used to obstruct attempts to bring security and stability to the region, to impede the mobility of NATO and Afghan troops whilst enhancing their own, as a show of strength in the face of overwhelming firepower and to protect key locations.[44] The Taliban tend to think in terms of mobility and counter-mobility, in the same way as a combat engineer. The ground is carefully studied in order to identify friendly force key terrain, likely avenues of approach, choke points and mobility corridors. These areas are subsequently sewn[45] with improvised Victim Operated (VO) or command detonated devices.[46] In keeping with military doctrine, the Taliban will often seek to bring effective fire to bear on friendly forces caught in such *mined* areas. They are highly skilled in the art of guerrilla warfare. During the war against the Soviets, the Mujahideen destroyed 1,191 vehicles and killed 1,995 troops using commercially manufactured and improvised mines;[47] such lessons have not been lost on the Taliban.

IEDs are also planted in order to exert psychological pressure. They demoralise troops, frighten the local populace and influence political leaders. The extremist often seeks to capitalise on the success of an IED attack by recording the incident on video camera and uploading to the internet. Social media, with its global reach, inadvertently enhances the strategic impact of the operation and the footage becomes a marketing tool for the extremist grouping. Such videos also serve to inspire and educate others.

Innovation and the IED system

Without question, extremists can be highly innovative[48] – but what is innovation? This chapter will draw heavily on the work of O'Sullivan and Dooley.[49] O'Sullivan and Dooley define innovation in a commercial sense as:

> The process of making changes, large and small, radical and incremental, to products, processes, and services that results in the introduction of something new for the organisation that adds value to customers and contributes to the knowledge store of the organisation.

A common misconception is that a terrorist innovation has to be highly sophisticated. The opposite is true. Very often, the innovative change actually simplifies an IED in terms of its initiation, its component parts or the manner in which it is to be tactically deployed. In the commercial world the term 'major innovation' is regularly associated with 'simplicity, low cost and effectiveness'.[50] The latter are principles that also inform IED design. According to innovation theorists in the commercial world a simple innovation may seek to 'minimise the amount of behavioral change required'.[51] Where IED deployment and initiation is out-sourced to a third party, simplicity in emplacement, arming and firing terms is critical. The Joint IED Defeat Organisation (JIEDDO) acknowledges that one of the most common IED threats to 'both mounted and dismounted troops' in Afghanistan is the 'simple, but non-discriminatory' Victim-Operated IED.[52] The Taliban VOIED threat has rapidly evolved in response to security force TTPs and technology. Initially the device consisted of metal saw blades in a sandwich arrangement,[53] wired into an electrical circuit. Anyone standing on the blades would cause them to come into contact with each other and act as a firing switch. Over time, small, incremental changes in VOIED design, reduced the metal content of the switch in order to thwart detection efforts. The Taliban have now perfected an extremely simple, *no metal content* variant of the VOIED.[54]

In Northern Ireland, PIRA produced a *radical innovation* in the form of an adapted radar speed gun to trigger an IED in response to ECM measures. The grouping also made small, incremental changes to its mortar bomb fuzes that were informed by testing and post-incident analysis. The commercial definition proposed by O'Sullivan and Dooley also discusses, 'products, processes, and services' which can be substituted for 'IED system'. Product innovation relates to 'making beneficial changes to physical products'.[55] In this instance the physical product would be the IED itself and changes might be an additional safe to arm switch within the device or changes to the liner of an improvised anti-armour munition to increase penetration. The product might be a vehicle used to smuggle IED components, the addition of secret compartments being an innovative enhancement. Process innovation is concerned with changes to the 'processes that produce products or services'.[56] The way in which an IED is fabricated is a process, whether the mixing of bulk HME, the assembly of printed circuit boards or improvised mortars. If a terrorist grouping found a way to reduce HME mixing time whilst enhancing safety this would be a 'process innovation'. Both FARC and PIRA standardised the large-scale fabrication of improvised munitions, something which can also be described as a process innovation. Finally, service innovation refers to 'making beneficial changes to services that customers use'.[57] Activities on the periphery of the IED system, such as fundraising for IED operations, is a form of service. Funds can be obtained in numerous ways, through extortion, sales of counterfeit goods such as DVDs, cigarette smuggling and internet scams. The list is endless but a consistent feature of such activities is usually sophistication and innovation.[58] The actual delivery and emplacement of an IED can be out-sourced in order to maintain a firewall between the bomb-maker/engineering department and the expendable bomb layer. In Iraq and Afghanistan one could argue that a service industry was created around bomb delivery with people being paid to deploy devices. Sometimes financial reward would be linked to the success of the operation, an innovative way of improving emplacement reliability.

To clarify then, prior to an IED being initiated and subsequently exploited a series of interrelated activities have to take place. Taken as a whole, the 'networks of nodes and linkages' that connect these activities are known as the IED system.[59] Also known as the overarching 'IED network', the various activities can be broken down into three distinct clusters of actions[60] with innovation occurring throughout:

- ***Resourcing and planning.*** Activities such as formulating ideas/method of attack concepts, obtaining funding, extensive target reconnaissance, enlisting local support, procuring IED components, smuggling, IED fabrication, arranging transport;

- ***Execution.*** Dry run, counter-surveillance, vehicle preparation, movement plan, emplacement, timings and initiation, counter-forensics and withdrawal plan;
- ***Exploitation.*** Post blast evaluation, feedback loop, media plan.

So what type of innovation might occur? Whilst *resourcing and planning* the terrorist may, for example. steal a vehicle for use as a VBIED and fit it with *ringer* plates. This was a common PIRA tactic to thwart police or traffic camera surveillance.[61] If the stolen vehicle was a white Ford Transit van, PIRA would simply look out for another white Ford Transit van, note its registration plate, make new plates with that number and screw them onto the stolen van. Any police check would, at first glance, simply show that the number coincided with that type of vehicle. Vehicle ringing is now a major criminal enterprise and it is likely that there could be some collaboration between terrorist and criminal networks in order to acquire the ringer vehicle. Reconnaissance can be physical, virtual or both. Extremists are increasingly using free, online geospatial resources for 'mission planning'[62] as was the case prior to the attack in Mumbai. This is likely to increase as IT systems evolve. Sophisticated Dual Tone Multi-Frequency (DTMF) Radio controlled receivers were recovered in Afghanistan in 2002 with intelligence officers reporting that they were made in Pakistan and smuggled over the border.[63] In order to aid the smuggling operation the spider devices, as they were known, were covered in high quality printed labels marked '12 volt battery' or 'Toshiba', simple but effective. Although not directly involved with bomb initiation, it is important to note that the individual who makes the number plates or prints the labels for spider devices,becomes part of the IED system.

In Afghanistan during the *execution* phase, the Taliban employed innovative methods of timing device initiation. An extensive system of 'dickers'[64] was used to observe friendly forces prior to an IED attack. They would tell the trigger man, hidden from view, when to initiate the device. This allowed the trigger man to make good his escape, often with the triggering device which could be reused.[65] Even if the dicker was apprehended there was little the troops could do, in the absence of any evidence.[66] Weapons Intelligence Teams (WIT) are routinely dispatched to the scene of an IED incident, in a bid to capture technical, tactical and forensic/biometric (FABINT) intelligence. The intelligence collected on scene will enhance the threat picture, confirm enemy TTPs, drive tactical and technical countermeasure development and potentially highlight innovative changes within the IED system. Naturally the enemy is becoming increasingly forensically aware and will take innovative steps to minimise and destroy technical intelligence and FABINT.[67] An incendiary device, placed in the back of the mortar baseplate van used for the PIRA attack on Downing street in 1991, was timed to initiate once the mortars had deployed in a bid to destroy forensics.[68]

Lastly, when the enemy seeks to *exploit* the IED incident the media plan will often be highly innovative, particularly where a video camera is used to document an IED attack. Throughout the world there are a plethora of sophisticated extremist media outlets.[69] The insurgent media campaign in Iraq continues to inspire extremists battling for control of the electronic frontier.[70]

Taking innovation a stage further, O'Sullivan and Dooley acknowledge that for something to be innovative it has to be more then simply novel, which would be classed an invention. If defence scientists produced a piece of equipment to counter a non-existent threat[71] or even one that had yet to be perceived, it could hardly be termed an innovation. An innovation has to 'add value' or result in an 'improved experience',[72] in terms of an existing product, process or service. In the current economic climate where funding and resources are scarce it is unlikely that a company would waste time developing a new process or product without identifying a specific demand or requirement. Within an extremist organisation, finances and resources can be particularly scarce and speed is crucial if it is to retain the upper hand. Therefore it will rarely attempt to innovate unless there is a specific demand or problem to address and they believe that the 'change' will result in some form of added value. If the 'innovation' does not subsequently add value it is likely that the grouping will revert to tried and tested methods or attempt to refine the changes. How quickly this takes place will depend upon the effectiveness of feedback loops within the organisation during the 'Exploitation' phase.[73] For example, it may not always be immediately apparent if the IED attack was successful or how effective the device was. Incidentally, from a security force perspective, it is critical to control the release of information, particularly to the media, following an incident to hinder terrorist bomb damage assessment (BDA).[74] Friendly force information operations (IO) also have an important part to play at this stage.

The importance of learning[75] from mistakes and failure[76] is recognised by O'Sullivan and Dooley. They argue that 'destroying poor ideas can be as important as nurturing good ones'[77] if valuable resources are to be effectively re-allocated, quickly. In the 1970s PIRA produced the 'Bombard', an improvised anti-armour weapon. It was initially effective against British Army armoured vehicles; however, 'up-armouring' rendered it ineffective and it was subsequently abandoned. The organisation went on to perfect more efficient methods of armour attack, including the Improvised Anti-Armour Grenade (IAAG). This was a small improvised munition, fashioned out of a tin can with a wooden or plastic handle and semtex main charge. Once a good idea has been identified it often has be conveyed to other cells within the terrorist organisation in a manner that does not compromise OPSEC. Sometimes concern for OPSEC and even paranoia will stifle the organisation's ability to disseminate word of the innovation to its members. In 2003, Israeli commandos boarded an

Egyptian fishing boat called the Abu Hassan and a significant amount of munitions were recovered. In addition, Hezbollah bomb expert, Hamad Masalem Mussa Abu Amra was arrested onboard. He had 36 detailed bomb-making CDs in his possession allegedly destined for Palestinians in the Gaza Strip.[78] The CDs contained various lessons outlining how to fabricate improvised munitions and effectively deploy suicide bombers. Crucially, the instructions were based on tried and tested designs that consisted of easily obtainable resources. The instructions were clear, unambiguous and broken down, step by step, in order that they could be readily digested. Israeli sources believed that it was an attempt to mitigate the elimination of key Palestinian bomb-makers. It was also an innovative way of disseminating bomb-making know-how and attempting to rejuvenate the Palestinian 'knowledge store'. Terrorist organisations are becoming increasingly wary of relying on one or two master bomb-makers and they will seek to disperse knowledge as widely as possible without compromising OPSEC.[79] The way in which the knowledge is dispersed is often innovative.

An organisation's 'knowledge store' is its repository of good ideas and lessons identified. When something fails, O'Sullivan and Dooley acknowledge that the 'learning gained can be a valuable asset for the organisation'.[80] This is something that is often overlooked by the counter-terrorism practitioner and academic. In an IED context we tend to focus on successful attacks and spend less time analysing device failure. In some instances of course, we may not even be aware of device failure which is where the likes of signals intelligence (SIGINT) and human intelligence (HUMINT) can fill the void. We need to consider what the implications of the perception[81] of device failure will be for device design and TTPs within the IED system.[82] The depth of an extremist organisation's 'knowledge reservoir'[83] will, in tandem with commercial thinking, 'determine the type and level of innovation possible'.[84] If it frequently carries out a post-attack analysis and retains the lessons identified – what worked, what did not – it will be in a much stronger position to innovate and avoid costly mistakes. However, the manner in which the knowledge is stored has to be balanced with the need to avoid compromising OPSEC. Terrorist training camps[85] are increasingly being pitched in the virtual world, due to UAV (Unmanned Aerial Vehicle) strikes and other disruptive operations. According to JIEDDO, 'social media websites, web-based video conferencing and virtual applications'[86] increasingly allow the extremist to obtain the necessary knowledge and share ideas. Encrypted messaging technology[87] is also providing the terrorist with new communication channels and methods of accessing information. Cloud computing, darknets and peer-to-peer networks are also areas of concern.[88] This is yet another element of the IED system, where the extremist is constantly innovating. The British Army is also aware of the need to 'institutionalise counter-IED knowledge and experience gained in Iraq and Afghanistan',[89] if it is to avoid costly mistakes and take best practice forward into the next campaign. Speaking from Afghanistan, Lt Col Martin Gliniecki, a bomb disposal expert, acknowledged the importance of a corporate memory for NATO as a whole, so that 'any future NATO deployment can utilise all those lessons identified and then subsequently those lessons learnt'.[90]

The previous paragraphs highlight that the definition of innovation supplied by O'Sullivan and Dooley has considerable utility in a counter-IED context. It is therefore proposed that the following modified definition be adopted, as it clearly outlines what innovation is, where it takes place and the purpose of innovation:

> The process of making changes, large and small, radical and incremental, to an element or elements of the IED system, that results in the introduction of something new and of value to the extremist organisation and contributes to its knowledge store.

In the commercial world O'Sullivan and Dooley state that there are four primary drivers of innovation[91] and they are all applicable to the IED system:

- ***Emerging technology.***[92,93] Smaller, cheaper electronic components, for example.
- ***Competitor actions.*** In this context, competitor equates to security force actions.
- ***New ideas from customers, strategic partners and employees.*** Bomb-making know-how from the virtual world, training camps, learning by doing. Strategic partners equates to state-sponsorship, i.e. Iranian EFP charges.
- ***Emerging changes in the external environment.*** Political/change of ideology, reduction of external support.[94]

This chapter further argues that extremist groupings usually innovate in order to keep the security forces off-balance and to remain unpredictable. Innovation in conventional warfare is also critical if one is to avoid becoming predictable and therefore vulnerable.[95] The chapter concurs with the work of Jackson,[96] in that extremists do not tend to innovate 'for the sake of innovating'. More often than not, the primary driver in this context is the evolving security environment in which they operate and the innovation itself will usually be of a tactical or technological nature. For example, increasingly hostile security force surveillance tactics may force the grouping to innovate tactically in terms of its IED component-part procurement strategy or caching tactics. In Northern Ireland and on the UK mainland PIRA developed sophisticated and innovative hides in order to ensure that components and weapons were handled or in transit for the least amount of time prior to an operation. Sometimes command wires were buried months in advance of an operation to reduce ground-sign in an attempt to defeat aerial surveillance and high-risk search procedures. In Afghanistan roadside IEDs were also buried for significant periods, with the Taliban simply waiting for a target of opportunity. Prior to the operation, against a logistics convoy for example, a battery pack would be rapidly attached to the buried main charge and concealed. This would reduce the potential exposure of the bomb layer to detection by UAV, and minimise his time on the ground and detection signatures.

In terms of technological innovation a bomb designer may deliberately select novel, off-radar components and materials with which to build the bomb in order to outwit the security forces and reduce the likelihood of suspicious members of the public raising the alarm. A tip-off by an alert member of staff at a self-storage unit in north-west London effectively foiled a major terrorist operation[97] in 2004. Remarkably, however, failed bomber Muktar Said Ibrahim was able to purchase 52 gallons of high-concentration hydrogen peroxide from a chemist in London without arousing suspicions. CCTV shows him wheeling the boxes out of the store in July 2005.[98] More importantly, it is widely acknowledged that there is a constant measure, countermeasure battle between the security forces and extremist groupings which drives innovation on both sides.[99] The use of Electronic Counter-Measures (ECM) by friendly forces is a classic example where the terrorist will look for a gap in the electro-magnetic spectrum and build a device accordingly. In numerous campaigns of violence extremists and security forces have vied for control of the spectrum. It was widely believed that Iran provided sophisticated telemetry modules[100] used to arm the firing switch within Explosively Formed Projectile (EFP)[101] devices fielded in Iraq.[102]

Whilst there are tactical and technical reasons for deviating from established IED designs the requirement to preserve OPSEC sometimes produces what may appear to be an unusual departure from device design and tactics, particularly if the extremist grouping considers itself to be compromised by informants. Sometimes device failure is attributed to informant activity as opposed to a technical problem. The departure will not always involve a new design, the grouping may simply delve into its knowledge store and reintroduce an historical method of attack.

In 1978 military intelligence officers fighting PIRA in Northern Ireland stated that changes in weapon design would be based on a number of factors, including 'the evolutionary improvement of existing homemade devices'. Furthermore they argued that the terrorist had a 'tendency to resort to methods that have been successful in the past'.[103] These observations, made well before terrorism studies became fashionable, are still valid today. If a particular type of bomb design is seen to be effective, there is usually little reason to modify it. What is 'effective'? A device would be deemed to be effective if it functioned as intended and circumvented security force counter/mitigation measures. An example would be an anti-armour weapon, as long as the device successfully breached or immobilised the target vehicle, as intended, there would be little or no reason to change the design. Sometimes an improvised weapon may simply be deployed to induce panic or fear, to reinforce a message or demonstrate capability and intent. In 1994, PIRA attacked Heathrow airport with MK6 improvised mortars.[104] None of the mortars exploded, but that may well have been the intention of the terrorist team. In any event the attack was highly effective, the weapon system breached the security of one of the most important airports in the world and prompted a massive security operation. The incident was quickly covered by the media and the story received international attention.

The MK6 mortar was a tried and tested improvised device,[105] so much so that the PIRA were confident enough to utilise it on the British mainland, against a prestige target. What was highly innovative was the tactical emplacement/deployment of the device. One mortar baseplate (launch platform) was buried alongside the perimeter fence of Heathrow airport and the mortars were launched by electronic time delay firing switch. Another baseplate was concealed on the backseat of a Nissan Micra parked in a nearby car park. One of the tubes in the baseplate was empty apart from a small 'window breaker' charge. This fired first clearing the back windscreen and allowing the mortars to deploy without obstruction.

Counter-IED doctrine and the IED system

Given that the terrorist will innovate in response to the actions of the security forces it is necessary to look briefly at what those actions are and where they have an impact on the IED system. The overall UK counter-terrorism strategy (CONTEST)[106] seeks to:

- ***Pursue.*** Detect, investigate, disrupt terrorist operations before they occur and prosecute individuals.
- ***Prevent.*** Tackle radicalisation and inhibit support for terrorism.
- ***Protect.*** Identify vulnerabilities within border security, transport networks, crowded places, critical national infrastructure and safeguard hazardous materials.
- ***Prepare.*** Lessen the impact of a terrorist attack by training, testing and equipping response and recovery agencies.

More precisely, at the C-IED level the strategy is based on key operational activities (KOA):[107,108]

- ***Deter.*** Attack the 'intent' of a potential bomb-maker, or bomb layer. IO campaign, to hinder recruitment for example. Counter-Terrorism Act 2008, risk of prosecution for obtaining IED related information from the internet.

- ***Predict.*** Analysis of previous IED attacks, understand IED TTPs and mindset, the operating environment, human terrain, ideology and likely target set.
- ***Prevent.*** Primarily an offensive operation to target key elements of the IED system in order to prevent an IED being constructed, delivered and emplaced. Largely based on all-source intelligence.[109]
- ***Detect.*** IMINT (Imagery Intelligence) platforms, reporting from local population, route clearance and search operations.
- ***Neutralise.*** IEDD teams,[110] use of ECM, deter a suicide bomber[111] through use of obstacles, force.
- ***Mitigate.*** Technical mitigation, reduce effects of explosion,[112] via blast walls, body armour, vehicle armour. Tactical, minimum man risk[113] and other TTPs. Information mitigation, disrupt extremist attempts to capitalise on incident, through social media, for example, and disrupt BDA activities.
- ***Exploit.*** Critical element of counter-IED KOA. Provides actionable intelligence for offensive targeting operations, identifies source of components, bomb-maker technical signatures, forensic and biometric intelligence (FABINT) and provides opportunities to predict future terrorist intent. A function of Weapons Technical Intelligence (WTI).

When a KOA applies sufficient pressure to the IED system, the extremist will attempt/be forced to innovate. The cycle can be reversed, whenever the IED system applies sufficient pressure to C-IED KOA, the security forces will be also attempt/be forced to innovate and so it goes on. Both sides have to be mindful of how their actions will influence either threat or countermeasure. Sometimes a countermeasure innovation can lead to unintended consequences, for example, forcing the extremist to follow a more lethal path or one where the materials for the device are harder to control or are more freely available. Equally, on the extremists' side, if they field an innovation that is particularly unusual it may enable the security forces to produce actionable intelligence following exploitation.

Conclusion

The chapter has clarified the meaning of the term innovation within an IED context. It has demonstrated that innovative activity takes place throughout the IED system. Furthermore it has illustrated what form that activity may take and why, using examples from previous campaigns of violence. The study should be regarded as a template or starting point for the analysis of current, IED-based activity in places like Nigeria, the Yemen, Pakistan and Somalia. It may also be used as a predictive methodology in order to identify where the extremist may go next or what his trajectory might be, particularly if one is aware of the future intentions of the security forces in terms of TTPs and technology. Useful indicators and warnings may also be developed. Returning to Europe it might be used to re-examine the Madrid bombings, London 7/7 or the Anders Breivik case in Norway. A starting point would be to identify the boundaries of the IED system in each case.

The Provisional IRA has featured heavily in this report and they have renounced violence, but before the guns fell silent the organisation had a truly international reach. Its IED system was carefully designed to defeat the security *modus operandi* of a European nation and in lethal fashion it often did. Its members were able to wage a guerrilla war in bandit country,[114] planting roadside bombs and booby-traps to target the security forces one minute and bring terror to UK cities the next. The group's ideology may have been different to the new breed of terrorist but its TTPs are still relevant today.

Looking to the future, there is much discussion about the migration of the threat from Iraq, Afghanistan, Pakistan and other conflict zones to Western shores. More work needs to be done to clearly identify the transferable skills. In such theatres of operation the human terrain, security, resource, target and ideological environment is very different to our own. In addition sensationalist journalism must not be allowed to drive our train of thought.

That said, key areas for concern and innovative activity amongst extremist groupings might be:

- increasing awareness of FABINT countermeasures as weapons intelligence procedures mature;
- further development of novel explosives using off-radar materials in places like Yemen and the dissemination of detailed preparation instructions using on and off-line means;
- innovative hostile vehicle attack methods, designed to counter mitigation measures;
- evolving suicide attack TTPs designed to defeat European security procedures;
- improved interactive and secure online learning environments, increasingly replacing the physical training camp;
- emerging off-radar technology being used to fabricate novel initiation systems.

One thing is certain, the extremist will continue to innovate and we need to maintain a listening watch if we are to stay one step ahead. In addition we must be aware of the cyclical nature of the measure, countermeasure battle. What is valid today may be null and void tomorrow. We must also step back from the actual device and consider where innovation may occur within the IED system as a whole. The printing press in Pakistan that produces the labels for spider devices leads to the spider.

It is important to stress that this chapter is not intended to be an exhaustive study, rather it is designed to stimulate debate within the academic community and promote new lines of enquiry within the field of IED innovation. The chapter has deconstructed the IED system and highlighted some of the areas where innovation may occur, from contemplation to detonation and beyond.

Notes

1 Brian Jackson, 'Technology Acquisition by Terrorist Groups: Threat Assessment Informed by Lessons from Private Sector Technology Adoption', *Studies in Conflict and Terrorism*, 24 (2001), pp. 183–213. Jackson appears to have initiated the study of innovation in this arena.
2 Maria. J. Rasmussen and Mohammed M. Hafez, *Terrorist Innovations in Weapons of Mass Effect: Preconditions, Causes and Predictive Indicators, Workshop Report*, Defense Threat Reduction Agency (October 2010), p. 1.
3 For example Brian A. Jackson, 'Aptitude for Destruction: Volume 1. Organisational Learning in Terrorist Groups and its Implications for Combating Terrorism', (Santa Monica: Rand Corporation, 2005).
4 Adam Dolnik, *Understanding Terrorist Innovation: Technology, Tactics and Global Trends* (London and New York: Routledge, 2007).
5 Terrogence workshop, C-IED, London 2012. Referred to use of RCIED (Radio Controlled IED) pack in suicide vest.
6 Thijs Turel, 'Learning Terrorist Organisations', Master's Thesis, International Relations (16 September 2005), p. 2.
7 Nick Hopkins and Sarah Hall, 'David Copeland: A Quiet Introvert, Obsessed with Hitler and Bombs', *Guardian*, 30 June 2000. Online. Available at www.guardian.co.uk/uk/2000/jun/30/uksecurity.sarahhall, accessed 14 September 2012.
8 Rasmussen and Hafez, 'Terrorist Innovations in Weapons of Mass Effect' (October 2010) and Maria. J. Rasmussen and Mohammed M. Hafez, 'Terrorist Innovations in Weapons of Mass Effect: Phase 2', Defense Threat Reduction Agency (January 2012).
9 Rasmussen and Hafez, 'Terrorist Innovations in Weapons of Mass Effect' (January 2012), p. 3.
10 Colombian Military Press Office (2011) 'Hallados 13 cilindros bomba en el departamento de Cesar'. Online. Available at: www.escom.mil.co/index.php?idcategoria=281400, accessed 13 September 2012.
11 Theodore W. Karasik, Toxic Warfare (Santa Monica: Rand Corporation, 2002), p. 27.
12 Rasmussen and Hafez, 'Terrorist Innovations in Weapons of Mass Effect' (January 2012), p. 3.
13 David McKittrick, Don Macintyre and Ian Mackinnon, 'Loyalists Attack Sinn Fein HQ: Car Bomb in Belfast but Republicans are Adamant that Peace Process will not be Halted', *Independent*, 5 September 1994. Online. Available at www.independent.co.uk/news/loyalists-attack-sinn-fein-hq-car-bomb-in-belfast-but-republicans-are-adamant-that-peace-process-will-not-be-halted-1446817.html, accessed 13 September 2012.
14 Bruce Cochrane, 'Explosive Legacy', Intersec (1 July 2012).
15 Brian McCaffrey, 'Loyalists Hold onto Weaponry', *Irish News*, 6 October 2005. Online. Available at www.nuzhound.com/articles/irish_news/arts2005/oct6_Loyalists_hold_on_to_weapons.php, accessed 14 September 2012.

16 Mumbai style small-arms attack key concern, see *CONTEST: The United Kingdom's Strategy for Countering Terrorism*, London: Stationery Office, July 2011, p. 97. Online. Available at www.homeoffice.gov.uk/publications/counter-terrorism/counter-terrorism-strategy/strategy-contest?view=Binary, accessed 14 September 2012.
17 Rasmussen and Hafez, 'Terrorist Innovations in Weapons of Mass Effect' (January 2012), p. 6.
18 Rasmussen and Hafez, 'Terrorist Innovations in Weapons of Mass Effect' (January 2012), p. 16.
19 Toby Harnden, 'Three Iranian Factories 'Mass-Produce Bombs to Kill British in Iraq', *Daily Telegraph*, 20 August 2006. Online. Available at www.telegraph.co.uk/news/1526809/Three-Iranian-factories-mass-produce-bombs-to-kill-British-in-Iraq.html, accessed 15 September 2012.
20 Dolnik, *Understanding Terrorist Innovation*, p. 56.
21 See Mike Dolomore in Rohan Guneratna (ed.), 'Terrorism in the Asia Pacific: Threat and Response' (Eastern University, 2003), p. 105.
22 NATO doctrine.
23 The Oxford online dictionary defines a device as 'a thing made or adapted for a particular purpose, especially a piece of mechanical or electronic equipment'. Available at http://oxforddictionaries.com/definition/english/device, accessed 15 September 2012.
24 Allied Administrative Publication (AAP)-6, *NATO Glossary of Terms and Definitions*, 2010.
25 See image 'Incendiary Device', Imperial War Museum, London. Available at www.iwm.org.uk/collections/item/object/30019741, accessed 15 September 2012.
26 Duncan Gardham, 'Glasgow Bombing Jury Sees Effects of Exploding Gas Canisters', *Daily Telegraph*, 7 November 2008. Online. Available at www.telegraph.co.uk/news/uknews/3400759/Glasgow-bombing-jury-sees-effects-of-exploding-gas-canisters.html#, accessed 15 September 2012.
27 'Marines Adapt, Improvise, Overcome', YouTube. Online. Available at www.youtube.com/watch?v=gyPI438nGjI, accessed 15 September 2012.
28 J. Edwards, J. Fletcher, C. Bain and S. Dennis, 'It's Just a Little Hobby of Mine; Mardi Gra Bomber; Blackmailer Made Bomb in his Kitchen', *Daily Mirror*, 8 April 1999, p. 9.
29 Simplicity is a key element of innovation and will be examined in more detail in following paragraphs.
30 Resources, such as equipment, manpower, knowhow, finance.
31 UK Forces Afghanistan blog post, 'Canadian Cowboy Trains British Soldiers to Spot IEDs', 29 July 2010. Online. Available at http://ukforcesafghanistan.wordpress.com/2010/07/29/cowboy, accessed 15 September 2012.
32 Excellent example of improvised munitions at 'Libya Rebels Make Weapons from Scraps', *Al Jazeera English*, uploaded 14 June 2011, YouTube. Online. Available at www.youtube.com/watch?feature=player_embedded&v=jCyRBUAEneM, accessed 15 September 2012.
33 FSA IEDs at 'Improvised Explosives Prepared by the Free Syrian Army are Pictured in Front of a House in Al-Ezaa district in Aleppo'. Online. Available at www.trust.org/alertnet/multimedia/pictures/detail.dot?mediaInode=6a76c937–9f0e-441c-8449-e5dea2f19fc1, accessed 15 September 2012.
34 Possible FSA directional charge being fabricated, uploaded 4 July 2012, YouTube. Online. Available at www.youtube.com/watch?feature=player_embedded&v=00YZhtOG-Gk#!, accessed 15 September 2012.

35 C. J. Chivers, 'Mao's Rockets and Modern War, Part 3', *New York Times*, 19 December 2011. Online. Available at http://atwar.blogs.nytimes.com/2011/12/19/more-on-maos-rockets-and-modern-war-part-iii, accessed 16 September 2012.
36 US Department of State.
37 Toby Harnden, 'Provisional Leadership 'behind US Gun Maze', *Daily Telegraph*, 14 June 2000. Online. Available at www.telegraph.co.uk/news/worldnews/northamerica/usa/1342843/Provisional-leadership-behind-US-gun-maze.html#, accessed 16 September 2012.
38 Loyalist paramilitaries modified blank firing pistols to fire live ammunition, for use in CQA. The smooth-bore barrel hampered forensic follow-up.
39 Bill Roggio, 'Mahdi Army Uses "Flying IEDs" in Baghdad', *The Long War Journal*, blog post, 5 June 2008. Online. Available at www.longwarjournal.org/archives/2008/06/mahdi_army_uses_flyi.php, accessed 16 September 2012.
40 Rob Olver, BFBS, 'Soldier Inventions in Afghanistan', 11 April 2012. Online. Available at www.bfbs.com/news/media/video/56398/20415, accessed 16 September 2012.
41 Bedford Museum, 2005, 'Aniseed Balls and the Limpet Mine', 6 July 2005. Online. Available at www.bbc.co.uk/history/ww2peopleswar/stories/53/a4376153.shtml, accessed 16 September 2012.
42 Press Association, 'Asim Kausar Jailed over Ricin Recipe and Bomb-Making Instructions', *Guardian*, 27 January 2012. Online. Available at www.guardian.co.uk/uk/2012/jan/27/asim-kausar-jailed-ricin-recipe?INTCMP=SRCH, accessed 16 September 2012.
43 Interview with IEDD operator, 16 September 2012.
44 Bomb factories, for example.
45 Also laid in defensive belts, in keeping with mine warfare doctrine.
46 Or in some cases a mixture of the two.
47 C. J. Dick, *Mujahideen Tactics in the Soviet–Afghan War*, Conflict Studies Research Centre, January 2002, p. 4.
48 Home Office, *Protecting Crowded Places: Design and Technical Issues* (January 2012), p. 5.
49 David O'Sullivan and Lawrence Dooley, *Applying Innovation* (London: Sage, 2009).
50 Universidade de Aveiro, 'Researchers from IEETA/UA Develop Fall and Posture Detector', 7 August 2012. Online. Available at http://uaonline.ua.pt/detail.asp?lg=en&c=24674, accessed 16 September 2012.
51 HBS Managing Innovation 2012 course, 'Simplicity in Innovation – Getting to "Just Right"', 12 February 2012. Online. Available at http://blog.hbs.edu/hbsinov8/?p=1284, accessed 17 September 2012.
52 Cdr Cedric Richardson, 'JIEDDO's Robotics Programs, Military Robotics Summit', JIEDDO, 29 August 2012, p. 3.
53 Separated by an insulator.
54 Pierre Claude Nolin, 'Countering the Afghan Insurgency: Low-tech Threats, High-Tech Solutions', NATO, October 2001, p. 6.
55 O'Sullivan and Dooley, *Applying Innovation*, p. 13.
56 O'Sullivan and Dooley, *Applying Innovation*, p. 14.
57 O'Sullivan and Dooley, *Applying Innovation*, p. 14.
58 'Terror group "cloning cards at petrol stations"', *Daily Telegraph*, 23 April 2007. Online. Available at www.telegraph.co.uk/news/uknews/1549409/Terror-group-cloning-cards-at-petrol-stations.html, accessed 17 September 2012.
59 NATO doctrine.
60 NATO doctrine.
61 David Connett and Martin Whitfield, 'Heathrow Bombing: IRA Exposes Airport's Vulnerability. Perimeter Protection "Almost Impossible". Delay in

Closing Runway Defended', *Independent*, 11 March 1994. Online. Available at www.independent.co.uk/news/uk/heathrow-bombing-ira-exposes-airports-vulnerability-perimeter-protection-almost-impossible–delay-in-closing-runway-defended-1428295.html, accessed 17 September 2012.

62 Army Sgt 1st Class Tyrone C. Marshall Jr, 'Future Holds Danger, Uncertainty in Complex World, General Says', American Forces Press Service, 15 May 2012. Online. Available at www.jcs.mil/newsarticle.aspx?ID=912, accessed 17 September 2012.

63 Rick Atkinson, 'You Can't Armor Your Way out of This Problem', *Washington Post*, 2 October 2007. Online. Available at www.washingtonpost.com/wp-dyn/content/article/2007/10/01/AR2007100101760_4.html?sid=ST2007092900754, accessed 17 September 2012,

64 Members of the local populace who observe and report troop/target movement.

65 For example a Motorola radio, car alarm fob trigger or battery pack.

66 Dickers were often unarmed children.

67 During the preparation and execution phase.

68 Lt Col S.E. Snedden, Defence Research Paper, *Northern Ireland: A British Military Success or a Purely Political Outcome?* Advanced Command and Staff Course, Number 10, September 2006 – July 2007, p. 9.

69 Evan Kohlmann, NEFA, Nine Eleven Finding Answers, senior investigator, 'Prominent Jihad Media Organisations in Central Asia, 2009', *Counterterrorism Blog*, 4 March 2009. Online. Available at http://counterterrorismblog.org/2009/03/new_nefa_chart_prominent_jihad.php, accessed 18 September 2012.

70 Hanna Rogan, Abu, 'Reuters and the E-Jihad: Virtual Battlefronts from Iraq to the Horn of Africa', Culture and Society, *Georgetown Journal of International Affairs*, Summer/Fall 2007, p. 91.

71 This could be likened to manufacturing a new household appliance for which there was no obvious demand.

72 O'Sullivan and Dooley, *Applying Innovation*, p. 5.

73 If the extremist grouping has access to a safe testing ground, limited feedback can be acquired earlier on in the process which is preferable to 'testing on the job'.

74 All too often, court reporting provides the terrorist with useful BDA-related information. PIRA made a point of scrutinising court documents, for the purpose of BDA and to clarify security force TTPs.

75 Hoax devices are often used by extremist groupings in order to test the security force response, a form of learning activity. All too often incidents involving hoax devices have been ignored, by practitioners and academics alike. In actual fact this type of incident should be carefully scrutinised as it may provide clues as to future intent. In effect, a hoax device is a way of 'testing the marketplace'.

76 A safe haven can be used for IED testing/development activities, which enhances/hastens learning and reduces the likelihood of device failure . Hence the importance of the fact that, for example, Anders Breivic had access to a remote testing location.

77 O'Sullivan and Dooley, *Applying Innovation*, p. 9.

78 Akiva J. Lorenz, 'The Threat of Maritime Terrorism to Israel', IICC, 1 October 2007, p. 21.

79 abcLocal, 'Al Qaeda's Bomb Maker may be Training Others', 9 May 2012. Online. Available at http://abclocal.go.com/kabc/story?section=news/national_world&id=8654911, accessed 17 September 2012.

80 O'Sullivan and Dooley, *Applying Innovation*, p. 11.

81 Sometimes the extremist grouping may perceive that the device has failed when in actual fact it was effective.
82 There are potential opportunities to shape threat trajectory through IO at this stage.
83 Its corporate memory.
84 O'Sullivan and Dooley, *Applying Innovation*, p. 11.
85 The traditional learning environment, where best practice would be disseminated and new ideas discussed.
86 JIEDDO, Lt Gen. Michael D.Barbero, *C-IED Beyond 2014*, 12 December 2011, p. 5.
87 Blackberry Instant Messaging, for example, used to great effect during the London riots, 2011.
88 *CONTEST*, p. 34. Online. Available at www.homeoffice.gov.uk/publications/counter-terrorism/counter-terrorism-strategy/strategy-contest?view=Binary, accessed 17 September 2012.
89 Lt Col Bob Seddon, C-IED conference, London, 2012.
90 Lt Col Martin Gliniecki, IQPC, Counter-IED 2011, Transcript. Online. Available at www.iqpc.com/uploadedFiles/EventRedesign/UK/2012/June/11590006/Assets/LTCMartinGliniecki_transcript_proofed.pdf, accessed 19 September 2012.
91 O'Sullivan and Dooley, *Applying Innovation*, p. 12.
92 JIEDDO, nano technology, etc.
93 Some terrorist groups may maintain a 'listening technology watch' for items that could enhance the effectiveness of their IED operations, for example by using news aggregators. Security forces may be able to interdict these efforts or at least monitor them.
94 Support comes in different forms: safe-havens, components, advice/training, finance. Some external supporters will want to distance themselves from the extremist organisation by providing indirect support.
95 Air Vice-Marshal R. A. Mason, 'Innovation and the Military Mind', January–February 1986. Online. Available at www.au.af.mil/au/awc/awcgate/au24–196.htm, accessed 19 September 2012.
96 Jackson, 'Technology Acquisition by Terrorist Groups', p. 11.
97 Duncan Gardham and Bonnie Malkin, 'Fertiliser Bombers Jailed for at Least 95 Years', *Daily Telegraph*, 30 April 2007. Online. Available at www.telegraph.co.uk/news/uknews/1550112/Fertiliser-bombers-jailed-for-at-least-95-years.html, accessed 19 September 2012.
98 Sandra Laville, 'Terror Suspects Shown on CCTV Buying Chemical', *Guardian*, 20 January 2007. Online. Available at www.guardian.co.uk/uk/2007/jan/20/terrorism.world, accessed 20 September 2012.
99 Hagai Segal, 2007, 'Cats Chasing Deadly Mice', Insight, *South China Morning Post*, 6 February 2007. Online. Available at www.hagaisegal.com/pdfs/SCMP_15_Feb_6_07_UK_kidnap_plot.pdf, p. 13, accessed 20 September 2012.
100 An attempt to defeat ECM by identifying a frequency that was not being jammed.
101 It is alleged that Iran supplied the insurgents with the technical know-how, equipment and components needed to fabricate EFP devices.
102 Atkinson, 'You can't Armor Your Way out of this Problem'.
103 Only a Northern Song blog, 'The Glover Report', 9 May 2012. Online. Available at http://northernsong.wordpress.com/2012/05/09/the-glover-report.
104 Andrew Gliniecki, Steve Boggan, James Cusick and David McKittrick, 'Heathrow: Flight Chaos after IRA Threats: Heathrow and Gatwick Shut in New Alerts', *Independent*, 14 March 1994. Online. Available at www.independent.

co.uk/news/heathrow-flight-chaos-after-ira-threats-heathrow-and-gatwick-shut-in-new-alerts-1428941.html, accessed 21 September 2012.

105 Minor, incremental changes had been made to its design over the years.

106 *CONTEST*, p. 34. Available at www.homeoffice.gov.uk/publications/counter-terrorism/counter-terrorism-strategy/strategy-contest?view=Binary, accessed 17 September 2012.

107 New Zealand Army, 'Familiarise Yourself with CIED Measures', 18 November 2010. Online. Available at www.army.mil.nz/at-a-glance/news/army-news-archive/415/fywciedm.htm, accessed 20 September 2012.

108 Rob Hyde-Bales and Edith Wilkinson, 'How Can Information Sharing between the UK Military and Civilian Organisations be Improved?', *Defence Procurement International*, Summer 2012, p. 2.

109 HUMINT, SIGINT, IMINT, FABINT, TECHINT, for example.

110 With the preservation of forensics being critical to exploitation.

111 If the bomber thinks that the attack won't succeed he may seek an alternative target or may call off the operation, mission success being the overriding factor as opposed to fear of death.

112 Blast energy, fragmentation.

113 Disperse personnel as widely as possible to avoid mass casualties.

114 South Armagh/rural border country.

4 The dynamics of terrorist innovation

Adam Dolnik

Introduction

This chapter revisits some of the conclusions reached in this author's earlier work, *Understanding Terrorist Innovation: Technology, Tactics and Global Trends* (London: Routledge, 2007). The chapter first focuses on making some observations about the general trends in tactical and technological innovation in terrorist campaigns, followed by a focus on the factors and conditions responsible for the differences between innovative tendencies of different terrorist organizations. Finally, the conclusion will identify implications for the future, especially in relation to innovative patterns of al-Qaeda and its affiliates.

First, some conceptual issues need to be clarified. While the topic of innovation in terrorist campaigns has until recently not been widely researched, the body of strategic studies literature focusing on state-level military innovations is considerably more extensive, and offers several important conceptual observations. The first one concerns the definition of "innovation" itself. Interestingly, most authors in the field of strategic studies have neglected to define the term, choosing instead to treat it as synonymous with "major military change."[1] Some strategic studies scholars such as Farrel and Terriff have criticized this approach, arguing that innovation constitutes only one of three pathways whereby military change occurs, the other two being adaptation and emulation. According to the definitional distinction of these authors, innovation involves "developing new military technologies, strategies, tactics and structures." Adaptation then constitutes the "adjusting [of] existing military means and methods." Last, emulation involves "importing new tools and ways of war through imitation of other military organizations."[2] Another useful angle has been provided by Evangelista, who defined *technological* innovation as "development of a new military technology that leads to significant changes – for example in the realm of strategy, in the organization of military forces, or in the distribution of resources among services." According to Evangelista, the term does not refer to "incremental improvements in characteristics of weapons that arguably constitute the main activity of military research and development."[3]

As we can see from these examples, the few authors that have chosen to define innovation have focused mainly on a very narrow aspect of military change, concentrating on innovations that are indeed completely new at the global level. In the realm of terrorist organizations, however, such a narrow definitional approach is hardly applicable, as we would be hard pressed to find a *single* terrorist invention that would fit either Evangelista's or Farrel and Terriff's definitions. Terrorist innovation would simply fall into the realm of emulation and adaptation, as technologies used by terrorists have never been completely new. In this sense revolution in terrorist technology has essentially consisted of attempts to emulate or adapt an existing capability, but has never really opened up a capability never before possessed. Terrorists have effectively engaged in what some authors have labeled evolutionary innovation (incremental) as opposed to revolutionary (radical) innovation.[4] For the purposes of this chapter, innovation will be defined as "an act of introduction of a new method or technology or the improvement of an already existing capability."

Historical trends in terrorist innovation

An exhaustive survey of all recorded terrorist incidents over the last 60 years reveals a clear pattern – contrary to popular perception, the scope of terrorist tactics and technologies is relatively limited and remarkably unchanging. In fact when one surveys the list of terrorist operations case by case, very few incidents strike the observer as creative *in any way*. This seems to confirm the observation previously made by Hoffman that terrorists are conservative by nature.[5] The main advances that we have seen in both tactics and technologies have consisted of incremental innovation, in the sense that terrorists have arguably improved in using traditional tactics and have made a better use of already existing dual-use technologies. In this sense technological innovation in terrorist campaigns has been a direct result of what Rosen terms "technology push" – the situation in which advances in civilian technologies drive military innovations, as opposed to the "demand pull" – the scenario in which military innovations drive advancement in civilian technologies.[6] Additional points of incremental innovation have included the increasing range of home-made artillery, miniaturization and improvements in detonation of explosive devices, and the greater incorporation of commercially available technologies such as internet applications, smart phones, computers and GPS into terrorist operations at the planning and execution phases. Another important tactical trend has involved the synchronization of attacks and the incorporation of various tactics into a single operation. This is especially evident in the relatively recent proliferation of "Mumbai style" suicide operations, which have become the model for many of today's most spectacular attacks.[7] Somewhat surprising may be the observation of another emerging trend, and that is the increasing emphasis of modern terrorism on *technologically crude* modes of attack, as well as an increasing understanding among the terrorists of the hurdles involved in successful weaponization of chemical, biological, radiological and nuclear (CBRN) agents. Contrary to popular perception, the current wave of global terrorism is not moving in the direction of high technology, confirming Hoffman's earlier prediction that "terrorist devices will be innovative in their simplicity."[8]

Several important implications stem from these prevalent trends in terrorist innovation. First, the common "business approach" to understanding innovation, which assumes that terrorists always seek new technologies in order to boost the ever-increasing lethality of their attacks, is simply not supported by empirical evidence. At the same time, however, the more empirically grounded observation that terrorists are essentially conservative by nature, is negated by the highly innovative approaches of prominent groups like Aum Shinrikyo, al-Qaeda (AQ), Popular Front for the Liberation of Palestine – General Command (PFLP-GC), or Liberation Tigers of Tamil Eelam (LTTE). It must therefore be recognized that there is limited value in treating the issues of "terrorist innovation" as an abstract concept, without losing a great deal of accuracy and representativeness in the process. It seems more fruitful to focus on the specific factors and conditions that distinguish the rare but influential innovative terrorist organizations from the more common but operationally conservative groups, in order to attempt to predict innovation before it occurs. The exploration of the reasons behind the differences in the approaches of different terrorist groups to innovation will be the objective of the next section.

Exploring variance between innovative tendencies of individual groups

In the original study, eleven variables were tested on four different case studies – Aum Shinrikyo, Popular Front for the Liberation of Palestine – General Command (PFLP-GC), Riyadus Shaliheen Suicide Batalion (RAS) and November 17th (17N) in an attempt to identify the level of correlation of these variables with the level of innovativeness demonstrated by the respective groups.[9] The goal of this section is to provide an analytical summary of the findings, in order to attempt to explain the circumstances and characteristics that determine the level of a group's involvement in tactical and/or technological innovation.

Role of ideology and strategy

It has been argued that this variable's importance lies in the fact that it is in essence an organization's ideological foundation that frames the worldview of its members and thus provides a sense of collective identity. Ideology is also instrumental in identifying the enemy, while also providing the necessary explanation and justification for targeting. Moreover, ideology determines a group's core objectives and the strategy for how and by what means these objectives are to be achieved. And finally, ideology is a critical component in determining a group's ambitions, as well as the overall perception of urgency for armed action in order to fulfill the given aspirations. At the operational level then, the group's core strategy translates into the frequency and intensity of its military operations. This is where ideology, strategy and innovation meet, in the sense that terrorists' innovation had been hypothesized to be driven by the need to achieve the capability necessary for reaching and sustaining the level of intensity preferred by the group.

Undoubtedly, understanding the ideology and strategic objectives of a group is a key component of threat assessment, as both of these elements provide us with an insight into the logic behind a particular group's decision to launch a specific type of operation at a given time. But while these aspects of a group's ideology can tell us a lot about where a terrorist group might be headed in terms of innovation, we must be aware of the limitations in the reliability of such an analysis.

First, in some cases ideology seems disassociated from strategy, in the sense that it merely seems to serve as a retrospective justification for violence, rather than the pre-step toward formulating a strategy.[10] This suggests that our analysis must go deeper than simple ideological compartmentalization, if we are to accurately predict the potential of a given group in engaging in innovative operational methods.

Second, we must be careful to make a distinction between ideology and operational practicality. So while some ideologies such as the "cosmically-scientific" millennialism of Aum Shinrikyo may signal a clear ideological predisposition of a group to becoming technologically innovative, tremendous danger lies in applying a reverse generalization to ideologically conservative groups. In other words, the fact that a given group despises modernity does not necessary mean that it will be reluctant to use modern means to their advantage. For instance, the AQ belief system, while heavily conservative and retrospective in nature, with its desire to return to the time of the Prophet and the rejection of modernity and globalization, has not constrained the group from reaping the benefits that modernity brings for operational purposes. In fact, the group derives significant pleasure from using the enemy's own tools against it, as documented by the words of a former AQ member who stated that "[the 9–11 attack] was like taking [your enemy's] finger and poking [it] into his own eye."[11] Similarly, according to Osama's close aid Abu Ubeid al Qurashi: "The Westerners' rage increased once it became clear to them that we could use the same computers that they did without espousing the same values. Against all their assessments, our culture cannot be shattered by technology."[12] AQ's practicality in these matters can further be documented on the instructions given to their operatives to help them avoid detection in the West such as drinking alcohol, shaving their beards, watching television and avoiding usual prayer times and places.[13] This level of ideological pragmatism is likely to be present among innovative groups that embrace conservative ideologies, as they will at times need to possess the will to act in conflict with their own belief system in order to succeed operationally. Contrary to popular perception, however, very few conservative groups have actually demonstrated such qualities, providing one possible explanation for why innovation has been rare.

Overall, the original hypothesis that organizations whose ideology identifies an ideal outcome with regards to definite objectives, and which prescribes a time-frame and a specific course of action for reaching those objectives, are likely to demonstrate a higher level of tactical and/or technological innovation than organizations with vaguely defined goals, a low sense of urgency and a low level of strategic planning has been confirmed in all of the four case studies and can be supported by plentiful additional evidence. The principal reason for this finding might be the fact that organizations with a greater sense of urgency for attaining their goals within their lifetime are generally more inclined to continually escalate in order to provide an "instant decisive formula" for victory, which can sometimes trigger a decision to pursue innovative means. However, while many terrorist groups encounter a need to escalate at some point throughout their lifespan, this step usually involves only the increased frequency and intensity of attacks as well as the broadening of targeting categories, as opposed to radical changes to the *modus operandi* per se. As a result, it appears that an increased perception of urgency can serve as a reliable indicator of the potential to *escalate*, but not necessarily a reliable predictor of the decision to *innovate*. Whether the escalation step will coincide with innovation will apparently be determined by other factors that lay outside of the scope of this variable. Also relevant in this regard may be the observation that all of the innovative groups examined in the book have shared a high level of ideological and strategic inconsistency and fluctuation. Given the fact that innovative tendencies of these groups have generally corresponded to the shifts in ideology and strategy, it seems to provide additional proof that it may not necessarily be the ideology itself, but rather the frequency of shifts in the ideological and strategic outlook that determine the innovativeness of a group. In this sense it seems safe to conclude that ideologically flexible and strategically adaptive groups are likely to be more innovative than groups whose ideology and strategy remains constant throughout their lifespan.

To conclude, terrorist ideologies and strategies have some predictive value with regards to understanding a group's motivation to innovate, but by itself this factor is not sufficient in explaining what sets aside innovative groups from conservative ones. More so than the type of ideology itself, the importance seems to lay in the ideological and strategic flexibility and adaptability – groups whose perception of reality changes in the direction of an increased perception of urgency can be expected to escalate, possibly employing innovative means if other suitable conditions and factors are present.

Dynamics of the struggle

Another variable that seems relevant to terrorist innovation patterns are the dynamics of the struggle. In this sense, tremendous differences were hypothesized to exist among organizations that are equipped with an area in which they can operate freely, and urban guerilla organizations that have to rely on safe houses and training grounds located in the urban setting. While the former have the option to freely conduct research and training without the immediate fear of detection and obviation, the latter have to take tremendous security precautions to ensure that their experiments and training do not arouse suspicion among the ever-present strangers. Further, the nature of the struggle also reflects on its frequency and intensity, having a profound impact on both the decision to innovate, as well as the likelihood of success in the case of a positive attitude toward such a decision. First, guerilla organizations in territorial control of a safe haven usually engage the enemy on a larger scale and with greater intensity and are therefore in more of a need of sophisticated weaponry to use in the field. Second, when groups that are involved in reciprocal clashes with the government decide to innovate, their greater overall combat exposure not only translates into more experience with handling weaponry, but also provides more ample opportunity to battle-test the new innovations in the field. Thus, greater frequency of attacks is likely to have a profound impact on both the desire as well as the capability of terrorists to innovate.

Overall, the hypothesis that organizations with guerilla characteristics, such as frequent reciprocal clashes with the enemy armed forces and control of a territorial stronghold, are likely to be both more willing and more capable of innovation than urban terror groups that are confined in their training and operations to the municipal setting has been partially confirmed in all four case studies. But while some aspects of the "dynamics of the struggle" were apparently relevant, this variable alone fails to provide a viable explanation for why the innovation practices of Aum, PFLP-GC and RAS were so radically different from other groups that have operated in very similar working conditions. Looking at other examples, some innovative groups that operated in safe havens like the AQ, LTTE, or the Revolutionary Armed Forces of Colombia (FARC) show a high level of relevance of this variable; on the other hand why did urban groups such as Provisional Irish Republican Army (PIRA) or Red Army Faction (RAF) engage in technologically more innovative practices than groups with guerilla characteristics such as the Armed Islamic Group (GIA) or the Sendero Luminoso (SL)? The answer apparently lies outside of the scope of this specific variable, and therefore it seems that while favorable "dynamics of the struggle" can provide a group with the working conditions that can generally increase the chances for successful innovation, the decision to engage in innovative practices in the first place is driven by other considerations.

Countermeasures

Another variable that has been credited with a possible causal role in driving terrorist innovation have been the security countermeasures introduced by a state to provide protection against specific tactics used by terrorists in the past. According to this hypothesis, target hardening efforts by government security forces that render the tactics previously used by terrorists ineffective frequently provide the critical push for a group' decision to innovate in order to overcome these countermeasures. Examples of this phenomenon in the history of terrorism are abundant, and it is thus not surprising that the hypothesis has proven to be relevant in three of the four examined case studies.

However, although the countermeasures variable can be credited with providing the impetus to both tactical and technological innovation in many concrete instances, the original hypothesis that organizations whose *modi operandi* are frequently countered by the adversary will demonstrate a greater innovative drive than organization whose tactics are not effectively countered is not applicable *universally*. Still, most groups that have demonstrated innovative tendencies in the past have done so partly because of the countermeasures variable. And while the majority of groups have responded to the introduction of security obstructions by switching to less challenging targets as opposed to choosing to innovate in order to defeat these, what sets the innovative groups aside is a desire to specifically overcome the enemy countermeasures as a symbolic gesture. Alternatively, new devices have sometimes been the product of a lack of access to modern weapons, as opposed to a forward-looking innovation. This suggests that while in our minds unconventional weapons typically represent an escalation, in reality they can also be employed out of desperation or a simple lack of conventional options. Such causes of technological innovation however, are unlikely to lead into a progression to a mass destruction capability.

Overall, while in many cases terrorist innovation has indeed been the product of an "arms race" between the terrorists and their enemies, terrorist groups are not necessarily limited to reactive strategies. Additional sources of the motivational drive to innovate, other than the need to adjust the *modus operandi* in order to survive in a more stringent security environment, apparently exist.

Targeting logic

Targeting logic is another factor that can have a strong effect on the level of innovation demonstrated by a particular group. At the most basic level, it has been hypothesized that terrorists identify the scope of their targets, and then seek to attain the capability to attack these targets at the desired scale. In other words, groups that embrace a very narrow and discriminate targeting logic will depend on a *modus operandi* that will allow such targeting, while organizations that embrace a highly indiscriminate targeting logic are more likely to engage in the process of innovation in order to obtain adequately destructive means of attack.

Another relevant dimension of a group's targeting logic has been the level of rigidity with which the given group approaches the issue. Organizations that have a highly rigid approach to their targeting throughout the entire period of their existence have been hypothesized to be less likely to demonstrate technologically innovative tendencies than organizations whose targeting logic is flexible in terms of frequency and extent of targeting shifts.

Overall, the assertion that the less discriminate and the more lethal the targeting logic of the group under scrutiny, the greater the organization's propensity to innovation has been confirmed in all of the case studies. Further, at the level of flexibility of targeting logic, the hypothesis that organizations whose targeting preferences shift over time are more likely to be innovative than groups embracing a rigid targeting logic throughout their lifespan has been confirmed as well. This particular phenomenon is closely associated with the point made earlier about the ideological flexibility and adaptability of innovative groups, as modifications in the strategic department are typically accompanied by shifts in the targeting scope as well.

However, it is also important to note that indiscriminate and highly lethal targeting is not always achieved through innovative means. For instance, some of the historically most lethal terrorist groups such as Armed Islamic Group (GIA) or the Lord's Resistance Army (LRA) rank among the operationally least innovative ones, while comparatively discriminate groups such as PIRA have engaged in an infinitely higher level of innovation. In fact, many highly discriminate operations might require a much higher level of operational talent than indiscriminate massacres, as in the example of highly focused assassinations of well protected persons.[14] For instance, the RAF's six assassinations of heavily guarded Germans all involved highly sophisticated weapons that included state-of-the-art remote-controlled bombs triggered by a light beam and high-powered rifles.[15] Similarly the ETA assassination of Louis Blanco involved the casing of the Admiral's travel route for over 12 months, the digging of a 20-foot tunnel from a basement under the road and the placing of three large devices under the road timed to detonate in intervals of one-tenth of a second to match the speed of the car.[16] Both of these groups were certainly more sophisticated in their approach than the GIA during its village massacres in the 1990s. As a result, it seems that even highly discriminate and highly rigid organizations can sometimes be driven to innovate in order to maintain their ability to conduct "surgical strikes" against their desired targets, even after these have been hardened. This point makes it clear that lethality cannot automatically be equated with innovation, just as innovation is frequently not associated with indiscriminate targeting.

In sum, while radical innovation in the category of mass-casualty CBRN attacks is likely to be pursued by organizations embracing a highly lethal and indiscriminate targeting logic, this observation does not necessarily apply to the realm of incremental or tactical innovation in the same way. Nevertheless, for the purposes of this analysis we are concerned primarily with the potential of terrorist groups to innovate in the direction of achieving mass-casualty capable weapons, and in this context the fluctuation of a group's "targeting logic" provides a useable indicator.

Attachment to weaponry

A particular organization's preferences in terms of *modus operandi* and weapons selection can also sometimes be driven more by non-rational factors rather than purely strategic or cost-benefit considerations. In fact, this non-rational attachment to a particular tactic or weapon has been found to be one of the key determinants of what type of *modus operandi* a group can be expected to adopt.[17] Examples of the importance of the "attachment to weaponry" variable as a determining factor behind a group's drive to pursue particular means of attack are abundant. For instance, the first terror groups to ever use dynamite – the Russian Narodnaya Volya, the transnational Anarchists and the Irish Fennians – all reportedly worshiped this explosive as the ultimate revolutionary weapon of the people that would topple the old world order and bring about the new secular millennium through its scientific, humane and even mystical powers.[18] Perhaps the most glaring example in this regard is the expressive nature of the 1995 Oklahoma City bombing, in which Timothy McVeigh meticulously followed the scenario described in the *Turner Diaries*, the William Pierce novel that has become the bible of right wing militias in the US.[19] Not only did the target, tactic and explosive mixture correspond precisely to the scenario described in the book – even the size of the explosive device was imitated. As such the *Turner Diaries* are a crucial text, in the sense that the book takes great expressive pride in the ability of the "patriots" to manufacture improvised weapons and devices from ordinary household items, because sophisticated weaponry in the novel's scenario is unavailable. Given the obsession of the American militia movement with this book, tactical improvisation with dual-use items is a much more likely scenario for the militia groups than a quest for highly novel technologies. In the same way, in AQ's case it has been the expressive emphasis on martyrdom as the principal jihadi vanguard and as the best way to achieve a victory of Islam that has been one of the key determining factors behind the group's *modus operandi.* Given the prominence of martyrdom in seminal AQ texts such as the "Declaration of Jihad against the Jews and Crusaders" and Ayman al Zawahiri's last will, "Knights under the Prophet's Banner," is it any surprise that since 1998, every single attack launched directly by AQ had involved suicide delivery? Another contemporary aspect of terrorist tactics that deserves close attention is the beheading practice used by Islamist groups in Bosnia, Algeria, Tajikistan, Abkhazia, Chechnya, Pakistan, Iraq, Syria and elsewhere. Here too we can trace the origin of this tactic to the literal interpretation of Allah's pronouncement: "When you encounter those [infidels] who deny [the Truth = Islam] then strike [their] necks."[20] While this quote is certainly taken out of context when used to support the contemporary practice of beheadings, for groups like Zarqawi's *Jama'at al-Tawhid wal Jihad* whose logo depicted a *mujaheed* holding a blood-soaked sword, it certainly represented a strict observance of the Qur'an in the purest sense. Interestingly, the slitting of throats has also been practiced by non-Islamist groups for punishment purposes, such as those of SL. Here again the expressive element was present, this time however deriving from Andean mysticism, where a person who is killed in this way cannot be saved because his or her soul cannot escape from the mouth.[21] As we can see from these examples, the attachment to a particular tactic or weapon is one of the key determinants of what type of *modus operandi* a group can be expected to adopt.

Overall, the groups that have been driven to tactically and/or technologically innovate differ from conservative organizations by the presence of a significant expressive or symbolic attachment to a particular weapon, tactic or the process of innovation itself, or in the sophistication needed to achieve this capability, to be more precise. While conservative groups often embrace a similar attachment, its fulfillment either does not require significant deviations from the group's present capability or the emphasis is on different values than innovation. Of all of the variables examined in the book the attachment to a particular weaponry or tactic seems to have the strongest predictive value with regards to providing an indicator of a group's attraction to using innovative means. As mentioned, other factors such as ideology and strategy, or countermeasures can have a profound impact on the modifications in a group's operations, but whether this process is accompanied by innovation or not will be determined directly by this variable.

Group dynamics

The first component of this variable was the background, the value system and the authority of the leader as a key determinant of the motivation of such a figure to instigate radical innovation, as well as his or her ability to successfully impose such a decision on the rest of the group. In this sense, the group structure is also extremely important. First, the structure shapes the decision-making dynamics, determining whether major operational decisions are based on a consensus of all members, or are rather a product of a top-down approach with the group's penultimate leadership making the decision and passing it on to operational sub-units or cells for execution. At the final level, ideational, operational or power disputes within a group can also sometimes trigger escalation or innovation as a reconciliation tool that will help the group overcome the differences and unite by channeling their energy into a major effort to strike the enemy.

Overall, the lessons of the group dynamics factor show that organizations led by an uncontested leader that provided a strong drive toward innovation were the most likely candidates to take this path and to complete it successfully. In contrast, incremental innovation and the operational evolution of groups over time has been most frequently a bottom-up process. However, as in the case of other variables tested throughout the book, there was the repeated problem of specificity. What seems to be the most important implication here is that radical innovation is likely to be driven by a group's leadership, whose decision is likely to be influenced by the presence of other critical variables, namely the attachment to a particular weapon, tactic, or the process of innovation itself.

Relationship with other organizations

Throughout history, various organizations have also modified their operational preferences as a result of both cooperation and competition with other terrorist groups. In the first instance, cooperation has allowed groups to get access to operational and technological know-how of others while, in the second instance, competition has played a role in a group's decision to innovate in order to achieve greater prominence over their competitors. What is important to highlight is that both levels of interaction among groups have not necessarily been driven by ideology, and that most organizations with similar ideologies have in fact historically engaged more often in competition, rather than cooperation. The reason for this is relatively simple: the closer the ideologies and aims of a given number of groups in the same operational theater, the greater the need on part of these groups to differentiate themselves from one another in order to gain a monopoly on the "dream" that these groups all claim to strive for. As a result, when two or more organizations with similar ideological foundations appear in the same theater, they are more likely to engage in violent competition rather than in cooperation. This is especially true in contexts where the terrorist activity is generally seen as a pre-step to taking over governance upon the event of victory in the respective struggle, and the claiming of rights to the eventual triumph is naturally taken into consideration even in the pre-victory phase.

Resources

The intuitive hypothesis that large organizations with hefty budgets, outside sponsors and highly qualified membership are more likely to demonstrate an inclination toward innovation with respect to both motivation and capability, than smaller groups with limited financial and logistical resources, was confirmed in all of the examined case studies. Still, some modifications to this hypothesis had to be made. First, the argument that the groups that are most likely to innovate were state-sponsored entities needed to be revised to take into consideration one of the most significant contemporary trends in terrorism – the dramatic decline of state sponsorship following the end of the Cold War. Contemporary terrorist organizations have learned to adapt to the reality of self-financing, in some cases so well that their budgets even exceeded that of many state-sponsored entities. Thus in today's world, the lack of state sponsorship would serve as a poor pre-indicator of a group's lack of capacity to innovate for threat assessment purposes. Second, the assertion that innovativeness of a group would positively correlate with its size also cannot be confirmed. In this sense it is not necessarily the size of the group, but rather the qualitative attributes of the cadres that will determine its innovation potential. In fact, while financial resources are only a supporting factor facilitating innovation, the availability of quality human resources is absolutely key to a group's motivation and ability to innovate. One of the key implications of this finding is the fact that a group's decision to innovate can represent a relatively sudden shift, if for instance a group finds itself in the possession of a key human resource with the idea or the skill to introduce a new method. This underscores that luck and coincidence can also play a considerable role in determining terrorists' *modus operandi*. Whether a group will make the leap toward taking advantage of a surprise acquisition of a resource will to a great extent depend on the following variable – the openness to new ideas.

Openness to new ideas

The level of openness to new ideas is another variable likely to be closely associated with terrorist innovation. At the first level of this variable are group decision-making dynamics. In highly autocratic organizations where members are closely watched and controlled and where dissent is not tolerated, the likelihood of innovation proposals coming from individual group members is lower than in the case of groups that base their decisions on a democratic vote. At this level it has been hypothesized that in order to facilitate innovation, a group's leadership has to be open to suggestions from below and individual members must not be afraid to put forward their proposals for adopting new methods. However, open communication links have not proven to be a critical variable for providing the push for the decision to innovate.

The second component of this variable was the technological awareness of a group. In this regard, organizations whose members are in everyday contact with modern technologies such as cell phones, computers or biological culture collections and laboratory equipment are more likely to incorporate them into their operations than groups whose members are secluded from the rest of the world. However, this hypothesis, while intuitive, again did not show universal applicability to the case studies.

The third important component of the "openness to new ideas" variable is the positive attitude toward risk taking, both at the level of the risk of failure and the physical risks associated with conducting experiments with unfamiliar weaponry. At the basic level of physical risk, virtually any terrorist is willing to die for his or her cause. However, while some organizations have shown a great keenness to sacrifice the lives of some of their members during suicide operations, others have gone out of their way to avoid as many physical dangers as possible, even sacrificing the effectiveness of their operations in order to see their operatives fight another day. It has thus been hypothesized earlier that innovative groups are likely to demonstrate less fear with regard to operational failure, as well as a greater willingness to sacrifice their members in the process of attack preparation and delivery.

However, the hypothesis that innovative groups could be identified beforehand by their higher tolerance to the risks of failure as well as the physical risks associated with experimentation did not show a universal applicability. Further, it should be stressed here that a group's attitude toward danger is a dynamically evolving phenomenon that changes throughout the life span of a group. Based on the principle of "risky shift" associated with the psychological processes of group decision making known as "groupthink," terrorist groups can be expected to take greater risks over time.[22]

Overall, the original hypothesis that groups that are in a regular contact with modern technologies, possess a positive attitude toward physical and operational risks, and embrace democratic elements in their decision-making process are more likely to demonstrate a high level of innovation than socially secluded, risk-averse and autocratically ruled groups, has shown a limited applicability for threat assessment purposes. On the one hand, while a high tolerance toward the risks of failure and physical risks can be highly important as a precondition to *radical* technological innovation, such as the adoption of CBRN weapons, in some cases it is the *low* tolerance toward physical risks that can drive groups to innovate incrementally, precisely in order to improve the safety of its members during operations. Further, we have also seen that informational seclusion is much more relevant to a group's lack of technological awareness than social or physical isolation, and that when an organization's leadership is in favor of innovation, the presence of open communication links and democratically based decision making is not a precondition to the initiation of this process.

Durability

The durability of an organization was hypothesized to be another key factor, based on the simple logic that organizations that exist longer have more time to progress in terms of their motivation to innovate, as well as the opportunity to gather enough experience to facilitate success of the innovation process. And while groups whose existence exceeds only several months may in some cases be significantly motivated to innovate, their ability to succeed is likely to decrease the shorter the duration of their life span.

In none of the examined case studies this hypothesis proved relevant, as all of the four organizations had reached their peak of capability relatively quickly from the outset of their existence. Overall, while we have seen the general improvement of many organizations over time as they have gathered experience and learned from past mistakes, such as the RAF whose study of past court cases had led to innovations such as a special ointment designed to prevent finger prints, we certainly cannot confirm the hypothesis that the length of life span of a terrorist group will be positively correlated with its demonstrated level of radical innovation.[23] The key implication of this finding is that the emergence of an innovative group can be rather sudden, and we would thus be wrong to expect a "superterrorist" group to necessarily be detected due to acts of lower level violence prior to their first mass casualty strike.

Nature of the technology

Once the decision to innovate has been made, additional factors that will determine the success of the innovation process will also come into play. These variables include some of the above stated factors such as the level of outside support, availability of financial and material resources, expertise, time, human resources, quality of membership, intensity of the struggle and the security environment in which the group operates. The one obvious variable that is likely to have the strongest impact on the level of innovational success, however, is the nature of the technology or tactic in question. Quite simply, the more complicated the new *modus operandi*, the less likely it is to be adopted successfully.

Overall, the nature of the technology has proven to be significant when it comes to the success of terrorist innovation. In this regard, the fact that most terrorist organizations have suffered significant setbacks and failures with even relatively primitive technologies may be one of the main reasons why radical innovation in terrorist campaigns has been a rare phenomenon. In terms of threat assessment, the nature of the technology pursued by a given group and the accuracy of its own assessments about the challenges of mastering this technology are among the most useful indicators of whether we can expect the organization to succeed in its innovative endeavors.

Conclusions and implications for AQ

As stated in the introduction the goal of this research was to identify distinct characteristics of especially innovative terrorist organizations, with the ambition of contributing to our ability to conduct predictive threat assessment of future terrorist violence. As we have seen throughout this chapter, terrorist innovation is a highly complex process, and only few of the individual variables proposed have demonstrated applicability in terms of being able to universally explain what distinguishes an innovative group from a conservative one. However, the presence of a certain combination of these variables does allow us to make some predictive judgments about future radical innovators.[24] In this conclusion, some of these scenarios will be identified, followed by the application of these lessons to AQ, an entity that is commonly associated with a high level of innovation.[25]

Innovation essentially occurs in the presence of an overlap between a terrorist organization's decision to innovate, and the capability to do so successfully. With respect to the motivational component, there are four distinct reasons that serve as possible triggers to innovation. The first such reason is the presence of an inherent ideological pre-disposition to using modern technologies, or the need to innovate in order to obtain the capability to match the level of violence associated with the respective ideological and strategic preferences. These innovative tendencies can be expected to occur right from the outset of the organization's existence, if the given group is founded on an apocalyptic ideology or in the event that it is a splinter group of another organization that was deemed not radical or innovative enough. In other cases, the points of innovation toward seeking more destructive means will be associated with points of escalatory shifts in strategy, which will likely be accompanied by frequent shifts in ideology and a continually radicalizing targeting logic. While newly formed apocalyptic organizations can remain under the radar screen of intelligence agencies if they do not pursue violence overtly, in the case of traditional terrorist groups the changing patterns in their statements and actions can provide early warning signals that the group might be moving toward the direction of radical innovation. While apocalyptic organizations can be detected by the ideological notion of destroying the world in order to save it, the key indicators in the latter case will include an escalatory pattern of violence, the broadening of target categories, radicalization of statements, a decreasing concern for constituency or international opinion, increased influence of an allied group or state sponsor, a demonstrated high level of tolerance toward risks and, most importantly, an expressive fascination with a particular weapon or technology or the process of innovation itself. At the same time, this reality may not apply to the scenario in which the group's decision to innovate constitutes a sudden shift triggered by the perception of a grave threat to the group, leading to the belief that there is no choice but a large-scale decisive strike as a last desperate attempt to save the group's very existence. However, the likelihood of success in such a scenario is relatively low, given the fact that organizations under such pressure are unlikely to have enough patience and calm to radically alter its operational methods with a positive outcome.

With respect to innovative tendencies of AQ, this scenario shows a high level of relevance. AQ embodies many of the alarming characteristics mentioned above, such as the apocalyptic dimensions to its ideology, highly indiscriminate targeting logic, disregard for the possibly negative reaction of constituencies and the international community to mass casualty attacks, a long-term preference for operational "quality over quantity" and emphasis on spectacular operations. It is thus no surprise that AQ's highly ambitious and innovative approach to terrorist operations is an integral part of the group's violent trajectory from its inception. This is unlikely to change in the future, as AQ has maintained a firm ideological and strategic line, with few shifts over its life span. On the other hand, with respect to the capacity to conduct innovative spectacular terrorist operations, AQ might become victim of its own success, in the sense that the 9/11 operation set such a high standard that AQ may never come anywhere near to matching. This unattainable nature of the goal of "out-doing" a 9/11 may serve as an impediment to innovation in the future.

The second situation in which we can expect a group to innovate is associated with the emergence of competition with other organizations operating in the same operational theater. Whether innovation will become a product of this rivalry will depend on a number of factors, among them the density of the operational theatre, ideological similarity among these groups, and clashes between leaders at an interpersonal level. The greater overlap in the goals, ideologies and ambitions of these groups, the greater the likelihood of innovative tactics and/or technologies being introduced by at last one of these groups. Most importantly, the occurrence of innovation will again depend on the presence of an expressive attachment to particular *modus operandi* or an emphasis on spectacular operations as a source of pride and prestige in the respective organization's value system. A group's condescending reactions to the operations of other rival groups, a lack of ideological distinctiveness, a demonstrated high level of tolerance toward operational failure, as well as the presence of an expressive emphasis on its own operational superiority will provide good early warning signs of a group's propensity to innovate.

In the context of AQ this dynamic seems to have a growing importance. While in the early years AQ has managed to integrate ideologically aligned local groups under its global jihadist umbrella with a comparatively high level of unity, recent years have seen much discord among AQ's various factions. This has ranged from healthy competitions (e.g., Al Qaeda in the Islamic Maghreb (AQIM) and Al Qaeda in the Arabian Peninsula (AQAP)),[26] to overt confrontation leading to an internal split (e.g., Mokhtar Belmokhtar from AQIM),[27] to outright open war (e.g., the violent confrontation between Islamic State of Iraq and al Sham (ISIS) and the Jabhat al Nusra in Syria). These competitions are likely to carry themselves over into the operational realm, and can lead to innovation in the drive of each faction to outbid its competitors.[28] For instance, the spectacular 2013 hostage taking at the Algerian gas plant in In Amenas was a direct manifestation of this phenomenon, as the primary motive behind the attack was Belmokhtar's desire to demonstrate operational superiority in the Sahara over AQIM.

The third situation in which a group can be expected to alter its operational methods in a novel direction comes in the presence of government countermeasures, such as target hardening efforts that serve as a direct obstruction to the tactics used by terrorists in the past. While most groups can be expected to yield to this pressure and substitute targets, an innovative organization will refuse go down this path of least resistance in order to increase its probability of success. Instead, such a group will work to overcome these countermeasures by means that have not been accounted for by the enemy, often placing an emphasis on projecting an image of invincibility as well as mocking the state for failing to stop the attack despite all of its resources. Indicators of such characteristics are statements praising operational success in the face of difficult conditions, emphasis on the disproportion in resources between the "believers" and the enemy, repeated attempts to attack the same target even after having previously failed, a demonstrated high level of tolerance toward operational failure, and an expressive emphasis on military skill or the process of innovation itself. In most cases, however, the innovations that will be triggered by the desire to overcome specific countermeasures are likely to be of an incremental, as opposed to a radical nature.

This scenario is highly relevant to AQ, as it was precisely the countermeasures employed in the context AQ's preferred realm of aviation targets that have been responsible for driving most of the group's attempts at innovation. AQ's long-term obsession of synchronized targeting of commercial airliners can be traced from Ramzi Yousef's Operation Bojinka plan in the early 1990s, through the failed 2001 Richard Reid and Saajad Badaat shoe bomb attempt, to the 2006 disrupted liquid bombs plot, the 2009 Omar Farook Abdulmutallab's underwear attempt, followed by the 2010 printer cartridges plot, and then yet another version of the underwear plot. Given AQ's obsessive preference for targeting airplanes, as well as its persistence and pride with respect to revisiting the same targets, and the expressive emphasis on overcoming the enemy's defenses – these are all factors that have driven AQ'a innovation in the past, and can be further expected to play a role in the area of aviation terrorism in the future.

The last possible trigger to terrorist innovation is an incidental or unintended acquisition of a particular human or material resource. This scenario is also highly relevant in explaining the innovativeness of AQ, especially in the realm of human resources. In fact, AQ's innovations have so far been primarily driven by entrepreneurial individuals like Ramzi Yousef and Khalid Sheikh Mohammed in the early years,[29] or, more recently, people like Ibrahim Hassan al-Asiri, whose bomb-making skills have been the driver behind some of the most important innovations in the world of terrorism today.[30] In essence, it is these terrorist entrepreneurs and middle managers – who are less driven by organizational affiliation than by their desire to be good in what they do – that continue to play the most important role in driving AQ's innovation.[31]

To conclude, while the situations described above are likely to be encountered by most terrorist groups throughout their life span, only some can be expected to innovate. This will depend on a combination of factors, most importantly on the presence of the expressive fascination with a sophisticated weapon or technology, or the process of innovation itself. Without the presence of this expressive component, groups forced to change their strategy, adapt to countermeasures, or prevail in competition with other groups can be expected to react by modifying their *modus operandi*, broadening their target categories, or intensifying their attacks, but not necessarily by engaging in radical innovation. In this sense, the presence of the non-rational component, such as the expressive emphasis to innovation, overly high ambitions in the operational realm or ideological or expressive attachment to a particular type of weapon or technology, serves as the strongest and most universal pre-indicator of the propensity of a terrorist group to innovate.

Ominously, AQ represents many of the alarming characteristics of innovative groups that have been identified here: it is a popular organization that attracts many ambitious entrepreneurial recruits from all over the world (many Western educated and trained); possesses safe havens and ample opportunities for in-battle experimentation and training; features competition among different factions and affiliate groups that can drive the desire to innovate; and has an expressive focus on overcoming countermeasures and on delivering spectacular attacks via innovative means. Given all these factors, it is safe to assume that AQ's innovative drive is unlikely to go away anytime soon. On the other hand, the 9/11 attacks have set a standard that AQ is unlikely to ever match again, which may serve as an internal deterrent within the organization to pursuing radical innovation of this scale. Further, the innovations that had led to the success of the 9/11 attacks took place in a very different security environment, which no longer exists today. So while we can expect AQ to constantly come up with incremental small-scale innovations in operational tradecraft, emergence of a sudden radical innovation that would be a "game changer" in the group's campaign against the West is by no means an inevitability.

Notes

1 Farrel, Theo and Terriff, Terry, *The Sources of Military Change: Culture, Politics, Technology* (London: Lynne Reinner Publishers, 2002) p. 6.
2 Ibid., p. 6.
3 Evangelista, Mathew, *Innovation and the Arms Race: How the United States and the Soviet Union Develop New Military Technologies* (London: Cornell University Press, 1988) p. 51.
4 Buzan, Barry, *Strategic Studies: Military Technology and International Relations* (New York: St. Martin's Press, 1987) p. 18; Murray, Williamson, "Innovation: Past and Future," in Murray, Williamson R. and Millet, Allan R., *Military Innovation in the Interwar Period* (London: Cambridge University Press, 1998) p. 306.
5 Hoffman, Bruce, "Terrorist Targeting: Tactics, Trends, and Potentialities," *Terrorism and Political Violence*, 5: 2 (1993) p. 378.
6 Rosen, Stephen P., *Winning the Next War: Innovation and the Modern Military* (London: Cornell University Press, 1991) p. 52.
7 Dolnik, Adam, "Fighting to the Death," *The RUSI Journal*, 155: 2 (2010) pp. 60–68.
8 Hoffman, op. cit.
9 The credit for generating the initial list of variables that were further refined and operationalized in the book goes to Brian Jackson (Jackson, Brian A., "Technology Acquisition by Terrorist Groups: Threat Assessment Informed by Lessons from Private Sector Technology Adoption," *Studies in Conflict and Terrorism*, 24: 3 (2001).
10 Sloan, Stephen and Anderson, Sean, *Historical Dictionary of Terrorism*, 2nd edition (London: The Scarecrow Press, 2002) p. 45.
11 Cited in Gunaratna, Rohan, *Inside al Qaeda* (New York: Columbia University Press, 2002) p. 103.
12 Ventzke, Ben and Ibrahim, Aimee, *The al Qaeda Threat: An analytical guide to al Qaeda's Tactics and Targets* (Alexandria: Tempest Publishing, 2003) p. 37.
13 National Commission on Terrorist Attacks upon the United States (Philip Zelikow, Executive Director; Bonnie D. Jenkins, Counsel; Ernest R. May, Senior Advisor), *The 9/11 Commission Report* (New York: W.W. Norton, 2004) p. 232.
14 Bell, J. Bowyer, *The Dynamics of the Armed Struggle* (London: Frank Cass, 1990) p. 88.
15 Hoffman, op. cit., p. 378.
16 Dobson, Christopher and Payne, Ronald, *The Terrorists, Their Weapons, Leaders and Tactics* (New York: Facts on File, 1982) p. 134.
17 Dolnik, A., *Understanding Terrorist Innovation: Technology, Tactics and Global Trends* (London: Routledge, 2007).
18 Ronfeld, David and Sater, William, "The Mindset of High-Technology Terrorists: Future Implications from the Historical Analog," in Alexander, Yonah and Ebinger, Charles K., *Political Terrorism and Energy* (New York: Praeger Publishers, 1982) p. 22.
19 Written by the leader of the National Alliance, William Pierce.
20 Qur'an sura 47, verse 4. al-Mawardi Abu al-Hasan, *al-Ahkam as-Sultaniyyah* (London: Ta-Ha Publishers, 1996) p. 192.
21 Marks, Tom, "Making Revolution with Shining Path," in Palmer, David Scott ed., *Shining Path of Peru* (New York: St Martin's Press, 1994) p. 214.
22 Lebow, Richard Nel, *Between Peace and War* (Baltimore: Johns Hopkins University Press, 1981) pp. 101–119.
23 Hoffman, op. cit., p. 377.
24 Another aspect that must be pointed out is that some factors such as "targeting logic" or "openness to new ideas" have proven to be a double-edged sword,

having the capacity to contribute to both directions of the innovation scale. But since we are particularly concerned with the prospect of radical innovation possibly leading to mass-destruction level, such variables will mainly be considered in terms of their potential to bringing about an escalatory outcome.

25 Assaf Moghadam, "How Al Qaeda Innovates," *Security Studies,* 22: 3 (2013) pp. 466–497.

26 Celso, Anthony, *Al-Qaeda's Post-9/11 Devolution: The Failed Jihadist Struggle against the Near and Far Enemy* (London: Bloomsbury Academic, 2014).

27 Bakari Gueye, "Belmokhtar breaks away from AQIM," Magharebia Online http://magharebia.com/en_GB/articles/awi/features/2012/12/11/feature-03.

28 Mortada Radwan, "Syria: al-Nusra Front Declares War on ISIS," February 26, 2004, al-Akhbar Online. Available at http://english.al-akhbar.com/node/18785.

29 *The 9/11 Commission Report,* pp. 145–149.

30 Al-Qaeda bombmaker represents CIA's worst fears. *Washington Post.* Available at www.washingtonpost.com/world/middle_east/al-qaeda-bomb-maker-ibrahim-hassan-al-asiri-has-tried-to-attack-the-us-three-times-officials-say/2012/05/08/gIQA16pkBU_story.html.

31 For an excellent in depth discussion of this phenomenon, see Moghadam, op. cit.

5 Attacking the West

The influence of al-Qaeda's ideology and propaganda

Søren Hove

> Al-Awlaki was a motivator for this person.... Inspire Magazine, which al-Awlaki was the prime mover behind, was the magazine, or the article that [Quazi Mohammad Nafis] read that justified him in the killing of children, the killing of women.
>
> (New York police chief Ray Kelly, speaking at a press conference on 17 October 2012 on why Nafis was targeting the Federal Reserve Bank in New York)[1]

How does al-Qaeda (AQ) and its affiliates choose their targets and what effect does propaganda have on the target designation? This is a question that both counter-terrorism officers and terrorism researchers have been asking since the 1990s in order to understand the past and prevent attacks in the future. The underlying assumption is that al-Qaeda's propaganda and training manuals provide us with important information about the intentions of the group and this has, over the years, given rise to a whole community of scholars and intelligence analysts who study jihadi strategic writings and propaganda in order to retrieve information about al-Qaeda's strategy and make predictions about al-Qaeda's future targets.[2]

From this way of thinking it often follows that the al-Qaeda senior leadership (AQSL) make the master plan for specific ways to inflict damage on their enemies, where they first develop a strategy for which targets are legitimate and then reach out to selected audiences and affiliates through propaganda in order to direct them towards these specific targets in the West.

But how can we be sure that things always work in this way? Are we right to believe that there is a correlation between the AQ propaganda and the kind of terrorism that is actually being conducted and planned for?

In order to answer the question of how al-Qaeda's propaganda affects the organisation's terror acts against the West,[3] this article will compare al-Qaeda's ideological writings from the 1990s to present with the pattern of target designation of Western targets in al-Qaeda's plots and attacks. In that way the article's approach will be two-pronged: first, by reading al-Qaeda's propaganda over the years the article will analyse how tangibly the central and local leaderships have envisioned the battle against the West and how the target designation has developed in al-Qaeda's propaganda in terms of what targets should be hit and who should take them on. Second, by holding these findings up to the actual terrorist attacks or plots the article will try to outline what correlation might exist between al-Qaeda propaganda and al-Qaeda deeds.

This chapter will argue that al-Qaeda propaganda has clearly developed over the years in terms of which targets the organisation promotes and which audiences it tries to instigate to commit terrorist acts. Comparing the propaganda with attacks and plots across three periods (1998–2002, 2003–2008, and 2008–present) it is possible to see some shifting trends in the propaganda that seem to have affected target designation. The article's analysis thereby sheds light on the kinds of targets and methods al-Qaeda is currently promoting and on how the organisation got there.

In the final section the chapter also contributes to the broader research debate on what the study of propaganda can actually show us.[4] Because even though it is possible to identify trends and changes in the propaganda material that have affected target designation, the connection between propaganda and the actual terrorist act is not as solid as one might expect. The imperative, guiding and instructional tone in most of the official propaganda material gives the reader the impression that leaders and ideologues seemingly point out what targets are important in their propaganda, along with how they should be attacked. And at times words and ideology can be shown to have formed actions but at other times the causal link is unclear.

In this respect the question of how al-Qaeda propaganda affects target designation raises some fundamental questions about (1) whether al-Qaeda over the years can be said to have a coherent ideology and a coherent propaganda catalogue and (2) whether changes in target designation continuously come "top-down" from announced changes in the propaganda or new practices are invented "bottom-up" and subsequently change the propaganda? For al-Qaeda these dynamics – between words and actions, between leadership and "foot soldiers" – seem to have changed as the organisation has changed and scattered over time, and especially today we increasingly have to acknowledge that we cannot make an automatic assumption that there is a direct correlation between propaganda and actions.

Approach and criteria

In order to analyse and give an overview of both al-Qaeda's focus on Western targets in their propaganda and the actual terror attacks and plots since the 1990s, this chapter builds on previous research that has outlined chronologies of attacks and drawn together collections of al-Qaeda propaganda. But because of both the large volume of al-Qaeda propaganda and the many attacks and plots that have occurred over the years, the chapter cannot claim to give a complete account of al-Qaeda's propaganda or terrorism. It considers a selection of the largest attacks and the most significant propaganda texts.

Concerning the criteria for including propaganda material, this chapter focuses on the publications where al-Qaeda and its official affiliates speak about target designation. Even though the volume of al-Qaeda's propaganda is extensive, it has been surprising to see that al-Qaeda seldom speaks explicitly and tangibly about target selection. There are texts that do this – both propaganda texts and internal documents in which the strategy for targeting the West is not just presented but also bluntly discussed – but they are fewer than one might think.[5]

In outlining the most important attacks this chapter builds upon the chronologies that researchers like Nesser (2008), Jordan (2012) and Silber (2012) have produced for al-Qaeda's attacks or plots against the West. The ambition is to make a broader periodisation but not another all-embracing chronology. For that reason this article does not go through all of the cases and incidents but tries to summarise the characteristics of al-Qaeda's target designation in the West. This means focusing on attacks and plots where there exists hard evidence for al-Qaeda involvement in a plot that was planned and prepared.[6] It is also important to note that because al-Qaeda is a clandestine and secret organisation and because information about terror attacks is often engulfed in secrecy and sometimes misinformation – even when the reporting comes from intelligence agencies – we, as analysts, have to make do with what limited source material we have available on attacks and plots. It is especially difficult to assess whether plots were directly connected to al-Qaeda or just more loosely inspired by al-Qaeda, and there is most often little information available about how targets are decided upon. We have to acknowledge that sometimes important information is missing and that there are substantial sources of error in evaluating the circumstances around a terror attack, especially when one is trying to decide whether a specific target was promoted and "ordered" by al-Qaeda's central leadership or whether the target was chosen more or less randomly by the local perpetrators.[7] Regarding Nesser, Jordan and Silber, this chapter has likewise to weigh and evaluate the evidence that is available through newspaper articles, courtroom testimonies, the information that al-Qaeda gives through official propaganda and that unearthed in recovered internal correspondence.

An historical overview of al-Qaeda's targeting of the West – in propaganda and deed

Al-Qaeda developed so profoundly from the 1990s to the present in terms of its propaganda, its *modus operandi* concerning attacks, and in terms of organisation and centralisation. This chapter divides al-Qaeda's target designation into three periods. Even though the characteristics of these periods overlap somewhat it is argued – and not just for the sake of clarity – that the years 1998–2002 constitute the "old" al-Qaeda style of top-down target designation, which resulted in large coordinated attacks. In this period al-Qaeda's organisation was as close to a hierarchical and centralised organisation as it would ever be, and in this capacity had the best opportunities for matching propaganda with the proper actions and targets. The period 2003–2008 is one where leadership was weakened and had to scatter; it seemingly outsourced most of the planning to local affiliates but was still able to extend the concept of "the Western Enemy" to European countries in their propaganda and, at times, to point out specific targets such as the Danish Embassy in Islamabad which was attacked in 2008. From 2008 onwards the propaganda produced by the leadership seems to be only inspirational in most attacks and target designation. Still, an ideological focus on Western Muslims launching their own attacks on the West on their own initiative is clearly traceable in the ideological writings of these years. The following is meant as an overview of these three periods – not an exhaustive account of all al-Qaeda's propaganda or terror attacks.

1998–2002: large coordinated attacks on the West

Al-Qaeda began their ideological battle against the West in the mid 1990s by publishing texts and fatwas criticising the West for occupying Muslim lands. The goal was to attack and expel Western targets from Muslim lands. Propaganda material was sometimes distributed by fax by Usama bin Ladin's Advice and Reform Committee in London, at other times it was distributed locally through taped sermons on cassette tapes.[8] In the 1998 fatwa, "Declaration from the World Islamic Front for Jihad against the Jews and the Crusaders", al-Qaeda gave the battle against the West top priority, over that of fighting local regimes, and al-Qaeda's senior leaders also broadened their definition of the way the war could be fought by insisting that "to kill the Americans and their allies – civilians and military – is an individual duty for every Muslim who can do it, in any country in which it is possible to do it".[9]

The 1998 fatwa widened the range of targets (civilian and military) and the location of these targets (in any country) but it did not provide clear guidelines on specific targets and specific means. Nonetheless, there seems to be a clear linkage between this groundbreaking ideological declaration and the terror campaign that al-Qaeda carried out in the years 1998–2002. First, the organisation hit the two American embassies in Kenya and Tanzania in August 1998 then, in October 2000, it attacked the American naval destroyer USS *Cole* in Aden. Then came the 9/11 attacks on the World Trade Centre and the Pentagon in 2001 and finally, in November 2002, the double attack on two civilian Israeli targets in

Mombasa. All these attacks were large, coordinated attacks, which required long preparation and high levels of expertise from professional operatives. All attacks involved suicide bombers and, except for the USS *Cole* bombing, all attacks were carried out against large Western civilian targets.

In line with this strategy of larger attacks and the ambition to kill as many as possible it is understandable why al-Qaeda, in the 1990s and early 2000s, also showed some interest in producing and using weapons of mass destruction. Bin Ladin and al-Zawahiri personally spent two days with two Pakistani nuclear scientists in Kandahar in August 2001 discussing al-Qaeda's determination to obtain nuclear weapons.[10] But even though al-Qaeda tried unsuccessfully to acquire uranium and also tested toxic gases in their camps in Afghanistan, these endeavours never transformed into weapons or attacks.[11] There is still an on-going scholarly debate about al-Qaeda's ambitions and capability concerning WMD and how far they actually went to pursue this, but it is telling that the organisation – in spite of their intention – never did use them and it is also important to note that the rather basic experiments that were conducted happened in a period prior to 2001 when al-Qaeda had its own camps in Afghanistan and was at the height of its power.[12] This makes it evident that al-Qaeda was never successful in these WMD experiments.

In the period between 1998 and 2002 several al-Qaeda ideologues and leaders spoke about the importance of targeting the West but they seldom went into details on how to carry this out in practice and on which targets were most suitable. In some few cases guerrilla warfare was mentioned as a possibility but not as a tactic specifically concerning Western targets.[13] But we know from courtroom testimonies, confiscated internal letters between al-Qaeda leaders and from al-Qaeda members who later published their memoirs about the period, that there was planning of and internal discussion about which targets were to be attacked.[14] According to one of the key operatives in the 1998 embassy bombings and the 2002 Mombasa attacks (Fazul Abdullah Mohammed, who was in contact with the senior leadership up to at least 2003), al-Qaeda's terrorist attacks were to distinguish themselves from those of other terror groups by being special in scope and also by always being strictly in accordance with Islamic law. In his memoir Fazul criticises other jihadists who act in the name of al-Qaeda but are not in fact al-Qaeda – in this case the Bali bombers who he believes were not sanctioned by al-Qaeda – by writing:

> There are many movements around the world today that adopt the same ideology of jihad but carry out operations against Westerners without the planning of al-Qa'ida. Most of the operations that have been carried out following the 9/11 attacks lack the authorization of Sheikh Usama or the central leadership of al-Qa'ida. One can observe that selecting very random targets does not serve the jihadi enterprise. Al-Qa'ida's operations are very specific in nature (*naw'iyya*), planned

> by its leadership, and aim with precision against lawful enemy targets; namely, military, political and economic targets.[15]

In the period 1998–2002, which could be said to be al-Qaeda's heyday in terms of operational activities against Western targets, the organisation relied almost exclusively on field commanders/operatives who had gone through intensive training in al-Qaeda's own training camps in Afghanistan and who were under orders and sent on specific missions.[16] Al-Qaeda's leadership relied on their trained operatives in the field to carry out orders from leadership concerning attacks, but it seems that it was up to individual field commanders to plan which method would be most successful and in some cases which specific targets would be chosen. Usama bin Ladin's bodyguard from 1994 to 2000, Abu Jandal (Nasir al-Bahri) explains in an interview:

> Al Qaeda pursues a method or principle that calls for "centralization of decision and decentralization of execution". The decision was made centrally, but the method of attack and execution was the duty of the field commander.... The planning for the *Cole* operation was carried out by the people [on the ground]. The idea was formed and the target was set and then it was referred to a higher military control committee in al-Qaeda, which does not plan, but gives the green light.[17]

In the period 1998–2002 al-Qaeda's official outreach was limited to a broader critique of the West. Leaders mainly gave interviews or speeches to either newspaper journalists or to pan-Arab television channels such as al-Jazeera. According to the propaganda of this period, al-Qaeda saw itself as a militant vanguard and an instigator that should incite and mobilise the Muslim masses but it is very difficult to find specific directions in the propaganda of this period as to which Western targets should be preferred. Usama bin Ladin spoke only in vague terms about targets; for example, in his speeches and interviews he talked about draining the American economy but not about how to do it. Ayman al-Zawahiri only spends two pages discussing the choice of targets in the West in his 252-page book *Knight under the Prophet's Banner* (2001), serialised in *Ash-Sharq al-Awsat* where he states:

> 1. Be sure to inflict maximum casualties on the enemy, kill the greatest number of people, this is the only language understood by West, no matter how much time and effort such an operation takes. 2. Concentrate on martyrdom operations as the most successful way of inflicting damage on the opponent and the least costly to the mujahedeen in terms of casualties. 3. Choose our targets and weapons to reach the vulnerable spots of the enemy.[18]

Al-Zawahiri's directions seem to summarise the terrorist attacks of the period 1998–2002: trained al-Qaeda "professional" operatives plan (mainly martyrdom) operations in large and time-consuming attacks aimed at causing mass casualties. And just as one would suspect from a hierarchical organisation there was a correlation between messages in propaganda and the actual deeds.

2002–2008: outsourcing the fight against the West

Overall the period from 2002 to 2008 was characterised by a weakened central leadership in al-Qaeda and the loss of territory and experienced operatives, which made new ways of targeting the West necessary. As a consequence of Operation Enduring Freedom, al-Qaeda lost most of their safe houses and training camps in Afghanistan and both senior leaders and operational leaders had to flee from 2001/2002. Many were killed or were apprehended in the early period of the War on Terror.[19] At the same time the organisation was struggling to maintain financial support since many of its supporters had their assets frozen or monitored.[20] The organisation would never possess the same capability to conduct large attacks against the West as they had prior to 2001/2002, but other opportunities would soon present themselves and some of the leadership managed to reorganise a few years later in Pakistan.

The invasion of Iraq in 2003 gave al-Qaeda a new ideological focal point as it represented a prime example of Muslim suffering at the hands of Western military and it gave the organisation an opportunity to fight the West locally, with the proximity of Western soldiers in Iraq.[21]

Al-Qaeda's warfare in Iraq signified a turning point in many respects. First of all it illustrated how local and affiliated groups – such as Abu Musab al-Zarqawi's group Tawhid wal Jihad – began to take over the fight against both local and Western targets at a time when central leadership was weak and scattered. Other groups in Saudi Arabia and North Africa would soon follow and central leadership gave them their support and even made some of them official al-Qaeda groups.[22] This "franchising" model was most likely possible because of al-Qaeda's fame after the 1998 and 2001 attacks which positioned al-Qaeda as the unrivalled opponent of the West.

Spreading out gave al-Qaeda an opportunity to fight Western targets in many places but it also represented a strategic challenge. Since many of the new local field commanders had a weaker affiliation with "core al-Qaeda" – many had never been to the training camps – al-Qaeda's central leadership seems to have lost control over which attacks were conducted and how in this period. We know from internal letters that leadership tried to discipline local affiliates concerning which targets were allowed and which not, but to little effect. In some instances the central leadership even advised local affiliates to refrain from attacking Western targets outside their local area. Ayman al-Zawahiri and Atiyah Abd al-Rahman's letters to Abu Musab al-Zarqawi in Iraq in 2005 are examples of this.[23]

The loss of control and the rise of different affiliated groups paved the way for a broader spectrum of propaganda coming now both from central leadership and local affiliates. Local affiliates in particular published more specific texts about concrete targets and *modi operandi*. Written instruction manuals and "Jihad encyclopaedias", instruction videos, series and periodicals increasingly began to find wider distribution. First on specific internet sites in the early 2000s and from 2004/2005, they spread to a wider use of websites, message boards, e-groups, blogs, instant messaging etc. Especially vBulletin fora such as al-Ekhlass and al-Hesbah became popular from 2005/2006 since they gave any user the opportunity to download material or communicate with other users, and these fora later became the template for other al-Qaeda affiliated fora. Even though this process did mean that knowledge about how to make explosives became more accessible and user-friendly for a growing audience of al-Qaeda sympathisers, most intelligence experts and scholars argue, however, that the online training manuals have limits and that technical skills and operative tradecraft require more than web-based instructions.[24]

The period from 2002 to 2008 also witnessed the booming of a jihadi genre which, to borrow Thomas Hegghammer and Brynjar Lia's words, could be called "jihadi strategic studies".[25] While strategic jihadi studies had existed since the 1980s they came into prominence following 9/11 and especially after the Iraq invasion. Moreover, these more recent writers initiated a more direct way of discussing strategy, tactics and means, which only added to the fragmented and multipolar picture of al-Qaeda's strategy for targeting the West. Also, they often criticised both al-Qaeda leadership and local affiliates, either for being too soft, or for killing too many civilians and alienating other Muslims etc.[26] Most of these strategic writers had no formal position within al-Qaeda but nonetheless sympathised with the group and were considered ideologues. The harsh tone of this semi-internal/semi-official debate and the cacophony of voices and opinions of these many writers made it almost impossible to talk about a single al-Qaeda formula for target designation.

What is clear, however, is that more targets were mentioned and that even the central leadership of al-Qaeda in this period began to mention more and more countries as legitimate targets, in contrast to the propaganda from 1998 to 2002 that focused on America and Israel. Individual European countries were mentioned more often. An al-Zawahiri speech from October 2004 is typical of this tendency:

> We shouldn't wait for the American, English, French, Jewish, Hungarian, Polish and South Korean forces to invade Egypt, the Arabian

> Peninsula, Yemen, and Algeria and then start the resistance after the occupier had already invaded us. We should start now. The interests of America, Britain, Australia, France, Norway, Poland, South Korea, and Japan are everywhere. All of them participated in the invasion of Afghanistan, Iraq, and Chechnya; they also facilitated Israel's existence.[27]

Many things could explain this extension of the Western enemy. Maybe it was a consequence of the fact that broader coalitions invaded Afghanistan and Iraq and therefore these individual countries became visible to al-Qaeda. Maybe it was because of the emotional debate about Islam in Europe which gave rise to discourses that al-Qaeda could use – e.g. the cartoons of the Prophet Muhammad, the Dutch 'Submission' film – which mobilized not only al-Qaeda sympathisers but large parts of Muslim communities living in Europe for public debates about freedom of speech in Europe, limits of secularisation, blasphemy etc.[28] This clearly happened when Usama bin Ladin in a video in November 2007 criticised European leaders "who flock to the White House" or in May 2008 when, in his speech "May our Mothers be bereaved of us if we fail to help our Prophet", he threatened the EU and especially Denmark for reprinting the newspaper *Jyllands-Posten's* cartoons of the Prophet Muhammad.[29] By picking up on grievances and discourses that were already established in Western/European communities and directly speaking to Muslims in the West and threatening European countries, AQSL tried to reach out to audiences in the West through propaganda and to mobilise Muslims in the West. By speaking more often about broader Western topics such as discrimination, poverty in the West and the Cartoon Crisis, al-Qaeda broadened the range of possible targets and made European countries and Australia more explicit targets for al-Qaeda.

If we compare this broader focus and the tendency to outsource and franchise the struggle with the actual attack pattern in the period 2002–2008, it becomes clear that al-Qaeda was successful at both mobilising and inspiring a variety of different people in Europe and in hitting European targets. Whereas Europe before 2003/2004 had been an arena for both local and global jihadists to live and operate in, Europe now had also become a target of global jihad.[30] Both the Madrid bombings in 2004 and the London bombings in 2005 showed that al-Qaeda sympathisers were able to conduct larger, coordinated attacks in Europe and against European targets. Even though connections between the train bombers in Madrid and al-Qaeda have not been uncovered, the London attack clearly showed that Muhammad Sidique Khan and at least one of his accomplices – contrary to initial reports – actually were in contact with al-Qaeda operatives in Pakistan before the attack and did receive training in manufacturing explosives in Pakistan.[31] The same was true of at least two attacks – one foiled and one successful – against Danish targets which were related to the cartoon crisis. Both Hammad Khürshid (who planned to bomb an undefined Danish target in Denmark in 2007) and the Saudi suicide bomber Abu Ghareeb al-Makki (who carried out the car bomb attack on the Danish embassy in Islamabad on 2 June 2008) had received training from al-Qaeda operatives in Pakistan and explicitly referred to al-Qaeda's senior leadership's propaganda.[32]

In this respect the Danish cartoon crisis and the following bombing of the Danish Embassy in Islamabad in June 2008 represented a big win for al-Qaeda. For the senior leadership in particular, the reprinting of the cartoons in 2008 and the following events gave them an opportunity – maybe for the last time in the organisation's history – to project an ideal image of al-Qaeda as an organisation with a strong and powerful leadership that could choose particular Western targets in their ideological proclamations and then make their affiliates follow this up with terrorist attacks against these specific targets.

As clearly illustrated by the narrative in al-Qaeda's as-Sahab produced video "The word is the word of the sword"[33] that came out after the Islamabad bombing: Usama bin Ladin had responded in March 2008 to the reprinting of the cartoons by publishing the speech "May our Mothers be bereaved … if we fail to help our Prophet"[34] in which he threatens those who mock the prophet. After this speech and Ayman al-Zawahri's call to go for specifically Danish targets in April 2008,[35] the local affiliates answered these calls by bombing the Danish Embassy and by directly referring to Usama bin Ladin's speech.

It is understandable why the narrative in the video is so appealing to the senior leadership because it illustrates how leadership can expound the ideology and through propaganda promote certain targets. In this particular case it seems that targeting was in fact a process that went from the top layers of the organisation down to foot soldiers like the Saudi suicide bomber Abu Ghareeb al-Makki at the bottom.

As Petter Nesser (2008) shows in his chronology of jihadism in Western Europe, the number of incidents increased steadily up to 2008. Targets were beginning to become European as well as American, which seems to reflect AQSL's broadening of the Western enemy to include Europe in the propaganda. But it is also clear – which is supported by Silber's (2012) analysis – that the majority of these attacks or foiled attacks against the West were only al-Qaeda *endorsed* plots or al-Qaeda *inspired* attacks and not al-Qaeda orchestrated 'command and control' attacks such as almost all the attacks between 1998 and 2002.[36] And besides the Madrid and the London bombings – that were both larger coordinated attacks aimed at producing mass casualties – most other attacks were also much smaller in both scale and targets. This could be seen as a consequence of the scattering of central al-Qaeda that no longer controlled their own training camps and after 2001 were put under increasing pressure which made it difficult to plan and prepare for larger operations.

The tendency in the period 2002–2008 towards smaller attacks, increasingly European targets and AQSL's gradual loss of control of the attack planning was to become even more evident in the period after 2008.

2008–2013: targeted communication and individual attacks

Since the attacks in London in 2005, al-Qaeda senior leadership has not been able to take responsibility for any large successful attack in the West and the attack on the Danish Embassy in Islamabad in 2008 was the last successful large attack on a Western target that al-Qaeda's senior leaders have taken responsibility for anywhere in the world (outside of Afghanistan and Iraq). Still, the period from 2008/2009 to the present has seen many smaller attacks against Western targets and many innovative changes to their *modus operandi*.

The tendency towards smaller and al-Qaeda endorsed or inspired attacks rather than al-Qaeda command and control plots continued. Al-Qaeda and their affiliated groups could still train some few people, who were then sent on missions against the West, but most of these people had been living in the West, were radicalised in the West and it was they themselves who offered their services to al-Qaeda. People such as Najibullah Zazi and his associates in the 2009 plot against the New York subway, the underwear bomber Umar Farouk Abdulmutallab who tried to detonate explosives on an airliner on 25 December 2009, and the Times Square bomber of May 2010, Faisal Shahzad, had all come from the West and sought out al-Qaeda themselves. When they had received training – which after 2006 mostly happened in Pakistan and was almost always given by affiliated groups and not al-Qaeda core members – they went back home to perform their deeds.

The fact that most of the al-Qaeda endorsed or inspired plots starting from around 2005 were carried out by people who had some form of "belonging to" and attachment to the West has given rise to the broader discussion of "homegrown terrorism", and the fact some of these people were acting independently and without association to foreign groups has promoted the term "lone wolf terrorism". But both terms "homegrown" and "lone wolf" are problematic: partly because it is difficult to define or set the boundary to "belonging to the West", partly because most terrorists often have some form of association with other terror groups apart from orders in the shape of inspiration, indirect communication etc.[37]

But it is also problematic since it makes us blind to the new role of al-Qaeda's propaganda, which since 2007/2008 has been directly targeting specific audiences in the West and directly instigating them to carry out specific acts against specific targets. Compared to earlier propaganda before 2007/2008 where al-Qaeda leaders occasionally spoke to Western audiences in broader political terms in order to intimidate the Western enemy and to awaken Muslims in the West, both al-Qaeda senior leadership and the affiliated groups started publishing very specific propaganda aimed at particular groups in the West – communicating specifically with young people, female sympathisers and, first and foremost, recruiting foreign fighters – a tendency that has only increased after 2012 with al-Qaeda's efforts to attract European foreign fighters to Syria.

Al-Qaeda propaganda has always sought to promote recruitment of foreign fighters, but since 2008 there has been a thematic shift to a much more personal approach. In order to recruit and mobilise Muslims in the West, al-Qaeda and its affiliates have begun to use spokespeople who have a special understanding of the West and most often speak in Western languages and use Western themes and topics. Al-Qaeda in the Arabian Peninsula (AQAP) used Anwar al-Awlaki and Samir Khan to produce their English language magazine *Inspire*, al-Shabaab distributed Omar Hammani's Western "raps", al-Qaeda's senior leadership have used the American Adam Gadahn, and both al-Qaeda and the Islamic Movement of Uzbekistan have benefited from German-speaking members such as Bekkay Harrach etc.[38] In 2013 Abu Khattab, a prominent Danish foreign fighter in Syria affiliated with the Islamic State of Iraq and the Levant (ISIL, the former al-Qaeda in Iraq), even uploaded Danish videos from Syria to YouTube in order to encourage young Danes to either come to the Syrian battlefield or to perform jihad in Denmark against Danes like Kurt Westergaard or NATO's secretary general, Anders Fogh Rasmussen.[39]

Magazines like *Inspire* or Omar Hammami's up-front and street-smart attitude seems to be a turning point compared to al-Qaeda's earlier propaganda. Distributed through social media like Twitter and YouTube, using slick slogans often coated in a snappy layout, these new types of propaganda have very little in common with bin Ladin's formal speeches or the typical al-Qaeda footage from the battlefield. It seems much more in tune with modern Western youth culture and commercial advertising.

As has been the case in many other places it seems that not only do foreign fighters attract more foreign fighters[40] but also that they have given al-Qaeda better opportunities to fight the West in the homelands of the recruited people. The prospect of targeting the West using Westerners as perpetrators has not been lost on al-Qaeda leaders and strategists. Contrary to earlier propaganda, al-Qaeda propaganda now also speaks about staying home to conduct attacks. As a writer in *Inspire*, Mukhtar Hassan (who is probably the now deceased Samir Khan), asks:

> Why not the West? If you are coming from the West, you might be asked by the mujahidin why you did not partake in the jihad inside your country. If you tell them, "to help the mujahidin," many will tell you that attacking the enemy in their backyard is one of the best ways to help the jihad ... I strongly recommend all the brothers and sisters coming from the West to consider attacking the West in its own backyard.[41]

Khan's advice is symptomatic of much of what *Inspire* Magazine and also persons from al-Qaeda senior leadership have published the last few years: it encourages sympathisers to stay home, and it gives specific advice on how to plan and conduct attacks alone without traveling to al-Qaeda training camps.[42] Target- and method-wise these types of publication also represent something new since they encourage people to conduct terrorism on the basis of the available materials. "How to build a bomb in your mom's kitchen" is one of the low-tech articles in *Inspire's* so-called "open source" section. In other sections of *Inspire*, attacks with lighter weapons such as firearms, knives or using your car as the "ultimate moving machine" are promoted.

Lighter measures also often result in smaller attacks. The tendency to choose lighter weapons and smaller targets is also evident both in the propaganda and in the actual attacks since 2008. Whereas there is still a fascination with homemade explosives and targeting aviation – e.g. AQAP have recently been trying to conceal bombs in packages on cargo planes (October 2010) and in underwear (the case of Umar Farouk Abdulmutallab, December 2009) – this new al-Qaeda propaganda also argues for targeting individuals, which is something that has not been associated with jihadi terrorism waged in the West before. The first volume of *Inspire* Magazine published a hit list with the names of specific persons who had mocked the Prophet and in 2013 Abu Khattab video from Syria ends with target shooting at pictures of Danish anti-Islamist opinion shapers. The statistical material that Jordan (2010) has collected on attacks in the West shows that since 2004 there has been a considerable rise in attacks in retaliation for alleged offences caused to Islam by the target: e.g. film director Theo Van Gogh and politician and activist Ayanna Hirsi Ali in Holland, as well as various cartoonists whose caricatures are considered offensive to Mohammed.

Most of these attacks were not directly linked to al-Qaeda operatives but were the work of Western militant Islamist networks or lone perpetrators who just sympathised with al-Qaeda. Targets designation came from the perpetrators themselves. The cases of Algerian-French shooter Mohammed Merrah in 2011, the Iraqi-born Swedish suicide bomber Taimour Abdulwahab al-Abdaly in 2010, American Nidal Malik Hassan's Fort Hood shooting in 2009, and Tamerlan and Dzhokhar Tsarnaev attack on the Boston Marathon in 2013 seem to support this tendency. The Fort Hood shooting in which Hassan in 2009 killed 13 people and wounded 29 using a normal semi-automatic pistol bought in a local gun shop is illustrative for the weak correlation between al-Qaeda's propaganda and terrorist actions in the period after 2008. Even though Hassan did correspond with Anwar al-Awlaki in Yemen prior to the shooting, nothing suggests that the assault-style method and the target designation came from al-Awlaki. It was Hassan's idea and he sought permission to carry out some sort of attack.[43] Only after this incident – and other incidents in 2009 – did al-Awlaki, and later on al-Qaeda senior leadership, begin to promote the so-called "lone wolf" attacks. In March 2010 Adam Gadahn praised Nidal Hassan as a "role model" and in another statement from October 2011 Gadahn talks about "the door" that Nidal opened.[44] Abu Musab al-Suri had indeed written about individual jihad and how it could be used against the West earlier, but it was not until after the Fort Hood shooting that the AQSL's and official propaganda changed in this direction.

Even though it is difficult to identify clear tendencies in the period 2008 to the present it is, however, clear that the al-Qaeda central leadership have had limited control over most of the perpetrators in attacks against Western targets in the period since 2008 and that most targets have been chosen locally by individuals who have not, in the large, been able to conduct successful terrorism. Even though local perpetrators might have sought out al-Qaeda the lack of direct orders in most cases means that what – was left of – AQSL naturally lost control over these operations. But it does not mean that al-Qaeda has not influenced these people. Since 2008 al-Qaeda propaganda – especially from affiliated groups but also from central leadership – has been directed towards specific audiences in the West and contains much more tangible and concrete advice on methods and targets in America as well as Europe. But what the Fort Hood shooting and the following propaganda also show us, is that the direction of the correlation between propaganda and terrorism to some extent has been twisted. The direction of the cause and effect relationship can now also go the other way: Nidal Hassan's actions changed the propaganda of AQSL and AQAP.

Concluding remarks – the relationship between al-Qaeda propaganda and al-Qaeda's target designation

This chapter's historical analysis of al-Qaeda's propaganda and al-Qaeda's terrorism does reveal some transformation and evolution in the targeting of the West and some correlation between propaganda and target designation. The chapter has also shown that the relationship between propaganda and targets designation clearly changed over the three periods. As the organisation developed so the role of propaganda and the nature of al-Qaeda's terrorism changed.

The link between propaganda and deed was strongest in the period 1998–2002 when the organisation was centralised and hierarchical. Even though target designation was rather unspecific in the propaganda AQSL was able to follow up their anti-American propaganda with large coordinated attacks and mass causalities. After the flight from Afghanistan AQSL was only occasionally able to point out specific targets (such as Denmark in the Muhammad Cartoon Crisis). It had to rely on local affiliates to act, but did manage to broaden the concept of the "Western Enemy" from America to include European countries and European persons that somehow had offended Islam. The broadening of the Enemy was visible in both propaganda and target designation. Even though al-Qaeda's propaganda after 2008 became much more professionalised and much more targeted towards specific Western audiences, AQSL lost control of the plots and attacks which increasingly were carried out by individuals who were picking their own targets based not on orders or directions but on inspiration from al-Qaeda.

Over the years the correlation between word and action, between leadership and sympathisers, between thinkers and doers, became much less straightforward than it was earlier. These developments have broader implications for how we today analyse the relationship between al-Qaeda's propaganda and terrorism and in this regard at least three questions could be raised.

1 Do words always precede actions?

The question of what effect al-Qaeda's propaganda has on target designations often has the implicit assumption that not only is there a connection between al-Qaeda's ideological texts and the actual terrorism; but also that there is a particular direction in the cause and effect relationship: propaganda – and broader ideology – is used as an explanation for and a cause of terrorism. The analysis of the attacks after 2002 show us that this was not always case and that the causal direction might even have been reversed in these later years.

Al-Qaeda's handling of the Muhammad Cartoon Crisis showed that al-Qaeda up to 2008 occasionally had the ability to promote specific targets in their propaganda and make their affiliates follow this up with attacks against these specific targets. But after 2008 cases such as Nidal Hassan's Fort Hood shooting indicate that actions came before words. It is an example of how an attack subsequently influenced and changed al-Qaeda's ideology and propaganda.

The tendency to act without direct incitement in the propaganda is increasing after 2008. Ex post facto rationalisation of targets is evident if one looks at how al-Qaeda and its affiliates, in these times of financial need, have begun to derive financial income from kidnappings, war booty and smuggling. Where al-Qaeda's writing has earlier been critical of kidnapping for money, looting and making money from smuggling, it has been interesting to see how al-Qaeda's senior leadership and affiliates such as AQIM, AQAP and ISIL have, these last few years, begun to give ideological/theological justifications for this.[45] In most of these cases it seems that practice has changed the ideology and not the other way around.

But even in the case of al-Qaeda's early ideological proclamations and attacks the causal direction is not clear. The most famous al-Qaeda fatwa of 1998 ("Declaration from the world Islamic front for jihad against the Jews and the Crusaders') could be seen as an explanation for the following attacks in the same year. The fatwa was published on 23 February and the attacks on the American Embassies were executed on 7 August. There is undoubtedly a clear linkage between the fatwa and the attack but if we look at Fazul Abdullah Mohammed's diary and the courtroom testimonies from the trial in 2001 about the 1998 Embassy Bombings, we see that planning for the attacks was most likely commenced years before the fatwa was ever published.[46]

Today it is increasingly difficult to prove the causal relation narrative "from words to actions" – whether ideology or propaganda is not a post-facto rationalisation instead of the cause of the events. Motivations for violent actions can be many and various – emotional, material, opportunistic etc. – and using propaganda as the single independent variable for either groups or individuals is too narrow an approach when it comes to al-Qaeda.

2 Is target designation a top-down process?

Both al-Qaeda senior leadership and scholars working with propaganda, such as myself, have an incentive to emphasise the importance of the official leadership propaganda. For scholars it is much more difficult to gain insights into the considerations and innovations from foot soldiers and operatives than it is to simply read the widely distributed and available propaganda. And the al-Qaeda leadership – like any leadership – of course wants to project an image of being in charge and setting the stage, also and maybe even more so when this is not the case.

But this chapter shows that after 2002 and especially after 2008 it seems doubtful whether the ideologues in the leadership have chosen the targets and means. AQSL's propaganda still inspires people and can – at times such under the Muhammad Cartoon crisis – point in a direction that seem to order and encourage actions from sympathizers. But Nidal Hassan and later attacks seem to indicate another process of how sympathisers engage in new and individual acts of terrorism that are later picked up by leadership and then promoted in ideological statements. Target designation and *modus operandi* seem here to be inspired from the bottom up.

3 Does al-Qaeda have coherent propaganda?

Comparing the three periods it is obvious that the idea of a single and coherent al-Qaeda with a well-defined propaganda profile and target designation becomes more and more problematic. Especially after 2002 al-Qaeda's propaganda material grew enormously in size. Even if one only focuses on the publications by AQSL one still ends up with a comprehensive body of text material containing many different opinions and directions concerning appropriate actions and legitimate targets in the West. One can find sanctioning of many different targets, depending on where in the text corpus one reads.

After 2002 it becomes increasingly clear in the propaganda – and also in internal correspondence of al-Qaeda members – that there are different opinions about what constitute the best and most legitimate targets in the West. Basic strategic considerations such as: "does it help our cause to kill Iraqi Shiites?"[47] or: "should we target Iran?". Others have to do with the theological legitimacy of certain targets: "are we allowed to kill children?". Fazul's writings are full of such considerations and discussions,[48] as are the internal letters between al-Qaeda's senior leaders found at Abbottabad, where bin Ladin in 2010 and 2011 warned about the killing of civilians.[49] And in December 2011 al-Zawahiri said in a public video that al-Qaeda should avoid killing civilians – even non-Muslims – in Egypt.[50] This could be understood as contradicting other propaganda messages that emphasised the need for large mass casualties.

Even though central leadership has a distinct voice in these debates there are rarely clear answers or specific targets and especially after 2008 there seems to be no absolute authority on several of these topics and, therefore, most of these questions keeps coming up. In other words: the idea of a single and coherent al-Qaeda propaganda with well-defined target designation is problematic since it does not exist anymore. It could be argued that this was the case before 2001/2002 where AQSL produced the majority of al-Qaeda's propaganda but even that can only be posited on the basis of selective interpretation of the material available. As has been shown AQSL was in 1998–2002 not very specific about target designation.

The ambiguous nature of al-Qaeda's propaganda is of course not a reason to avoid analysing those broader trends that we actually have identified in both the propaganda and internal letters. It just means that al-Qaeda's propaganda is not clear and therefore our findings too will reflect that and be conjecturable.

Notes

1 "Terror Plotting: Bravado or Sincerity?", CNN, 19 October 2012, http://edition.cnn.com/2012/10/18/justice/new-york-federal-reserve-plot/index.html?iref=allsearch.

2 Many researchers such as Lia and Hegghammer (2004) have analysed al-Qaeda's ideological writings in order to explain strategy and attacks, but also terrorism analysts such as Katz (2007) have argued for the importance of analysing al-Qaeda's speeches and videos. See B. Lia and T. Hegghammer (2004) "Jihadi Strategic Studies: The Alleged Al Qaida Policy Study Preceding the Madrid Bombings", *Studies in Conflict and Terrorism*, 27, no. 5, September/October.

3 In order to limit my analysis I will focus on the specific ways in which AQ has mentioned targeting the West so I will exclude the rather vast material that deals with how local groups should fight their local or "near" enemy.

4 For different opinions on what al-Qaeda's propaganda can tell us and how al-Qaeda's use of the internet should be countered, see the testimonies by Williams McCants, Aaron Weisburd, Brian Jenkins and Evan Kohlman, Committee

on Homeland Security (2011) *Subcommittee Hearing: Jihadist Use of Social Media* (6 December 2011, Washington, D.C.), http://homeland.house.gov/hearing/subcommittee-hearing-jihadist-use-social-media-how-prevent-terrorism-and-preserve-innovation.

5 This chapter focuses on propaganda and internal documents from al-Qaeda senior leadership or al-Qaeda's official sub-groups (such as AQAP, AQIM, etc.). Internet preachers or independent sheiks with ambiguous affiliation to jihadism (e.g. Abu Muhammed al-Maqdisi, Abu Basir al-Tartusi) are not part of the study.

6 I define hard evidence as Nesser (2008: 926):

> Hard evidence might be video surveillance of a selected target, the discovery of bombmaking materials and/or written plans on computer files, so-called martyrdom videos or testaments, money transfers, testimonies, and released intelligence, such as that gathered from wire tapping, and so on.
>
> (Petter Nesser (2008) "Chronology of Jihadism in Western Europe 1994–2007: Planned, Prepared, and Executed Terrorist Attacks", *Studies in Conflict and Terrorism*, 31: 924–946)

7 Silber's (2012: 3–4) distinction between "Al Qaeda command and control", "Al Qaeda suggested/endorsed" and "Al Qaeda inspired" plots is to some extent useful for this article since it tries to define how much al-Qaeda was part of the plot or attack – and also concerns target selection. In this respect this chapter has been inspired by Silber's study of 16 plots, but more often than not essential data about the role of al-Qaeda's leadership or senior operatives seems to be missing or is hazy which makes categorization uncertain. See Mitchell D. Silber (2012) *The Al Qaeda Factor: Plots Against The West*, Philadelphia, University of Pennsylvania Press.

8 The most famous fatwa was Usama bin Ladin's 1996 fatwa, "Declaration of War against the Americans Occupying the Land of the Two Holy Places" that was sent as a telefax on 23 August to several Arab newspapers and published in *Al-Quds al-Arabi.* The propaganda material from Advice and Reform Committee (ARC) can be found in Combating Terrorism Center's Harmony Database, e.g. ARC (1994) "Saudi Arabia Unveils its War Against Islam and its Scholars" (Harmony documents, AFGP-2002–003345).

9 The Fatwa was published in *al-Quds al-Arabi* on 23 February 1998 after a meeting in Afghanistan where leaders of different militant Islamist organisations, including Usama bin Ladin and Ayman al-Zawahiri, met. For a discussion of why al-Qaeda made the shift from the near to the far enemy, see Magnus Ranstorp (1998) "Interpreting the Broader Context and Meaning of Bin-Laden's 'Fatwa'", *Studies in Conflict and Terrorism*, 21, October/December: 321–330 and Fawaz A. Gerges (2005) *The Far Enemy: Why Jihad Went Global*, London, Cambridge University Press.

10 Committee on Foreign Relations (2009) *Tora Bora Revisited: How We Failed to Get Bin Laden and Why It Matters Today* (November 30, Washington, D.C.), www.foreign.senate.gov/imo/media/doc/Tora_Bora_Report.pdf.

11 Peter Bergen (2006) *The Osama bin Ladin I Know*, New York, Free Press.

12 I am aware that the former intelligence officer Mowatt-Larsen (2010) argues that al-Qaeda still intends to develop WMDs and puts great effort into this, but I tend to agree with Stenersen's argument, that al-Qaeda's interest in using chemical, biological, radiological and nuclear (CBRN) terrorism to achieve their goals after 2001 appears lower than commonly anticipated and that al-Qaeda would have used the weapons and know-how by now had they had it. See Anne Stenersen (2008) *Al-Qaida's Quest for Weapons of Mass Destruction,*

Saarbrücken, VDM Verlag Dr. Müller and Rolf Mowatt-Larsen (2010) *Al Qaeda Weapons of Mass Destruction Threat: Hype or Reality?* Belfer Center for Science and International Affairs, Harvard Kennedy School.

13 Bin Ladin declared in his 1996 declaration of war that "due to the imbalance of power between our armed forces and the enemy forces, a suitable means of fighting must be adopted i.e. using fast moving, light forces that work under complete secrecy. In other words to initiate a guerrilla warfare." See Usama Bin Ladin (1996) *Declaration of Jihad against the Americans Occupying the Land of the Two Holiest Sites.* Translated by the CTC at West Point Military Academy; http://en.wikisource.org/wiki/Osama_bin_Laden%27s_Declaration_of_War.

14 For Khalid Sheikh Muhammed's role in picking out targets in the 9/11 operation, see "Verbatim Transpript of Combat Status Review Tribunal Hearing for Khalid Sheikh Muhammed", in T. McDermott and J. Meyer (2012) *The Hunt for KSM: Inside the Pursuit and Takedown of the Real 9/11 Mastermind, Khalid Sheikh Mohammed,* New York, Little, Brown: 292ff.

15 From Fazul's memoire *al-Harb 'ala al-Islam: Qissat Fadil Harun* (The War against Islam: Fadil Harun's Story), 1: 523. Posted on a jihadi website in February 2009. Translated and quoted in Nelly Lahoud (2012) *Beware of Imitators: Al-Qa'ida through the Lens of its Confidential Secretary,* CTC Report, West Point: 63, www.ctc.usma.edu/posts/beware-of-imitators-al-qaida-through-the-lens-of-its-confidential-secretary.

16 A pre-2002 internal al-Qaeda document (AFGP-2002–000078) sets out the elaborate list of qualifications one must possess in order to become a supervisor for al-Qaeda's special operations section, which indicates that pre-2002 there were strict rules for who could plan and train people for terrorism abroad.

17 "Bin Ladin's Former Bodyguard Interviewed", *al-Quds al-Arabi,* 3 August 2004 and 20 March–4 April. Quoted in Bergen (2006).

18 Quoted in Gilles Kepel, Jean-Pierre Milelli and Pascale Ghazaleh (2008) *Al Qaeda in its Own Words,* New York, Harvard University Press.

19 The apprehension of Khalid Sheikh Mohammed in 2003 was undoubtedly a big blow for al-Qaeda. Mohammed was the chief of al-Qaeda's out-of-area special operations unit and the mastermind of the 9/11 attacks.

20 Greg Bruno (2010) "Al-Qaeda's Financial Pressures", *Council on Foreign Relations,* www.cfr.org/terrorist-organizations/al-qaedas-financial-pressures/p21347#p3: "By 9/11, the CIA estimated al-Qaeda's annual operating budget to be about $30 million, raised primarily through donations." This financial network was hit hard with multiple international initiatives aimed at impeding al-Qaeda's ability to fund after 2001. The lack of funding was even visible in the propaganda as well. In the period after 2001 AQSL released numerous video appeals for donations.

21 Thomas Hegghammer (2006) "Global Jihadism after the Iraq War", *Middle East Journal,* 60, no. 1, Winter.

22 For development of Saudi-Arabian groups, such as al-Qaeda in the Arabian Peninsula in 2001–2003 and its linkages to al-Qaeda, see Thomas Hegghammer (2010) *Jihad in Saudi-Arabia,* London, Cambridge University Press. For GSPC's (Salafist Group for Preaching and Combat) morphing into AQIM in 2006, see Manuel R. Torres Soriano (2011) "The Road to Media Jihad: The Propaganda Actions of Al Qaeda in the Islamic Maghreb", *Terrorism and Political Violence,* 23: 72–88.

23 In December 2005 Atiyah warns Zarqawi: "Concerning foreign activities, my brother, that is to say, outside of the Iraq arena, I would advise you not to expand that way until you have reviewed it with the brothers."

24 See Hanna Rogan (2006) "Jihadism Online: A Study of How al-Qaida and Radical Islamist Groups Use the Internet for Terrorist Purposes", Kjeller:

Norwegian Defence Research Establishment and Anne Stenersen (2008) "The Internet: A Virtual Training Camp?", *Terrorism and Political Violence*, 20: 215–233.

25 Lia and Hegghammer (2004).

26 Strategic writers such as Yusuf al-Ayiri, Abu Musab al-Suri, Abu Umar al-Sayf, Abu Bakr al-Naji and Lewis Atiyatallah.

27 Ayman al-Zawahiri, aired on al-Jazeera on 1 October 2004, English transcript from CNN.

28 Roy (2004) and (2010) has in both books elaborated on the impact of these controversies. See Olivier Roy (2004) *Globalized Islam: The Search for a New Ummah*, New York, Columbia University Press and (2010) *Holy Ignorance: When Religion and Culture Part Ways*, New York, Columbia University Press.

29 See Usama bin Ladin (2007) *A Message from Shaykh Usama Bin Ladin to the People of Europe*, www1.nefafoundation.org/miscellaneous/FeaturedDocs/nefabinladen0308.pdf and (2008) *May Our Mothers be Bereaved of Us if We Fail to Help Our Prophet* (English translation), posted to various internet sites, www1.nefafoundation.org/miscellaneous/FeaturedDocs/nefabinladen0308.pdf.

30 Nesser (2008). See also Javier Jordan (2012) "Analysis of Jihadi Terrorism Incidents in Western Europe, 2001–2010", *Studies in Conflict and Terrorism*, 35, no. 5, 1 May.

31 On Khan's training, see Silber (2012: Ch. 7).

32 On Khürshid, see "Portræt af Hammad Khürshid", *Politiken*, 18 August 2008. On al-Makki, see al-Qaeda video, released on 4 September 2008, "The Word is the Word of the Sword" (SITE 2008090502).

33 See al-Qaida/as-Sahab (2008) *The Word is the Word of the Sword.*

34 Usama Bin Ladin (2008) *May Our Mothers be Bereaved of Us if We Fail to Help Our Prophet.*

35 Ayman al-Zawahiri (2008) *Selected Questions and Answers by Dr. Ayman al-Zawahiri*, released on different web fora 17 April 2008, www.nefafoundation.org/miscellaneous/FeaturedDocs/nefazawahiri0508–2.pdf.

36 Silber (2012: Ch. 3) considers Operation Overt in the UK in 2006 (where plans made by British militant Islamists against transatlantic airliners were discovered) to be the only case of an al-Qaeda "command and control plot" after 2001.

37 Mani Crone and Martin Harrow (2010) *Homegrown Terrorism in the West 1989–2008*, Copenhagen, DIIS Working Paper 2010: 30.

38 Samir Khan and Anwar al-Awlaki have contributed to many *Inspire* volumes since 2010, e.g. *Inspire* no. 2 (October 2010). Hammani's raps "Send Me a Cruise [Missile]" and "Make Jihad with Me" were uploaded in December 2011. Bakkay Harrach is featured in many IMU and as-Shabab videos, e.g. "Der Islam und die Finanzkrise" (2009).

39 "Danskere kæmper for al-Qaeda", Ekstrabladet, 4 October 2013.

40 As illustrated by Hegghammer's historical account of the Muslim foreign fighter phenomenon. See Thomas Hegghammer (2011) "The Rise of Muslim Foreign Fighters: Islam and the Globalization of Jihad", *International Security*, 35, no. 3, Winter.

41 *Inspire Magazine* no. 2, Fall 2010: 24.

42 Adam Gadahn, Ayman al-Zawahiri and Abu Yahya al-Libi (along with Gadahn) also praised "lone wolves" such as Fort Hood shooter Nidal Malik Hasan and Dutch filmmaker Theo van Gogh's killer Mohammed Bouyeri, and encouraged jihadists to strike at targets of opportunity in their Western Homelands in the 2011 video "Do Not Rely on Others, Take the Task Upon Yourself".

43 According to an interview on al-Jazeera in December 2009, Anwar al-Awlaki claimed that Hassan initiated the email correspondence with a message on 17

December 2008: "He asked whether killing American soldiers and officers is lawful or not." See "Transcript of interview with Al-Awlaki", NBC, 23 December 2009, www.msnbc.msn.com/id/34579438/ns/us_news-tragedy_at_fort_hood/t/transcript-interview-al-awlaki/#.T-NFxys5Iag.

44 Adam Gadahn, Ayman al-Zawahiri and Abu Yahya al-Libi (2011) *Do Not Rely on Others, Take the Task Upon Yourself*, as-Sahab Production, www.youtube.com.

45 AQAP has used the Islamic concept of "ghanimah" in some of their writings (e.g. *Inspire* no. 6) in order to justify looting as a way of making money.

46 See Lahoud (2012) and Vahid Brown (2008) *A profile of Abdullah Muhammed Fazul*, Combating Terrorism Center, www.ctc.usma.edu/posts/a-profile-of-abdullah-muhammad-fazul. For court testimonies, see U.S. (2001) *U.S.* v. *Usama Bin Laden et al.*, S(9) 98 Cr. 1023, S.D.N.Y., http://cryptome.org/usa-v-ubl-dt.htm.

47 Ayman al-Zawahiri and Atiyah Abd al-Rahman's letters to Abu Musab al-Zarqawi in Iraq in 2005 discuss this issue.

48 Regarding the internal discussion about targeting children, see the memories of al-Qaeda fighter Paulo Jose de Almeida Santos in Bergen (2006: 119). See also Fazul's critique of the Beslan operation by Islamic Chechen militants summarised in Lahoud (2012: 58). See Gadahn's (2011) letter to central leadership found in Abbottabad, criticising al-Qaeda in Iraq.

49 Bin Ladin stressed in his internal letter to both al-Shabaab leader Abu Zubeir and to an unknown Shaykh Mahmoud that killing civilians Muslims is damaging al-Qaeda and should be avoided. See Usama Bin Ladin (2010) *Letter to Shaykh Mahmoud*, SOCOM-2012–0000019, www.ctc.usma.edu/posts/socom-2012–0000019-english and (2010) *Letter to Mukhtar Abu al-Zubayr*, SOCOM-2012–0000005, www.ctc.usma.edu/posts/socom-2012–0000005-english.

50 Al-Zawahiri (2011):

> I do not call for fighting in Egypt at this moment, and I repeat and stress, at this moment ... the Egyptian people in general and the Islamic movement in particular must be peaceful and protect non-Muslims who do not attack Muslims.
>
> (Ayman al-Zawahiri (2011) A Message of Hope and Glad Tidings to Our People in Egypt (SITE 2011120101)

6 Jihadist innovation and learning by adapting to the "new" and "social media" Zeitgeist

Nico Prucha

The ideology of al-Qaeda (AQ) and its ability to propagate that ideology as a monopoly of truth through professional promotion and marketing material disseminated via modern communication technology has proven to be its most resilient foundation and greatest innovation. This jihadist media activism is evident and strengthens this resilience on a daily basis with new audio-visual and written propaganda uploaded from a number of conflict zones, in numerous languages, to a wide range of online social platforms and multimedia channels. This propaganda is underpinned by an ideological authority which has maintained an important and influential role, despite the killing of Osama bin Laden by SEAL Team 6 and the demise of AQ core leadership in drone strikes across the Afghan-Pakistani border area, events which have made it increasingly difficult to locate or define "al-Qaeda". Instead, AQ remains alive and has regained an active role in offshoot groups and individuals who subscribe to this monopoly of truth, the ideological parameters and interpretations of Islamic sources and the role model of the "proper" Sunni Muslim according to these principals of faith (*'aqida*) and religious methodology (*manhaj*). The doctrine of AQ provides a clear identity of what it should mean to be a "true" monotheistic Muslim (*muwahhid*) and how to profess the divine laws of God (*shari'a*) as based on the authoritarian interpretations and guidelines freely available on the internet. Fighters and clerics likewise and sometimes in rotating roles relate the written ideology into actions and fierce emotional sermons conveyed by the extremists' most powerful platform: Twitter and YouTube, jihadi forums and dedicated websites. These role models, the corpus of texts and, most importantly, the AQ-affiliated iconography most prominently represented and identifiable by the black banner with the imprinted shahada,[1] has successfully manifested within majority Sunni Islamic societies, and is even more visible within the Arab public space amid the chaos and aftermath of the so-called "Arab Spring".[2] Perhaps even more so in times of upheaval and (re-)spawning of militant groups throughout the countries affected by the "Arab Spring", first and foremost Syria, the emphasis of jihadist media advisors, strategists and contributors is set to undermine their own role and importance:

> We are indeed convinced that the battlefields of the media jihad are of the most important streams of jihad and an elementary front thereof in this raging war with our enemies. Therefore, we [media workers] commit ourselves just as any jihadi brigade, working correspondingly with our brothers in the fields of war, directly encountering the enemies.[3]

The self-esteem of the media battalions is further expressed in a strategy paper published in 2012 outlining the ambition of the jihadi media as:

- Resetting the focus to the interior of the Arab countries affected by the transformation processes; as the counter-terrorism and police forces are in a weakened position. Referring in particular to the intensifying civil war in Syria but also post-Gaddafi Libya where local jihadi militias are active, the online media workers define this as "the interior media departments".[4] The strategy is clear: to infiltrate or embed ideological adherents with al-Qaeda media workers to indoctrinate fellow fighters and to frame group actions accordingly. This, however, is not restricted to the new media. For pre-internet means are to be exploited by "relying on local radio and television stations, magazines and newspapers"[5] where the Mujahideen are in control over territory.
- Not neglecting the media work on a global level, emphasizing the role of the media outside of the actual conflict zones, defined as "international media departments". Jihad videos advertising and sanctioning violence while providing the theological basis for the active Mujahideen are broadcast, "depending almost entirely on the Internet and social media networks. [This work allows us to] maintain the anonymity and high professional security standards by the professional media workers, extending on the global level, by the grace of God."[6]

Or, as AQ leader Ayman al-Zawahiri summarized in his decree of "General Guidelines for Jihad" in September 2013: "It is well known to the brothers that our work in this phase consists of two sections: 1. Military; 2. Propagating."[7]

The value of the internet for strategic communication

The media in principle is used on a tactical level, not only to incite and recruit individuals worldwide, but to create ideological denominators and operational paradigms with policy guidelines for various jihadist groups. While "jihad" is first and foremost agreed as being implemented by militant and violent means in general or, in short, linked to "combat" (*qital*), the media thereof follows the rationale of propagating Islam (*da'wa*) and the proper religious methodology (*manhaj*). This propagation thus is part of the overall objective to "re-establish the rightly guided Caliphate" while uniting various Sunni fighting groups and individuals under this particular propagated *manhaj*.[8]

As a consequence of the strategic value of jihadist media, the perception has shifted to a new role model: the "media Mujahid" and the "media martyr". For any media worker embedded with fighting Mujahideen or committed in secret to develop and disseminate jihadist media worldwide has in the meantime been equated to any "regular" Mujahid or even proactive martyrdom seeker. The jihadist media departments are as decentralized as the internet itself. Subsequently, this role model is backed by the call to take the fight onto a greater level on all channels online as issued by al-Fajr in their response to the killing of bin Laden:

> The Internet is a battlefield for jihad, a place for missionary work, a field of confronting the enemies of God. It is upon any individual to consider himself as a media-mujahid, dedicating himself, his wealth and his time for God.[9]

For the 'media mujahid' there is no need to physically join the real-life militant groups to participate in the divine obligations of combat, as his means of propagation are functional and of greater value far from the battlefield. This divine obligation in jihadist terms consists of incitement, inspiring (*tahrid*) the targeted audience to follow the actions of the Mujahideen. AQ's senior leadership has emphasized this in publications over the last years and made direct statements in recent as-Sahab videos. The Libyan ideologue 'Attiyatullah al-Libi (d. 2012) had been quoted in a video entitled "You are only accountable for yourself", a direct quote from the Qur'an (4:84), equating the urge to join the Mujahideen with the propaganda work that will inspire individuals to engage in

> missionary calls to each and every Muslim for jihad; to join the caravan of jihad (*qafilat al-jihad*) under any circumstances. This means without preconditions or limitations on the basis of the call to arms (*nafir*) towards this or that front – and the incitement to jihad is a *shari'a*-law obligation. God, exalted is He, says: "Prophet, urge the believers to fight…",[10] and: "Urge the believers on. God may well curb the power of the disbelievers."[11,12]

The jihadist endeavours online are part of a global culture dependent on modern technology to broadcast themselves and their messages to the world. This modern agitprop is not restricted to jihadists and is used likewise by other militant and hate groups. However, the jihadi use of the internet is unprecedented in terms of the vast quantity of various categories of propaganda material. Especially the audio-visual productions are perhaps the most powerful tool to

> lure consumers into their specific interpretation or world perception, trying to gain sympathy through modern, pop-cultural elements. Jihadists employ *nasheed* ["praiseful hymns"] and *huda* ["encouraging battle songs"] in combination with appealing training, combat, or everyday aspects of the Mujahideen and use a powerful yet comprehensible rhetoric. The rhetoric is inseparable from the (audio-) visual content and enforces key elements such as grievances and the need to respond, usually framed as a "call" to arms (*nafir*).[13]

The jihadist media departments, the most visible and public side of groups such as al-Qaeda, have developed, maintained and seriously improved a massive virtual footprint since the attacks on September 11, 2001. As Magnus Ranstorp outlines, AQ has created a "virtual sanctuary" from where the ideological parameters seem untouchable and freely roam the internet. In a move from static websites and classical online forums, jihadist media has nowadays infested the contemporary social media platforms following the same similar strategic incentive of the past "virtual sanctuary". The growth of jihadist propaganda outreach has boosted the network's

> polymorphic structure or design with multiplicity of nodes or pods swarming towards a mission or resurrecting shortly before or after an operation. More fundamentally it allows survivability through a constant virtual presence with no real or tangible physical centres of gravity and in constant stealth mode and ideological motion. Having simply an online presence confers a certain degree of legitimacy which they otherwise would not have. It also allows them to resurrect and reconfigure at any time.[14]

Online media platforms facilitate a blend of audio-visual media interspersed with writings that further sanctions and explains specific ideological dimensions of jihadi activity.[15] As the range of online platforms has expanded, so jihadist groups have increasingly used sites such as Twitter,[16] Facebook and YouTube,[17] with their use of Tumblr also gathering pace.[18] The role of the "media mujahid" has been approved,[19] sanctioned[20] and encouraged with the release of suggested strategies,[21] although not all have seen the move away from the traditional forum based interaction as positive.[22]

Jihadist agitprop online and new media strategies

This Zeitgeist reflects the virtual behaviour of aspiring real-deal real-life jihadists who are influenced not only by the interactive and inclusive electronic propaganda of jihadi groups, but are moved and affected by the real-world grievances of mainly Islamic countries. The growth of extremist agitprop and the influx of such output on social media platforms like Facebook and Twitter has led to a new generation of jihadists, who had been active within the virtual dominions of al-Qaeda, where exposure to extremist media has had an impact on their personal lives and understanding of religious conduct.[23] The term agitprop refers to the Soviet Department of Agitation and Propaganda that deployed lay actors to perform theatre plays for workers throughout the Soviet Union in the 1920s. Thus, "Agitprop theatre, unlike other forms of propaganda, was able to communicate directly to the masses; [with] one explicit purpose: to reach the working class directly."[24] The massive audio-visual content disseminated by jihadist media departments on multiple layers by various facets online is comparable, also in terms of its success, to agitprop theatre.

This trend reflects to great extent a specific Zeitgeist, a contemporary as well as generational shift from texts and scripts to a visual literacy. Ideology is presented by iconographic, habitual and, most importantly rhetorical, means. Elements shown in jihadi videos are thus most appealing to initiated consumers who can read and identify the greater ideology at work.[25]

The main AQ forums have picked up on this Zeitgeist and are active on Twitter, advocating their official accounts on the main forum pages.[26] Jihadi media activists and fighters on the ground use Twitter in the meantime on a regular basis to upload their personal pictures and videos, made on their cell phone cameras. This has furthered in general a renewed role model, namely that of the "media martyr", in reference to any killed media activist within the ranks of the regular Mujahideen.

This material enters the jihadist online sphere where it is instantly used and reused to strengthen the worldview of al-Qaeda and affiliated groups. As a result, new communication channels have emerged through which the new generation of activists and fighters, including those working for, or on behalf of, al-Andalus, al-Shabab and Jabhat-al-Nusra can interact with potential influence multipliers and sympathizers. Groups such as the Somalia-based Harakat al-Shabab al-Mujahideen (HSM) rely on Twitter to publish pictures, statements and links to YouTube jihadi videos primarily in Arabic and English. Twitter was used by HSM to update their followers regarding the failed French operation to free Denis Allex, who had been taken by the group in 2009.[27] Pictures of an alleged killed French special operative and his gear had been put on Twitter and Facebook,[28] the statement claiming the revenge execution of the hostage was exclusively claimed on Twitter by the "HSM Press Office (@HSMPress)".[29] The Afghan Taliban (@ABalkhi), the media department al-Andalus (@Andalus_Media) of al-Qaeda in the Islamic Maghreb (AQIM) and AQ's Iraqi al-Furqan media branch (@abo_al_hassa) have Twitter accounts and frequently publish and disseminate new and old content. The Yemeni branch, AQAP, for example, recently resorted to publishing and linking statements first on Twitter.[30] These "official" media channels, to facilitate an active communication with their sympathizers and followers, also use Twitter. The Twitter account of Al-Andalus Media was advertised for by the official and thus verified member-account on the jihadi forums.[31] Four days later, on 1 April 2013, an "open interview" was announced via the forums with "Shaykh Ahmad Abu 'Abd al-Ilaha, the head of the Media Board for al-Andalus", setting a time window to ask the questions on Twitter.[32] Only English and Arabic questions had been considered and answered respectively on 18 April.[33]

The "AQTweets" section of the eleventh edition of the *Inspire* magazine, which highlighted tweets regarding the Boston bombing, further elevates the use of Twitter by multilingual jihadist media activists.[34] These English-language tweets reflect a pop-cultural youth slang while the Tsarnaev brothers who committed the bombing are applauded as "only 2 soldiers of Allah", who by their operation "are lone wolves. That z wat ALQ wants, the #jihad project to be da ummah's prjct." The Boston bombing is furthermore a success within the framework of the sympathizers and supporters of al-Qaeda, for the Chechen-born brothers Tamerlan and Dzhokhar Tsarnaev had been residents of the United States. Furthermore, they used the bomb-making instructions of the English-language *Inspire* magazine on how to build the pressure-cooker bomb. The innovative jihadi media outlets, of which the *Inspire* magazine is the most prominent English substrate, celebrates the Tsarnaev brothers as the new role models who have been virtually guided by both the ideological as well as bomb-making parameters.

The classical jihadi discussion forums remain the vital hub for authoritative, cohesive and coherent propaganda disseminated via all channels of the internet. This coherent and cohesive ideology, that in past years has filled many gaps and open questions by responding to questions and criticism online, is difficult to counter, as no "single-narrative" of al-Qaeda or its underlying substrate is evident. Rather, on a multi-layered basis, emotional, theological, moral and other elements are triggered by its massive professional propaganda output combining the textual layer with the audio-visual one. However, jihadi media groups and activists have turned with increasing speed to social media and are fully committed on Twitter, where selected parts and the most popular aspects of this cohesive ideology are disseminated in a vociferous style.[35]

After all, as Jason Lanier puts it, "digital information is really just people in disguise"[36] and the internet is an intersection where content from the real world, our daily lives, enter a global platform in textual, visual and audio-visual form. As much as any of the virtual content, jihadism with its vast media material from the real world has created spheres of its own which attract various types of audience, who in turn are also able to participate in the "online jihad" by creating and publishing their own versions of fan-culture. Instead of following the mainstream parts of global and/or local culture, individuals hence are able to subscribe to a specific and nevertheless global lifestyle that is the essential and integral foundation of a worldwide active terror group, according to their own standards. As outlined in the jihadist media in general, and summarized by AQ leader Ayman al-Zawahiri in a speech eulogizing Abu Yahya al-Libi in September 2012:

> we do not tie our jihad to any organization, to any [specific] leader or leadership, not to any group, and not to any name or territory not even to the Holy Land. For the land does not hold anyone sacred but sanctifies one for one's deeds. Therefore, we support this form of divine service while we engage the enemies of God, fighting them wherever [God] enables us to, in Afghanistan, Pakistan, Somalia, Iraq, Algeria, Syria, and Libya.[37]

The jihadi culture on the internet is fed by authorized media material, which in the past years has more and more been re-disseminated based on increased fan activities. The fans and followers subscribe to AQ's monopoly of absolute truth and their work in turn is valued as part of the overall jihad. Thus, engaging and drawing in their audiences by the new media in general is part of the jihadists' motivation and tactic on the internet:

> as versatile, the converging medium, as wide is the range of different fan activities we find online. Beyond the "one-click fandom" on Facebook, the diversity of these practices necessitates the further qualification of how they cluster among different fan groups. Two definitions are thus required here: Firstly, the emphasis on fans' productivity in online spaces warrants further attention to the different types of productivity in "participatory culture" (Jenkins, 2006). John Fiske (1992) distinguishes between three types of productivity among fans: semiotic productivity – the act of reading/watching/listening and thereby meaning construction on behalf of fans in encounters with the object of fandom and texts surrounding an object of fandom; enunciative productivity – which describes the interactions between fans surrounding their fan object such as fan talk, gossip, banter and speculations; and textual productivity – found in fans' creation of new texts such as fan fiction, music remixes, or fanzines.[38]

The role of the "public relationship departments" of AQ, via the online forums, the electronic magazines where individuals can be drawn into and even participate by writing articles, has resulted in a "new martyr" and renewed role models in general, who had been engaged online and then managed to join jihadi detachments in the real world to die fighting "on the path of God". Such stories are on the rise, with individuals being portrayed having had their first point of contact on the internet and then moving on to the battlefields.

The intersection of offline acts and online visualization of these purported legitimate defensive and morally justified actions construe and fertilize the jihadists' frame of reference and the motivations that have consequently led to both individual as well as collective actions. The effect does not only resort to connecting, relating and directly bridging the offline real world with the "virtual sanctuary". Rather, it is more devastating as "this virtual domain simultaneously expands and de-territorializes the social interactions between the local and global jihadist milieus."[39] By using singular words or short phrases like the Islamic creed (*shahada*), "There is no God but God and Muhammad is His messenger", the jihadists seek to hijack Islam exclusively for their cause and interpretation, with the ambition of drawing the consumer into their virtual "network of knowledge" to further the understanding of this ideology that in the past has covered more and more niches. In combination with the assumed position of AQ ideologues and their countless pseudo-scholarly writings, ideologues as well as "regular" fighters or "martyrdom operatives" serve as guides for the consumers, leading them further into this online milieu of jihadism. By citing renowned jihadist figures or scholars the individual actor of jihadist videos relays the ideology in their corresponding dialect, language or emphasis, enabling the audience to read and expand their knowledge within the virtual worlds where countless religious writings and essays by historical as well as contemporary jihadi scholars can be easily accessed.

These religious authoritative writings serve as the backbone and as the intellectual motivation that finds its expression by mostly male actors[40] in the audio-visual propaganda – who openly address the audience with details of their motivation to conduct, for example, "martyrdom operations" or who had filmed their testimony (*wasiyya*) in case they were killed in "regular" combat. The videos are the mainframe of the intersection of online and offline worlds.

As a recent phenomenon, pro-jihadist but apparently non-militant groups, usually termed by Western media as "salafists", openly demonstrate in Western cities using the same iconography and slang of militant jihadists movements, with local language inspired by the jihadist propaganda at hand and in turn adding new files to the overall corpus. Arabic core files, writings, and the authoritative use of specific symbols are thus in a localized context and a local language used in a repetitive manner. This creates a coherent identity and group cohesion, while individuals receive guidance and can strengthen their notion of belonging within this community. According to Anthony Cohen, "people construct community symbolically, making it a resource and repository of meaning, and a referent of their identity."[41] Subsequently, a specific religious habitus is propagated as an audio-visual frame of reference that includes vital cornerstones of an ideology put into action with the perception of fighting for both justice and defence. The intersection of real life and online, virtual worlds – accessible for anyone unlike most real-life scenarios – lead to questioning the effects of the accessibility to the jihadist propaganda for consumers sometimes thousands of miles away, but in close emotional, spiritual and politicized proximity with the fear, anger and cognition of being "persecuted" because of ones faith and one's identity as "Muslim".

The new martyrs of the internet as innovative role models for the real-world: the death of AQ's second-in-command, Abu Yahya al-Libi

In a message broadcast by AQ's notorious as-Sahab media foundation in September 2012, Ayman al-Zawahiri eulogized the killing of his comrade and second-in-command, Abu Yahya al-Libi.[42] Al-Libi, whose real name was confirmed after his death by Ayman al-Zawahiri as Hasan Muhammad Qa'id, was killed in a drone strike in Pakistan in June 2012:[43]

> The death of al-Libi is a great loss for he had been a productive writer and somewhat of one of the remaining faces of as-Sahab. His comprising writings are hosted on Abu Muhammad al-Maqdisi's website tawheed.ws next to the prominence of jihadist scholars such as Osama Bin Laden, 'Abdallah 'Azzam, Sayyid Qutb, and many others.[44]

Thus al-Zawahiri elaborates on the role of al-Libi as a missionary worker for Islam (*da'i*) who was committed to "incite the umma" even more so after his episode of torture in the U.S. military prison in Baghram, Afghanistan.[45]

Abu Yahya al-Libi's importance for AQ's senior leadership, and al-Zawahiri in particular, is perhaps further evident as al-Zawahiri wrote an comprehensive introduction for al-Libi's main framework on dealing with Muslim spies, the "Guidance regarding the Ruling on Muslim Spies".[46] Abu Yahya al-Libi's framework for operationalizing shari'a law for jihadist groups was widely read and implemented by groups such as the Somali-based Harakat al-Shabab al-Mujahideen who sanction the execution of alleged spies in propaganda videos[47] and ideologues in AQ's Yemeni department such as Harith bin Ghazi al-Nazzari, warning imprisoned Muslims against interacting with their captors and thus be "turned" into spying on their brothers.[48] Grading this work as "judicial knowledge" (*al-i'lmiyya al-fiqhiyya*) and equating his 149-page text to a fatwa, a legal decree valid for all Muslims, al-Zawahiri further praises al-Libi who had:

> migrated [*hajir*], embedding himself spiritually and physically-military [*rabit*], conducting the jihad since the communist rule over Afghanistan [in the 1980s]. Since that time his commitment for jihad and the Mujahideen has never been revoked. Not even during his time in capture in Pakistan and Afghanistan in the prisons of the crusaders and their henchmen; he was – by the kindness [*munna*] of God – rescued from

> imprisonment despite his aforementioned captors and then re-joined the fields of jihad and military preparation [*al-i'dad*]. He then became the companion of the commander [Abu] al-Layth, the martyr – as we reckon him to be – Abu' l-Layth al-Libi – may God have mercy on him – who was assassinated by an American missile [fired by an unmanned drone] based on the treason of spies and the deception of the hypocrites. Abu Yahya in this time did not eat and was tirelessly seeking knowledge,[49] studying and lecturing, and sought further knowledge by travelling to Mauretania to master linguistic and juridical topics being instructed by the best scholars in the field.[50]

The death of al-Libi in 2012 and the eulogy by al-Zawahiri published in September of the same year provide an interesting insight into the overall concept of "martyrdom" of AQ's leader. For al-Zawahiri's perception of the martyrdom of al-Libi is focused on the latter's work and contribution to the media field, the tutoring and lecturing of Muslims worldwide, while having been a military field commander as well. Thus, one may argue, al-Libi had been a "hybrid" whose martyr story comprises both elements and is as such of great value to jihadi propaganda online, as al-Libi is a role model having professed his faith online and in the real world. His legacy online, consisting of various writings and videos, remains.

For the Mujahid is framed as an individual loyal to God in this and the next life, whereas in this world he is thus part of the group of the most pious and rightly guided believers, being someone who pursues death to be alive. For the reward of one's God will compensate in the next world for deeds in this, as outlined in the selective reading of verses of the Qur'an according to extremist interpretation: "for people desire life, fearful of death, while the martyr pursues death demanding life."[51] As Ayman al-Zawahiri elaborated in this eulogy:

> [al-Libi's] attained martyrdom will raise the attention of the people to his writings and his missionary work to propagate Islam, God willing. For the martyrdom is the profession of the words of life as the two martyrs – as we deem them to be – Sayyid Qutb and 'Abdallah 'Azzam – may God have mercy on them both – [have proven].
>
> This is exactly what America cannot cope with, who define the Mujahideen and AQ as their primary enemy; for they are ignorant of the reality that AQ is merely a message for the Muslim umma to strive, struggle, resist, and to fight the foreign Crusader-Zionist injustice and the corruption [*fasad*][52] inside [our countries]; for the attained martyrdom of the Mujahideen of AQ and the leadership only emphasizes the sincerity and promotes the message of AQ, for whenever our blood is shed for the sake of our articles of faith ['*aqida*], our words become alive among our umma. Whenever the martyrs fall, the *da'wa* [propagation of Islam] of jihad is filled with life.[53]

The message or the comprehensive narration by AQ is mainly alive on the internet where the martyrs and their stories remain present. Even if the use of the internet was in the beginning of the early days of jihadism online a unique phenomena with limited actors, it is nowadays unthinkable that most individuals who subscribed to AQ's worldview would not have profiles online or would be ignorant of the internet and the jihadist forums and social media channels. As al-Zawahiri further outlines, the internet is the safest zone for AQ's message, independent of the slain leaders, activists, or fighters in general and empowers jihadist actors in general and by principal to broadcast their "message of truth". As the jihadi media claims to provide information without hostile or enemy interference, it seeks to be an information platform for anyone, even for the Western public that is subjected to governmental censorship and information warfare, according to al-Zawahiri.

> For by these lies the Americans are deceived [by their own leadership] that the war against terror would be won against AQ by killing this or that person. It eludes their perception of reality and of their defeat in Iraq, and the coming defeat in Afghanistan. They have lost in Egypt, Tunisia, and Libya, and they miss the point that AQ has indeed divided its real tasks into various parts, consisting of the incitement of the *umma* [*tahrid*]. This is an omen of the defeat of America, God willing.[54]

For the extrajudicial execution of AQ personnel in general by drones and by other means, has no effect or impact on preventing the jihadist incitement of mainly Arab Muslims, especially in the aftermath of the Arab Spring.

Al-Zawahiri re-emphasizes the hybrid role of martyrs, who first of all have fulfilled their part either way in this world and are mediatized on the internet for their posthumous fame, that "our martyrs will enhance the message of jihad by greatest means possible, spreading and presenting this message as firm as a wall, for we are killed for our da'wa to be alive."[55]

The "Arab Spring" as a renaissance for AQ affiliates

Within the framework of the turmoil in the wake of the "Arab Spring", and especially with the conflict in Syria growing in intensity and scope, AQ has been able to re-emerge with two linked groups, Jabhat al-Nusra (JN) and the Islamic State of Iraq and al-Sham (ISIS). While JN has pledged allegiance (*ba'ya*) to Ayman al-Zawahiri, ISIS under the rule of Abu Bakr al-Baghdadi has morphed into the greatest success of al-Qaeda iconography and doctrine by adhering to the ideology without being subjected to its formal leadership.[56] Thus, the renaissance of al-Qaeda doctrine is largely based on its most powerful tool that allows the network to morph and spread in many directions: the professional and coherent decentralized use of the internet and the continuous and tireless deployment of media-workers and media-dedicated brigades embedded with fighting jihadi units in real-life battle arenas.

The new AQ considers itself as an avant-garde, an icon (*ramz*) and in parts in its rhetoric even as *awliya'* (friends/saints)[57] of God who are fighting against injustice, suppression, occupation, imprisonment, or the distortion of "Islam". Anyone is invited to tap into this freely available culture on the internet and be a part, either as passive or active consumer – by various means within the shared space of the inclusive online territories. This has led to the construction of a virtual sense of a real (global) community, where individuals may feel compelled to commit acts of terror by themselves out of a presumed personal responsibility for the "Islamic" collective in the West or by seeking to join the Mujahideen within Islamic territories deemed as occupied by Western nations or at war with non-Sunni armies in general, as is the case in Syria currently.

In this virtual realm, the principles of the al-Qaeda dominated ideology remain active and resilient, whereas in real-life, offline fighting groups, with either direct or indirect connections to the remnants of the AQ network, propagate the implementation of these principles framed as "*shari'a* law" by force. Groups adhering to the jihadist theology and mindset in general learned their lessons from past conflicts in Iraq and Algeria, where the indiscriminate killing of Muslims contributed to the demise of jihadi outlets, and emphasize positive elements in their audio-visual incitement material published frequently online. The culture of jihadi videos is the main launching pad to document the replication and translation of AQ ideology into visible action for jihadi camera lenses, sanctioning not only policy and policing measures, but also providing filmed materials of highly violent content alongside elements that can be graded as either winning the hearts and minds or state building measures. This mixture of "the book leading the right way" (*kitab yahdi*) and the "sword that assists" (*sayyf yansuf*)

> refers to a famous saying "the foundation of religion is a book leading the right way and a sword assisting" (*qawam al-din kitab yahdi wa-sayf yansur*) attributed to Ibn Taymiyya (1263–1328), the famous Hanbali scholar, which is often used by Jihadis to legitimize their fight.[58]

The slogan remains a popular and powerful element. A video published by the newly founded Islamic State of Iraq and al-Sham in July 2013 from North Syria, Aleppo Province, bears the title *Kitab yahdi wa-sayf yansur*. It was disseminated by the al-Furqan Media Foundation that sprang from al-Zarqawi's pioneer media work in Iraq.[59] With the widening of the Islamic State of Iraq into Syria and thus the declaration of ISIS, this slogan is used on a frequent basis for the video series *Nawafidh 'ala ard al-Malahim* published by the al-Furqan sub-department al-I'tisam, highlighting education, infrastructure and, in general, state building elements, while at the end of these videos attacks, executions, suicide bombings and the like conclude the "book leading the right way" by the "sword that assists". Such violence on display is subsequently embedded in sequences of religious tutoring and ideological speeches and sermons validating and sanctifying the depicted violence; underlying fundamentals are determined as being a divine obligation and the publication online is partially a media-campaign and, most important, a form of missionary work (*da'wa*). The majority of the videos are in Arabic, but nevertheless appeal and attract non-Arab Muslims to "join the caravan of jihad" as described by the German convert Abu Osama in the thirty-second instalment of the series. He joined ISIS in Syria to avenge the unjustified killing of Sunni Muslims by al-Assad's forces and is proud of the Taliban repelling the occupation of Afghanistan while considering his status as a migrant committed on the "path of God" as a guarantee to attain entry to paradise.[60] By portraying foreign fighters, the jihadi media seeks to display the extremist interpretation of "Islam" as an identity applicable to any individual – Western Muslim and non-Muslim alike – and frames militancy as righteous action sanctioned by God and as moral obligation, by calling on Islam and questioning the foreign policy of Western states. For example, French convert Abu 'Abd al-Rahman al-Fransi states in an interview with the al-Furqan Media Department to have committed himself to fight in Syria not only to avenge the raped Sunni women in prisons worldwide but also the demands of François Hollande to withdraw all troops from Mali.[61]

These materials can be highly appealing, particularly when sympathizers can interact in online forums and more so on social media platforms like Facebook and Twitter with proclaimed real-life Mujahideen and can ask for advice about what one would need prior to joining.[62] Subscribers to jihadist ideology and media strategists seek to initiate consumers by crafting a reformulation of Muslim identity of the "Islamic public" in general. The electronic incitement materials online serve to reinforce the offline and are made popular as a mirror of the real world, the real Mujahideen, the true Islamic implementation of *shari'a* law and so on. Global jihad provides a comprehensive religious ideology and a belief system, based on inclusion and exclusivity for anyone, within a specific frame of reference – rules, guidelines, guidance, conditions and belonging by coherent standards of identification.

This intersection between the offline and online construes and fertilizes the jihadists' frame of reference. This framework enables, for the sake of understanding individual actions and deeds, the motivations that have consequently led to both individual and collective actions. While "the question, however, is always about how 'Weltanschauung' and 'ideologies' translate into individual perception and interpretation, and how the acts of individuals are affected",[63] the jihadist media groups and battalions are in place to enforce guidance and to reproduce not only high-ranking ideologues but rather portray common fighters or suicide-bombers as much as renown jihadist figures. The written and filmed statements of jihadist groups "are given credibility by their social context [... and] the power of performative speech – vows and christenings – is rooted in social reality and is given currency by the laws and customs that stand behind it".[64]

Subsequently, a specific religious habitus is propagated as a filmed frame of reference that includes vital cornerstones of an ideology put into action with the perception of fighting for both justice and defence of mistreated Muslim civilians. The intersection of real-life and online, virtual worlds – accessible for anyone unlike the real-life scenarios – lead to questioning the effects of the accessibility to jihadist propaganda for consumers sometimes thousands of miles away, but in close emotional, spiritual and politicized proximity with the fear, anger and cognition of being "persecuted" because of ones faith. Thus, it would be deceptive to understand the intention as well as the desired effects of radicalizing and inspiring the audience as a sole process. By studying and monitoring written and filmed jihadist online propaganda, one must also find an "understanding of online practices and spaces not as distinct and isolated pursuits, but as closely linked with the everyday and offline milieu".[65] The jihadist online territories on the internet are public spaces, which on a first glimpse may seem to be a pure online feature, a fantasy. However, these online territories are

> a variety of intersections between the online and the offline. This includes virtual spaces that are anchored offline, or extensions of online entities, as well as those social territories that may emerge when online activities are "lived out" or re-enacted in other parts of everyday life.[66]

The jihadist community, physically and also virtually, exercised individually or in small groups, is a collective united by the absolute conviction of being the victims of a global conspiracy against Islam and in need of stars and heroes framed as the sole and legitimate defenders. The self-perception of the jihadi is essentially about the firm conviction of being part of the "soldiers of God" (*jund allah*), who in their death prove their loyalty to God. The lines of demarcation are clearly drawn, any Mujahid, any fighter striving on the "path of God" (*fi sabil l-llah*) is portrayed and propagated as a real believer adhering to the "truth" (*haqq*) unlike those who have fallen into a state of apostasy by their ignorance or open loyalty to anything defined by the jihadist reading of the Qur'an as "falsehood" (*batil*). This enables AQ, as well as other jihadist movements ideologically in close proximity to AQ, and its massive amount of religiously authoritative writings to defend, sanction, explain and justify in short any military operation.

Conclusion

Facilitating the internet as the prime and most effective (as well as cost-effective) communication facility to lure consumers into their specific interpretation or world perception is not restricted to the jihadi web. Militant and hate groups of all colours employ similar means to gain sympathy through modern and pop-cultural elements. However, the quantity as well as quality, not to mention the multi-lingual capacity, of jihadi media departments is unmatched and unprecedented. The rhetoric is inseparable from the (audio)visual content and enforces key elements while reaching out to the audience to get active, empower the *umma* by individual response.

With abundant jihadist material freely available online and with an ever present crowd of sympathizers at hand, the nature of jihadist spheres on the internet have changed, adapted, increased and simply synchronized with the technical developments of the world wide web over time. Independently of AQ's actual size and its operational status, the genre, corpus, and narrative of its online published jihadist materials remain both active and effective. Pseudo-AQ members and sympathizers can lead a double life, adhering to the jihadist narrative and iconography online while remaining – more or less – subjugated to their individual surrounding and social environment. This phenomenon, described as "homegrown radicalism" or "homegrown terrorist", if decisive enough to attempt to undertake actual operations in the real, offline, world, is certainly the next major burden for security agencies. As Jarret Brachman and Alex Levine put it, "there is a lower barrier for entry to becoming a practicing extremist on the Internet than there is in the real world".[67]

Social media such as Facebook and Twitter are the new platforms for jihadist propaganda to deliver statements and videos shortly after operations and attacks committed by militant groups, delivered by their media brigades and tactical teams. While the level of adaption to the "new" and "social media" in general has proven to be a fundamental part of the AQ ideology to maintain resilient, inspiring groups and individuals worldwide to adhere and subscribe to this frame of reference, it, on the other hand, may also pose a potential threat to jihadist groups and their physical networks. As the nature of social media consists of a constant information flow, groups such as Kata'ib al-Muhajirin bi Ard al-Sham, active in Syria, felt inclined to publish the pictures and names of every martyr on their Facebook group, linking to YouTube. This loud strategy, however, backfired and the Facebook page went silent. On 5 May 2013, the administrator posted an explanation of the group's media outage:

> Peace be upon you, the mercy of God and His blessing, we apologize to every member of [our] honorable page for our absence recently, not publishing news and other things [...]. [We] had been busy with

internal issues and when we had uploaded a couple of pictures [on our Facebook page] of those who we deem being martyrs, the local security forces responded in their respective countries, raiding their houses. We therefore stopped this to protect the families [of the martyrs]. We are very sorry and ask you for forgiveness for our failure to protect them.[68]

Notes

1 The black flag comprises a grey circle imprinted with (from down upwards) *Muhammad – rasul – allah*, a reference to the Islamic creed "There is no God (*allah*) but God and Muhammad is the messenger (*rasul*) of God".

2 The manifestation of pro-AQ styled jihadist groups, exercising violence or "soft-power", such as restoring basic infrastructure and handing out supplies to the Sunni population, is evident from the transnational appearance of the Ansar al-Shari'a network. The Ansar al-Shari'a groups propagate openly jihadist iconography and a similar rhetoric based on the fundamental teachings of AQ. The Libyan branch based in Bengazi and Derna as well as the chapter in Tunisia had been declared by the U.S. State Department as "foreign terrorist organizations" and "global terrorist entities" following the attack on the U.S. Consulate in Benghazi, 11 September 2012. Thomas Joscelyn, "State Department Designates 3 Ansar al Sharia Organizations, Leaders", 10 January 2013, www.longwarjournal.org/archives/2014/01/state_department_des_2.php.

3 Mu'assasat al-Ma'sada al-I'lamiyya/Fursan al-Balagh al-I'lam (eds), "Bayyan nusrat al-ansar li-ikhwanihim al-mujahideen al-abrar", 4, 10 October 2013, https://shamikh1.info/vb/showthread.php?t=212520. This statement is signed by the most influential contemporary jihadi media figures.

4 Mu'assassat al-Qayrawan al-I'lamiyya, Tunis al-Islam, "Al-Mujahid al-I'lami– al-Khatawat al-'ula li-ihtiraf al-jihad al- I'lami", 4, *Shabakat al-shumukh al-Islam*, 16 March 2012, www.shamikh1.info/vb/showthread.php?t=154861.

5 Ibid.

6 Ibid.

7 Ayman al-Zawahiri, "Tawjihat 'amma li l-'aml al-jihadi, as-Sahab Media", https://shamikh1.info/vb/showthread.php?t=211363.

8 Mu'assasat al-Ma'sada al-I'lamiyya/Fursan al-Balagh al-I'lam (eds), "Bayyan nusrat al-ansar li-ikhwanihim al-mujahideen al-abrar", 3–5.

9 Bayan min markaz al-Fajr li-l i'lam, "Ta'ziyya wa-tahani'a li-ummat al-Islam bi-munasiba istishhad al-shaykh al-asad Osama bin Laden", 1432 a.H.

10 Qur'an, first part of 8:65.

11 Qur'an, part of 4:84.

12 "La tukallafu ila nafsaka", as-Sahab Media Production, June 2011.

13 Ali Fisher and Nico Prucha, "Jihadi Twitter Activism, Part 2: Jabhat al-Nusra on the Twittersphere", *Jihadica*, 13 May 2013, www.jihadica.com/jihadi-twitter-activism-part-2-jabhat-al-nusra-on-the-twittersphere.

14 Magnus Ranstorp, "The Virtual Sanctuary of al-Qaida and Terrorism in an Age of Globalization", in J. Eriksson and G. Giacomello (eds), *International Relations and Security in the Digital Age*, London: Routledge, 2007.

15 The persistent as well as ideological cohesive online presence of jihadist propaganda, framed as authoritative rulings and determinations, has become an open sub-culture. The jihadist narrative, enforced by audio and visual elements strengthens in-group cohesion and affects mainstream Muslim culture, the main targeted audience.

16 For an introduction to the jihadist presence on Twitter as well as the social-media strategy deployed since the outbreak of violence in Syria, see Ali Fisher and Nico Prucha, "Jihadi Twitter Activism – Introduction", *Jihadica*, 27 April 2013, www.jihadica.com/jihadi-twitter-activism-introduction.

17 Cori E. Dauber, "YouTube War: Fighting in a World of Cameras in Every Cell Phone and Photoshop on Every Computer", Strategic Studies Institute, U.S. Army War College, 2009, www.strategicstudiesinstitute.army.mil/pubs/display.cfm?pubID=951.

18 Rüdiger Lohlker, "Tumbling along the Straight Path – Jihadis on tumblr.com", Research Notes, University of Vienna, 2012, www.univie.ac.at/jihadism/?p=391.

19 Mu'assasat al-Furqan and Markaz al-Yaqin, *Al-Manhajiyya fi tahsil al-khibra al-i'lamiyya*, part 1, May 2011.

20 The sanctioning of jihadi activity is related to the existing core fatwas, authoritative rulings, ideological decrees and so on. Thus any local jihadi AQ-affiliated action is placed under the virtual umbrella, increasing the appeal. Prem Mahadevan, "The Glocalisation of al-Qaedaism", Center for Security Studies (CSS), 22 March 2013, www.isn.ethz.ch/Digital-Library/Articles/Special-Feature/Detail/?lng=en&id=161729&tabid=1454211886&contextid774=161729&contextid775=161659.

21 Discussed in Fisher and Prucha, "Jihadi Twitter Activism – Introduction". Members of the Ansar al-Mujahideen forum and Shumukh al-Islam have posted advice encouraging fellow users to develop social media profiles to disseminate their message to a wider group of users; for example, "The Twitter Guide // the Most Important Jihadi Users and Support Accounts for Jihad and the Mujahideen on Twitter", https://shamikh1.info/vb/showthread.php?t=192509 and "Selected Tweets from Twitter", https://shamikh1.info/vb/showthread.php?t=192509.

22 The jihadi success of Twitter has had its impact on leading Jihadi writers such as Abu Sa'd al-'Amili to highlight the shift of "major [jihadi] writers and analysts", lamenting the general decline in participation in jihadi online forums. Cole Bunzel, "Are the Jihadi Forums Flagging? An Ideologue's Lament", *Jihadica*, 20 March 2013, www.jihadica.com/are-the-jihadi-forums-flagging-an-ideologue%E2%80%99s-lament.

23 Martyr biographies such as that of Abu Qasura al-Gharib (Muhammad al-Zulaytni), a 19-year old Libyan fighting for Jabhat al-Nusra, demonstrate how the internet and the regular consumption of ideological bearings have their impact on a new generation. According to this eulogy he used his iPad to improve his knowledge, embrace AQ ideologues and remain active online within jihadi spheres while he fought in Syria. The virtual footprint of real-life martyrs and their internet habits is part of the advocated role model. He was killed on 4 January 2012 (https://shamikh1.info/vb/showthread.php?t=193834).

24 Kevin Brown, "Agitprop in Soviet Russia", *Constructing the Past*, Vol. 14, Issue 1, 2013, 5–6, http://digitalcommons.iwu.edu/constructing/vol14/iss1/4.

25 On the strategy and tactics of how the primarily Arabic language ideology is conveyed by individual preachers and activists, in this case for Germany and Austria: Nico Prucha, "Die Vermittlung arabischer Jihadisten-Ideologie: Zur Rolle deutscher Aktivisten", in Guido Steinberg (ed.), *Jihadismus und Internet: Eine deutsche Perspektive*, Stiftung Wissenschaft und Politik, October 2012, 45–56, www.swp-berlin.org/de/publikationen/swp-studien-de/swp-studien-detail/article/jihadismus_und_internet.html.

26 The Ansar al-Mujahideen Network, a tier-one jihad forum, for example, advertises its Twitter account (@as_ansar) on its main page. The less renowned al-Minbar al-I'lami al-Jihadi forum (www.alplatformmedia.com/vb/index.php) does likewise (@alplatformmedia).

27 "Denis Allex: French Agent 'Killed' by Somalia al-Shabab", www.bbc.co.uk/news/world-africa-21056592.
28 The Facebook link to HSM's Arabic media account, "Wakala shahada al-ikhbariyya", circulated on Twitter with the intention of gaining traction to the Facebook account, www.facebook.com/photo.php?pid=1050347&l=f3ee046c97&id=357503417665041.
29 HSM Press Office, "PRESS RELEASE: Dennis [sic] Allex: An Agent Betrayed", published on Twitter as a link to TwitLonger, www.twitlonger.com/show/knkqp9.
30 At the end of January 2013, three statements by AQAP were only disseminated via Twitter. A high profile jihadi Twitter user, @STRATEAGY, 'retweeted' these. A member of the Ansar al-Mujahideen forum posted a screenshot of the subsequent 'retweets' onto the forums, to gain more traction to AQAP, "Urgent – Three New Statements by AQAP", http://as-ansar.com/vb/showthread.php?t=79875.
31 "Statement regarding the New Twitter Account", http://as-ansar.com/vb/showthread.php?t=85175.
32 "Now the Open Interview with Shaykh Ahmad Abu 'Abd al-Ilaha, the Head of the Media Board for al-Andalus", http://as-ansar.com/vb/showthread.php?t=85501&highlight=andalus_media.
33 English: http://ia801702.us.archive.org/5/items/Ansewrs.AQMI/answers.pdf; Arabic: http://ia801703.us.archive.org/11/items/answersI.1/answersI.pdf.
34 *Inspire*, al-Malahem Media, Special Issue, 11, Spring 2013, 17.
35 Ali Fisher and Nico Prucha, "Tweeting for the Caliphate: Twitter as the New Frontier for Jihadist Propaganda", *CTC Sentinel*, June 2013, West Point.
36 Jason Lanier, *Who Owns the Future?*, London: Penguin, 2013, 52. EBook edition.
37 Ayman al-Zawahiri, "Asad al-'ilm wa-l-jihad Abu Yahya al-Libi", as-Sahab Media, 2 September 2012, http://archive.org/details/Nosr.A.
38 Cornel Sandvoss, "Fans Online: Affective Media Consumption and Production in the Age of Convergence", in Miyase Christensen, André Jansson and C. Christensen, *Online Territories: Globalization, Mediated Practice and Social Space*, New York: Peter Lang Publishing, 2011, 50.
39 Ranstorp, "The Virtual Sanctuary of al-Qaida and Terrorism in an Age of Globalization".
40 Female martyr and corresponding biographies exist, for example: Yusuf al-Qaradawi, "Min qisas al-batula wa-l-fida', al-istishhadiyyat 'arafna al-dukhula ila l-hayyat", www.palestine-info.info/arabic/feda/2004/isteshhadeat.htm.
41 Anthony Cohen, *The Symbolic Construction of Community*, London: Tavistock, 1985, 118.
42 Ayman al-Zawahiri, "Asad al-'ilm wa-l-jihad Abu Yahya al-Libi".
43 Bill Rogio, "Abu Yahya al-Libi Killed in Latest Drone Strike, US Officials Say", *The Long War Journal*, 5 June 2012, www.longwarjournal.org/archives/2012/06/abu_yahya_al_libi_ru.php,.
44 "The folder of Abu Yahya al-Libi", *Minbar al-Tawhid wa-l-Jihad*, 14 June 2013, http://tawhed.ws/a?a=hv5znv47.
45 Al-Libi had managed to escape with three comrades and talked extensively about the escape and the torture in as-Sahab videos. Parts of the videos are re-used for al-Zawahiri's eulogy, "Asad al-'ilm wa-l-jihad Abu Yahya al-Libi".
46 Abu Yahya al-Libi, "Al-Mu'allim fi hukm al-jasus al-muslim" (Markaz al-Fajr li-l i'lam, 2009), February 2012 www.tawhed.ws/dl?i=y34deswq. For an analysis of al-Libi's writings justifying the execution of Muslim spies: Nico Prucha, "Kangaroo Trials: Justice in the Name of God", in Rüdiger Lohlker (ed.), *Jihadism: Online Discourses and Representations*, Göttingen: V & R Unipress, 2013, 141–207.

47 The Harakat al-Shabab al-Mujahideen in late 2008 published a video entitled "implementation of the rule of God on two spies" whereas in the video al-Libi's book is shown, quoted and the ultimate framework within which to execute two alleged Muslim spies.
48 His alleged real name is offered as Muhammad bin 'Abd al-Qadir al-Murshidi, as stated in Harith bin Ghazi al-Nazzari, *al-Wajiz fi ahkam al-asir al-Muslim*, al-Malahem, 2012.
49 *Talab al-'ilm* is a reference, with historical and traditional connotation, to those who seek to gain knowledge and understanding in religious practice and manners. It is an honorary and obligatory determination to actively seek knowledge on the proper orthopraxis and conduct of religious parameters for worldly circumstances.
50 Introduction by Ayman al-Zawahiri, in "Abu Yahya al-Libi: al-Mu'lim fi hukm al-jasus al-Muslim", Markaz al-Fajr li-l i'lam, 2009, 4–5.
51 Ayman al-Zawahiri, "Asad al-'ilm wa-l-jihad Abu Yahya al-Libi".
52 The term *fasad* is taken from the Qur'an but used by jihadists in the modern meaning of "corruption". By exploiting this as a main slogan, jihadists seek to further position themselves as righteous people who, just like any average Arab, are victims of the "interior corruption", referring to the widespread nepotism of Arab dictators.
53 Ayman al-Zawahiri, "Asad al-'ilm wa-l-jihad Abu Yahya al-Libi".
54 Ibid.
55 Ibid.
56 See Cole Bunzel, "The Islamic State of Disobedience: al-Baghdadi Triumphant", *Jihadica*, 5 October 2013, www.jihadica.com/the-islamic-state-of-disobedience-al-baghdadis-defiance.
57 The *awliya' allah*, "the friends of God", as stated in Quran 10:62 is an expression used by jihadists for describing and elevating their deeds for God by approximating themselves to God as representatives exercising His will in this world.
58 Rüdiger Lohlker, "Religion, Weapons, and Jihadism: Emblematic discourses", Manuscript, in Lohlker, *Jihadism*, 65–89.
59 "The Islamic State of Iraq and al-Sham: Video Publication: kitab yahdi.. wa-sayf yansur", 24 July 2013, https://shamikh1.info/vb/showthread.php?t=207606&p=1059462800#post1059462800. Simultaneously with the forum posting, the video was uploaded to YouTube, www.youtube.com/watch?v=OzoPa6Xjua4. It had been watched over 20,000 times by 31 July 2013.
60 Mu'assasat al-i'tisam, "Nawafidh 'ala ard al-malahim 32", 2 December 2013, https://shamikh1.info/vb/showthread.php?t=215518.
61 Mu'assasat al-Furqan, "Al-ladhina amanu wa-hajaru wa-jahadu", 10 July 2013, https://shamikh1.info/vb/showthread.php?t=206004.
62 As the following thread on the Shumukh al-Islam jihad forum highlights: http://shamikh1.info/vb/showthread.php?t=126040. Foremost, responding to the member 'Abd al-Hakim's question, (1) a sincere intention and (2) faithfulness are required. Other members, such as Osama Hakim, reply in length by copying-and-pasting a "methodology to construct a Mujahid-terrorist personality".
63 Sönke Neitzel and Harald Welzer, *Soldaten: Protokolle vom Kämpfen, Töten und Sterben*, Frankfurt am Main: S. Fischer Verlag, 2011, 17.
64 Mark Juergensmeyer, *Terror in the Mind of God: The Global Rise of Religious Violence*, Third edition, Berkeley and Los Angeles: University of California Press, 2003, 127. Juergensmeyer references Pierre Bourdieu, *Language and Symbolic Power*, Cambridge: Harvard University Press, 1991.
65 Christensen *et al.*, *Online Territories*, 2.
66 Ibid.

67 Jarret Brachman and Alex Levine, "You Too Can Be Awlaki!", http://jarret-brachman.net/?p=1157.

68 Facebook page of the Kata'ib al-Muhajirin bi Ard al-Sham, published and accessed 5 May 2013, www.facebook.com/permalink.php?story_fbid=473426306061452&id=443758765694873.

7 Learning terror

The evolving threat of overseas training to the West

Paul Cruickshank

Overview

Terrorist training matters. The four British suicide bombers who set off their explosives to such devastating effect on the London transport system on July 7, 2005 succeeded where so many other plotters failed because two of their group – Mohammed Siddique Khan and Shehzad Tanweer – received hands-on bomb-making training and instruction in terrorist tradecraft from al-Qaeda instructors in the tribal areas of Pakistan, during which time they test-detonated the devices they would later build to target London.[1]

How did this training make a difference? First, and most importantly, it greatly improved their chances of making a viable explosive device. Since 9/11, only rarely have Islamist terrorists in the West successfully completed the tricky task of building bombs from chemicals available in Western countries by downloading bomb-making instructions from the internet – a task not helped by the variable quality and reliability of online bomb recipes.

Second, Siddique Khan and Tanweer's training provided them with good terrorist tradecraft. Their instructors stressed the importance of secure communications and keeping as low a profile as possible, reducing the risk they would be caught. In the years after 9/11 poor operational security led to many "home-grown" terrorist cells in the West being easily intercepted by security agencies.[2]

Third, their training bolstered their commitment to carrying out a terrorist attack. While a significant number of Islamist extremists in the West "talked the talk" about launching attacks, few progressed to actually trying to carry them out. Unlike many terrorist wannabes, Siddique Khan and Tanweer's sense of belonging to al-Qaeda's cause was not restricted to online forums. Their handlers in Pakistan had assiduously instilled in them a sense of mission, religious duty, and – in having them record martyrdom tapes – the belief there could be no turning back.[3]

This chapter explores the danger posed by Western militants receiving training in Jihadist encampments overseas. As concern mounts over the emergence of a large terrorist safe-haven in Syria and Iraq, it seeks to answer the following two questions: to what degree does terrorist training better equip Western militants to carry out terrorist attacks, and in what ways? And does it make Western militants more likely to carry out such attacks, and if so how?

These questions are at the heart of the debate over the relative danger posed by lone wolf actors, home-grown terrorist cells, and plots that are either directed or encouraged by al-Qaeda and like-minded Jihadist terrorist groups. The answers should dictate the relative priority policy-makers give towards allocating resources for policing work at home and measures to reduce terrorist safe-havens overseas in Pakistan, Yemen, Somalia, North Africa, Iraq, Syria, and elsewhere.

There have been a number of attempts to quantify the impact of terrorist training. In an exhaustive survey of 60 global Jihadist terrorist plots in the West between 1993 and 2008 American researcher Marc Sageman found that formal training from al-Qaeda or its affiliates doubled the chance that a terrorist plot would succeed, with the rate of success going from 10 to 20 percent.[4]

While the loss of large-scale training facilities in Afghanistan after 9/11 reduced al-Qaeda and allied groups' ability to produce sophisticated and dangerous terrorists, this chapter will outline how the group subsequently set up smaller training encampments in the tribal areas of Pakistan and elsewhere, where they continued to offer bomb-making training and continued to be able to orchestrate terrorist plots against the West.

The emergence of terrorist safe havens in Syria and Iraq has seen the terrorist threat grow once again. Jihadist groups like the Islamic State of Iraq and al Sham (ISIS) and Jabhat al Nusra operate camps in Syria and Iraq on a scale last seen in Taliban-run Afghanistan.

Training and terrorism: the numbers

Over the last decade there has been a strong correlation between the most serious Islamist terrorist plots and training in Jihadist encampments overseas. According to a tally by this author, in the majority of the "serious" terrorist plots against the West between January 1, 2004 and November 1, 2014 one or more cell members received overseas training.[5] (See Table 7.1.) Serious plots were classified as those which killed or had the potential to kill ten or more people and saw plotters obtain weapons or bomb-making components without the assistance of undercover law enforcement agents.[6] There were 47 serious plots in total during this nearly 11-year period, with the most recent the October 2014 attack on the Canadian Parliament in Ottawa.[7] In 27 serious plots one or more cell members received terror training somewhere overseas.

In 20 serious terrorism plots that training took place in Pakistan (see Figure 7.1). In many cases this instruction was provided in the tribal areas

Table 7.1 Location of training in 47 serious jihadi terror plots

Serious plot	*Date*	*Overseas training (at least one member)*
1 Madrid Bombings	March 2004	None
2 Crevice UK	March 2004	Pakistan
3 Torrance, California	July 2005	None
4 7/7 London bombings	July 2005	Pakistan
5 7/21 London	July 2005	Pakistan
6 Pendennis, Australia	November 2005	Pakistan
7 Hofstad group, Netherlands	June 2003	Pakistan
8 Toronto	June 18, 2006	Pakistan
9 Cologne Commuter, Train	July 2006	None
10 Transatlantic Airline	August 2006	Pakistan
11 Vollsmose, Denmark	September 2006	None
12 Glasgow	June 2007	None
13 Fort Dix	May 2007	None
14 Glasvej, Denmark	September 2007	Pakistan
15 Sauerland, Germany	September 2007	Pakistan
16 Barcelona	January 2008	Pakistan
17 Andrew Ibrahim, Bristol	April 2008	None
18 Manchester	April 2009	Pakistan
19 Quantico, Virginia	July 2009	None
20 Najibullah Zazi, New York	September 2009	Pakistan
21 Nidal Hasan, Fort Hood	November 2009	None
22 Underwear, Detroit	December 2009	Yemen
23 Faisal Shahzad, Times Square	May 2010	Pakistan
24 Mikael Davud, Norway	July 2010	Pakistan
25 Hiva Alizadeh, Canada	August 2010	Pakistan
26 Printer Cargo Bomb Plot	October 2010	Yemen
27 Taimur al Abdaly, Stockholm	December 2010	Iraq
28 London Stock Exchange	December 2010	None
29 Mounir Dhahri, *Jyllands-Posten*	December 2010	Pakistan
30 Khalid Aldawsari, Texas	February 2011	None
31 Arid Uka, Frankfurt	March 2011	None
32 Düsseldorf	April 2011	Pakistan
33 Naser Jason Abdo, Fort Hood	July 2011	None
34 Birmingham	September 2011	Pakistan
35 Berlin	September 2011	None
36 Jose Pimmentel, New York	November 2011	None
37 Mohammed Merah, Toulouse	March 2012	Pakistan
38 Denmark	April 2012	None
39 AQAP, U.S. Plane	April 2012	Yemen
40 EDL, Yorkshire	July 2012	None
41 Southern Spain	August 2012	Pakistan
42 Cannes-Torcy	October 2012	None
43 Boston	April 2013	None
44 Cannes	February 2014	Syria
45 Brussels Jewish Museum	May 2014	Syria
46 London plot to shoot police	October 2014	Syria
47 Ottawa Parliament attack	October 2014	None

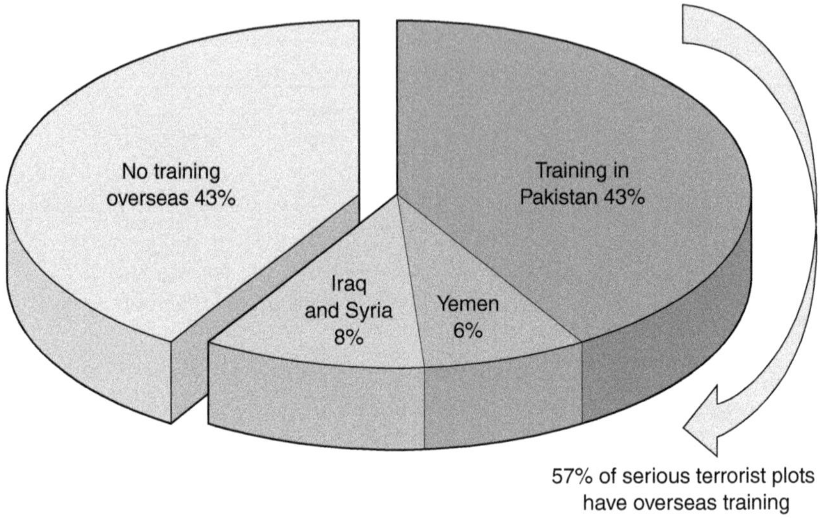

Figure 7.1 Location of training in serious terrorism plots on Western soil January 1, 2004–September 1, 2014 (source: Updated from Paul Cruickshank "The Militant Pipeline between the Afghanistan–Pakistan Border Region and the West," 2nd Edition, New America Foundation Counterterrorism Initiative Strategy Document, July 2011).

of the country, a region that was the epicenter of al-Qaeda plotting against the West in the years after 9/11 before an expansion of the U.S. drone campaign in the late 2000s began to degrade al-Qaeda's capabilities in the region. In four of the 47 serious plots extremists planned to carry out attacks in the West after training in Syria and Iraq, a number that will likely rise significantly in the years ahead.

Until recently the training camp threat to the West was almost entirely located in Pakistan. In the post 9/11 years Al-Qaeda "Central" was the only part of the broader network of affiliates that made attacking the West a central priority. A key driver was bin Laden; the Abbottabad documents made clear his determination to attack the West again. From 9/11 until 2014 only "core" al-Qaeda and the Pakistani Taliban offered recruits bomb-making training specifically tailored to launching attacks in the West. By contrast, al-Qaeda's affiliate in Yemen, despite plotting several attacks on U.S. aviation, only dedicated a fraction of its resources to plotting attacks on the West. Al-Qaeda in the Arabian Peninsula (AQAP) trained recruits in detonating already assembled bombs rather than making them.

The concern about terrorist training camps now centers on Syria and Iraq. On September 22, 2014 the United States launched 47 Tomahawk missiles against encampments to the west of Aleppo run by the Khorasan group, an outfit of al-Qaeda veterans from Afghanistan and Pakistan with ties to the Syrian al-Qaeda affiliate Jabhat al Nusra. U.S. officials said the group was in the final stages of plotting an attack on Western aviation and was also looking to send European recruits back home to launch attacks. They said they believed sophisticated bomb-making knowledge had been transferred to the group from al-Qaeda in Yemen.

The Cruise missile strikes were not assessed to have removed the threat from the group. Several operatives, including a key bomb-maker, survived, and as of November 2014 they continued to pose a significant threat. Khorasan had an unprecedented large potential pool of Western extremists it could recruit for attacks. According to European counter-terrorism officials, more than 3,000 Europeans had by then traveled to fight in Syria. The mix there of sophisticated explosive technology and Western passport holders was a perfect storm that was keeping Western officials up at night.[8]

As of November 2014 there was little evidence that ISIS had yet initiated a training program specifically tailored to attacking the homelands of Western countries. However, after the United States launched a campaign of air strikes agasint ISIS in August 2014, there was rising concern the group may retaliate with attacks in the West by providing such training to Western recruits.

British Prime Minister David Cameron has called ISIS a "greater and deeper threat" to Western security than any terror group before it. Despite U.S. and coalition air strikes in late 2014 the group controlled large swaths of territory in Iraq and Syria, had up to 1,000 European recruits in its ranks, operated several large training camps, and likely had tens of millions of dollars in cash reserves.[9]

This chapter is structured as follows.

Section one outlines how terrorist training encampments have evolved since 9/11.

Section two explains how terrorist training has made Western militants more capable by looking in turn at (1) explosives training, (2) weapons instruction, and (3) lessons in terrorist tradecraft.

1 Explosives training

Many plots against the West have involved bombs made from easily purchasable chemicals. Pre-prepared or military grade explosive is hard to acquire, in part because of stricter legal restrictions and enhanced international law enforcement cooperation, including Program Global Shield.

But making bombs out of chemicals purchasable in the West has been seen as tricky and dangerous by Western Islamist militants plotting terrorist attacks, because many of the components are volatile.

Few Islamist terrorists have succeeded in building a viable explosive device without receiving explosives instruction in a Jihadist encampment overseas. Downloading bomb-making instructions from the internet has often been a dead end. According to a leading explosives expert many of the instructions in online bomb manuals produced by pro-al-Qaeda Jihadists are of low quality.[10]

Between January 1, 2004 and November 1, 2014 there were just two Islamist terrorist plots in the West in which a cell member with no formal training in Jihadist encampments overseas successfully finished building a viable explosive device from everyday chemicals with the capacity to inflict serious damage.[11]

The first was Andrew Ibrahim – a Bristol-based "lone wolf" Islamist extremist – who successfully manufactured the high explosive HMTD (hexamethylene triperoxide diamine) by downloading instructions from the internet. Ibrahim put together a suicide vest before he was arrested by British police in April 2008, and was later convicted of a plot to bomb a Bristol shopping center with it.[12] It should be noted that the March 2004 Madrid bombers constructed bombs from dynamite stolen from a quarry – and so did not need to build explosives from scratch using easily purchasable chemicals.[13]

The second case was the April 15, 2013 Boston marathon bombing, which killed three and injured hundreds. Two brothers – Dzhokhar and Tamerlan Tsarnaev – allegedly constructed pressure cooker and pipe bombs from a recipe in AQAP's *Inspire* magazine "How to make a Bomb in Your Mom's Kitchen."[14] Four days later they threw several pipe bombs at police as they attempted to flee.[15]

Although Tamerlan Tsarnaev spent time in Dagestan in 2012 and developed connections to militants in the region no evidence has emerged that either he or his brother received any training overseas.[16] Despite receiving no formal training, they had a degree of technical ingenuity. The brothers allegedly set off the bombs using a makeshift remote control triggering mechanism from a toy car – an adaptation of the original instructions in *Inspire* magazine which called for the use of a clock timer.[17] Indeed the overall sophistication of the devices was high enough for investigators to initially suspect they may have received training overseas.[18]

Boston illustrated that lone wolf terrorists, despite the difficulties in making home-made bombs, could inflict significant damage. But the death toll would have been much higher if the bombs had been more powerful. The Boston devices contained low explosive powder sourced from fireworks which gave them significantly less destructive power than the high explosive devices al-Qaeda instructors have taught Western recruits to make in overseas training camps. The Boston devices did not technically "detonate" – a process in which chemicals extremely rapidly decompose releasing large amounts of energy. Instead the devices exploded the walls of the pressure cooker outwards because of the build up of gas in the confined space of the pressure cooker.[19]

If the Tsarnaev brothers, like the 7/7 bombers and other Western al-Qaeda recruits, had been taught by al-Qaeda instructors how to build a high explosive main charge out of hydrogen peroxide or ammonium nitrate (fertilizer) and how to detonate it with the high primary explosive HMTD or TATP (triacetone triperoxide), the attack could have been significantly more deadly.[20]

Only one self-taught terrorist in the West has carried out a bomb attack using high explosive since 9/11. In July 2011 Anders Brevik, an Islamophobic extremist, carried out the second most deadly terrorist attack in modern European history with no formal training. While most of his victims were killed during a shooting rampage on Utoeya Island, the most technically difficult part of the operation to pull off was the large ammonium nitrate-based device he exploded at government offices in Oslo, which killed eight. In contrast to Timothy McVeigh, the 1995 Oklahoma City bomber, who stole dynamite and blasting caps from a quarry for components of his fertilizer-based device, Breivik built his from scratch.[21]

Breivik is, however, arguably not an indicator of what lies in store from most cases of "lone wolf" Islamist terrorism in the future. As this chapter will outline, Breivik stood out from the majority of those who have been recruited into Islamist terrorism in a number of ways, including his deep financial pockets, extraordinary dedication, and access to a remote farm to tinker and test his device.

2 *Weapons training*

As its training encampments in the tribal areas of Pakistan have come under increasing pressure from drones, al-Qaeda has increasingly encouraged followers in the West to launch gun attacks. Compared to bomb plots, training in encampments overseas has not been as essential to the success of shooting plots in the West because Western Islamist militants can easily learn how to shoot a weapon on gun ranges in their home countries.

But even in shooting attacks, time spent in Jihadist encampments can make a difference. Toulouse terrorist shooter Mohammed Merah learned how to fire guns in the tribal areas of Pakistan just six months before he went on a killing spree.[22] Many of the Western militants who spent time in the region learned how to fire weapons in Jihadist encampments. A subset of these practiced how to storm enemy positions or experienced combat, bolstering their capability to carry out a gun attack. Alleged Brussels Jewish museum shooter Mehdi Nemmouche returned from Syria skilled in handling weapons.

3 *Tradecraft training*

In any kind of plot, terrorist tradecraft matters. Western Islamist militants who learn counter-surveillance techniques and how to communicate securely with each other or with handlers overseas are more likely – all other things being equal – to succeed. The 9/11 plot hijackers and the ringleaders of the July 7, 2005 London bombings received such training direct from al-Qaeda instructors. While al-Qaeda and its affiliates have attempted to replicate such instruction through do-it-yourself manuals in *Inspire* magazine, and other publications, most home-grown Islamist terrorist plots have displayed poor terrorist tradecraft. For example, the alleged Boston bombers succeeded despite serious lapses in operational security. Before the attack Dzhokhar Tsarnaev boasted to friends he knew how to build bombs, but none took him seriously.[23]

Section three examines the link between training and commitment. Spending time in Jihadist encampments overseas has not just boosted the capabilities of Western militants, it has also had an equally significant impact on their commitment to carrying out terrorist attacks in the West. In contrast to surfing Jihadist websites online, there appears to be something about spending time in the rarified air of the remote mountains of Waziristan or the Jihadist battlegrounds of Syria which makes militants feel they are part of a wider global cause and gives them a sense of mission. As this chapter will outline, al-Qaeda handlers have become expert in psychologically manipulating Western militants to return home to launch attacks by insisting such missions are a religious duty which will be rewarded in the afterlife.

Section four addresses how overseas training has affected target selection.

Section five examines the future of terrorist training.

The evolution of terrorist training

Afghanistan and Pakistan

The heyday of the Jihadist training camp was in Taliban-run Afghanistan in the years leading up to 9/11. Jihadists operated around two dozen training camps in the country, a significant number of which were funded by Osama bin Laden. Though they were large in scale and included exercise and shooting areas, most of the instruction in these camps was dedicated to the teaching of basic fire-arms skills. It was only the most promising militants that al-Qaeda cherry-picked for more advanced terrorist training at the hands of its experienced instructors in bomb-making, target reconnaissance, operational security, counter-surveillance, secure communications, and a variety of other aspects of terrorist tradecraft.[24] For this purpose al-Qaeda directly operated no more than two or three camps in the country.[25]

With ISIS and other Jihadist terrorist groups now operating training camps in Syria and Iraq on a scale last seen in Taliban-run Afghanistan, it is more relevant than ever to study how al-Qaeda organized training for recruits in these camps two decades ago.

One of the fullest accounts of the extensive and varied training undertaken by Western militants in al-Qaeda camps in Taliban-run Afghanistan was recently provided in court testimony in 2012 by Saajid Badat, a British al-Qaeda operative convicted of plotting to blow up a transatlantic airliner with a shoe bomb in December 2001. Badat testified that Jihadist recruits arriving in Afghanistan would typically receive about 35–40 days basic weapons training. He said he himself was trained in a camp run by Chinese Uyghur near Kabul for 4–6 weeks. After moving to a guest house, they might then graduate to a variety of other more specialized training courses.

After being recruited into al-Qaeda in the summer of 2000 and meeting bin Laden, Badat said he took several courses in several camps around Kabul, Jalalabad, and Kandahar. Each month al-Qaeda paid him a salary of 1,000 Pakistan rupees. By the end of his training he had been taught about light weapons, heavy weapons and navigation, according to his account, and learned how to make explosives so well that he himself became a bomb-making instructor at a camp run by al-Qaeda's Egyptian chief bomb-maker Abu Khabab al Misri.

When Abu Hafs al Masri and Saif al Adel tapped Badat for a possible mission to scope out Jewish targets in South Africa, he was placed on an intelligence course by the group. "We were taught about how to interact, how to present ourselves, how to form our appearance," he stated. After being entrusted personally by bin Laden with the shoe-bombing operation after 9/11, Badat recalled how he and fellow conspirator Richard Reid met with Khalid Sheikh Mohammed in Karachi who gave them final orders and specific operational advice on how to communicate with each other and plan their mission.[26]

After the fall of the Taliban, al-Qaeda lost its ability to train recruits until it began to establish a new safe haven in the tribal areas of Pakistan. As late as 2003 it appears only to have been loosely involved in setting up training for Western militants present in the Afghanistan-Pakistan border region. In the summer of 2003 a group of British militants later arrested in "Operation Crevice" for plotting to blow up fertilizer bombs around London, were forced to organize their own bomb-making training and recruit their own instructors.

During a period of several months the group – led by British-Pakistani Omar Khyam – tried to organize training by reaching out to a variety of contacts including Lashkar-e-Taiba, the Taliban, and al-Qaeda.[27]

In June 2003, several of the leading figures in the group, including plot ringleader Omar Khyam, received explosives training from a freelance Afghan bomb-maker, in an apartment they rented in Kohat.[28]

The following month the cell organized training in a makeshift camp in Malakand in the mountains of North-western Pakistan. Among their number was future 7/7 ringleader Mohammed Siddique Khan. The UK group offered a local Muslim scholar over £3,000 for use of a makeshift facility in the hills above his school, an encampment consisting of two tents and a small area which they used to fire weapons, and covered all the expenses of their training.[29]

Rather than being trained by al-Qaeda instructors, the group got its bomb-making lessons, training in military and weapons skills from the Islamic scholar's son. All this was a far cry from the large training complexes that had been at al-Qaeda's disposal in Afghanistan.[30]

At trial, one of those who attended the Malakand camp – Mohammed Junaid Babar – testified that Omar Khyam tricked members of the group by telling them that if they did well at the training camp they would be selected to fight in Afghanistan, when in fact Khyam's plan was to persuade them to return to attack the UK.[31]

Almost all the Western militants travelling to the Afghanistan–Pakistan (Af–Pak) border region – then and since – initially sought training so they could participate in Jihad against Western forces in Afghanistan, rather than because they wanted to carry out terrorist attacks back home. But their passage through the area gave al-Qaeda the opportunity to recruit them to attack the West. In contrast to the pre-9/11 period when they were able to cherry pick the most promising recruits from training camps in Afghanistan, al-Qaeda's pool of potential recruits was largely restricted to whichever militants happened to be travelling through Pakistan, seeking to fight in Afghanistan.

By the end of 2004 al-Qaeda was sufficiently established in the tribal areas of Pakistan to provide direct terrorist instruction to Western militants. Three waves of British recruits subsequently involved in the July 7, 2005 London bombings, the July 21, 2005 London bombings plot, and the August 2006 Transatlantic Airline plot trained there during this period.

The point person in all three plots was the British al-Qaeda operative Rashid Rauf, who in 2008 authored a 46-page account of how al-Qaeda planned the operations. The document was discovered in the possession of an al-Qaeda recruit in Berlin in the spring of 2011.[32]

Rauf outlined how in the last few months of 2004 he met the future ringleaders of the three plots in Pakistan and arranged for them – in turn – to travel to al-Qaeda encampments in the tribal areas of Pakistan. His account revealed al-Qaeda was by then taking on a much more direct role in terrorist instruction. He revealed that the ringleaders of each of the three cells were trained by explosives instructors overseen by Abu Ubaidah al Masri, a senior member of al-Qaeda's external operations unit. The training of several of those recruited into these conspiracies, he wrote, involved not only practice in making devices but also test-detonations of the bombs they had made. According to Rauf, several of the plotters were also taught in counter-surveillance and how to communicate securely with their al-Qaeda handlers in Pakistan once they returned home.[33]

Rauf's account suggests al-Qaeda took their time in training the ringleaders of the 7/7 plot. He himself spent a significant amount of time with them in the tribal areas of Pakistan during their training.[34] But he wrote that al-Qaeda decided to speed up training for recruits following the July 2005 London bombings. "We would now send back brothers quickly and not keep them for a long time," Rauf wrote. "This was largely due to the fact that we knew there would be intelligence monitoring of people travelling from the UK to Pakistan."[35]

In subsequent years several additional groups of Western militants were given explosives training by al-Qaeda or affiliated groups in small outdoor encampments in the Waziristan mountains. In 2008 the U.S. responded by intensifying its drone campaign. According to a tally by the New America Foundation there were 36 strikes in the Federally Administered Tribal Areas (FATA) that year compared to just nine in the previous four years.[36] This forced al-Qaeda to adapt by scattering its operatives into smaller groups across the tribal areas of Pakistan and conducting much of its instruction indoors in small mountain shacks away from the sightlines of the drones, reducing the amount of time recruits could spend outdoors either on weapon firing ranges or areas where they could test their devices.

Nevertheless these adaptations allowed al-Qaeda to continue to be able to teach terrorist tradecraft to recruits. Bryant Neal Vinas, an American recruit from Long Island and Walid Othmani, a French recruit, described a variety of classroom training courses offered by al-Qaeda in the first half of 2008 in the tribal areas of Pakistan, including explosives training, electronic circuitry, poisons, sniper shooting, and a special course set up by senior al-Qaeda operative Saleh al Somali on kidnapping and assassination, including instruction on the use of silencers and how to break into and enter a property.

In classes attended by 10–20 recruits, Vinas described being instructed how to handle a variety of explosives including TNT, RDX, Semtex, C3, and C4. Vinas stated that when they completed a particular training course, al-Qaeda instructors did a written evaluation of their performance to judge whether they could graduate to the next level. Instruction in the actual construction of bombs, he stated, was offered to al-Qaeda recruits who had become more advanced in their training.[37]

Najibullah Zazi, the 2009 New York Subway plot ringleader trained by al-Qaeda the same year as Vinas, described how a trio of instructors taught him how to make a variety of explosive devices in late 2008 in a small encampment in South Waziristan which they test-exploded in the mountains near their encampment.[38] His explosives training was compressed into a week, a shorter period of time than provided to several groups of recruits prior to that, indicating that by late 2008 drone strikes were putting increasing pressure on al-Qaeda.[39]

Furthermore, the decentralization of al-Qaeda's structures appears to have created some costs for recruits. Othmani and several other French and Belgian Jihadists part of his travel group described feeling increasingly cut off, bored, and fed up with the primitive living conditions in their mountain shacks. They often did not seem to know what their next orders would be or where their handlers would take them. They also described feeling deeply frustrated about being repeatedly given false promises that they would be able to fight in Afghanistan.[40] The Belgian-French group also described their frustration at having to pay around $1,800 each for training and equipment, reflective of al-Qaeda's growing cash flow problems.[41]

In 2009 as a result of the increasing pressure from drones, al-Qaeda operatives called for recruits to be trained as rapidly as possible. An internal al-Qaeda strategy document entitled "Future Works" found on an al-Qaeda suspect in Berlin in mid-2011 and thought to have been authored in 2009 – acknowledged the ease with which Western security services were thwarting plots and called for European Jihadists to be trained rapidly and sent home – rather than use them as fighters in Afghanistan and Pakistan – with instructions on how to keep in secret contact with their handlers.[42]

By 2010 the ever escalating drone campaign appears to have made it significantly more difficult for al-Qaeda to train recruits in the tribal areas of Pakistan. Not only were experienced instructors being killed but movement of fighters and trainers around the tribal areas became much more risky. According to a tally by the New America Foundation there were 54 drone strikes in the region in 2009 and a record 122 strikes in 2010.[43]

Rami Makanesi, a German recruit who trained in an al-Qaeda encampment that year described drone strikes as a constant danger to militants living in North Waziristan, significantly constraining their freedom of movement. Foreign militants were prohibited from leaving their residence unless they were given permission and were in the presence of a Pashtun guide out of concern they would be targeted for drone attacks by agents working for the United States, according to Makanesi. He described his training as rushed. It only consisted of two weeks of classroom sessions on weapons handling inside a residence and no bomb-making training. He recalled that the threat of drones meant he was not given shooting practice outdoors. And he claimed that his lack of funds meant he was unable to procure a weapon, illustrating the terrorist organization's continued cash shortages. At a certain point Makanesi was told by his handlers that the time was not yet right to provide him practical bomb-making training, suggesting such training had become more difficult for al-Qaeda and affiliated groups to organize.[44]

Makanesi also described the increasing paranoia al-Qaeda fighters had of drone strikes. They feared the drones were being guided by Western spies on the ground in FATA. That fear had made it more difficult for Western militants to connect with the terrorist group. Makanesi himself was interrogated by what he described as the group's intelligence branch, who deliberated for several weeks before he was allowed to join al-Qaeda ranks. His account painted a picture of chaotic conditions in the tribal areas of Pakistan with al-Qaeda having decentralized into as many as 40 subgroups, and lines becoming increasingly blurred between different Jihadist organizations.[45]

Such was the threat of drones that in October 2010 bin Laden in a letter to Atiyah abd al Rahman, a Libyan who was then al-Qaeda's head of operations in the tribal areas of Pakistan, instructed al-Qaeda operatives in Waziristan to keep a low profile or leave the area, and to avoid travelling around in cars.[46]

The pressure appears to have eased somewhat in the early part of the following year. In an April 2011 follow up, just days before his death, bin Laden revealed that al Rahman had told him there had been "an improvement in the security situation."[47] This was around the time two British militants received bomb-making training in the tribal areas of Pakistan, the last documented case of explosives training being provided to Westerners in the region when this book went to press in November 2014.

The two militants in question were Irfan Naseer and Irfan Khalid, the ringleaders of a conspiracy to launch attacks in the UK, which was thwarted by British police in September 2011. The men recorded martyrdom videos while in Pakistan and had previously received several days training in the country in 2009.[48]

Six men in total from the Birmingham area were charged in connection with a plan to conduct suicide bombings in crowded places in the UK, attempting to make home-made bombs, and discussing a plan to launch an attack with AK-47 assault rifles. In February 2013 Naseer and Khalid were among those convicted of the plot.[49]

In the weeks before they were arrested, Naseer and Khalid outlined their 2011 trip to Pakistan to others recruited into the conspiracy. The conversations were bugged by British police and produced in court. They said that after travelling to Pakistan in late 2010 they were initially provided 40 days' training by Harkat al Mujahideen, a Pakistani group closely affiliated with al-Qaeda inside the settled areas of Pakistan, during which time they were given weapons firing training. "You get the best training there because the government doesn't attack them," Naseer told his co-conspirators.

They then travelled to Waziristan where – in around May 2011 – they were trained by an Arab teacher and Pakistan al-Qaeda operative in explosives and poisons. Their conversations indicated they received ten or 11 days of instruction which took place inside a house rather than an outdoor training area. "W hasn't got no more camps now," Naseer explained to his co-conspirators. "The brothers in the mountains the drones just get them straight away, they just bomb the camps, so what they do is they stay, you know all our this what they taught us was inside houses."

Naseer and Khalid described drones as still posing a significant danger, despite the slight ease in strikes. When they drove 40 minutes away from their house to test detonate devices they spotted a drone above their location and had to take shelter under trees for more than four hours.[50]

Khalid described how the drones restricted movement for fighters in the region: "They were restricted to one place most of the time: One place to eat, sleep, go to toilet and do everything."[51]

The Birmingham case revealed that al-Qaeda had begun actively dissuading recruits from going to Pakistan. Instead al-Qaeda suggested the duo train operatives in the West to launch attacks. "They told me don't send anyone [more to us]," plot ringleader Naseer said in one of the recordings.

This was consistent with the message conveyed in an al-Qaeda video released a month after the duo trained in the tribal areas of Pakistan. Entitled "You are Only Responsible for Yourself" and featuring Ayman al Zawahiri it urged Western sympathizers to stay home and launch attacks.[52]

If, in early 2011, the security situation eased slightly for al-Qaeda in the tribal areas of Pakistan, the respite was short lived. Two key al-Qaeda commanders were killed by drone strikes in the summer of 2011 – Ilyas al Kashmiri in June and Atiyah abd al Rahman in August 2011 – rocking the terrorist organization further onto the back foot after the death of bin Laden. Although the overall number of drone strikes in the region dropped to 72 in 2011, the successful targeting of some of its most experienced commanders was a painful blow to the terrorist organization. In response, al-Qaeda and its allies in the tribal areas of Pakistan appear to have heeded the "rapid training" recommendation of the "Future Works" strategy document.

In the first half of September 2011, Toulouse terrorist shooter Mohammed Merah, by his own subsequent account, received less than half a day of training in the tribal areas of Pakistan during a two-week stay in the Miranshah area. U.S. and French intelligence agencies believe the instruction was provided by Jund al Khilafah (JaK) a small FATA-based "start-up group" affiliated with al-Qaeda focused on Kazakhstan.[53]

Merah claimed he was provided this training after first connecting with al-Qaeda elements in the tribal areas of Pakistan. He said his travel to the region was organized by a known Jihadist sympathizer he approached in a mosque in Islamabad, but that when he arrived in Waziristan, the al-Qaeda operatives he was introduced to had no idea who he was, and as a security measure confined him to house arrest for a time until he was able to win their trust.[54]

Merah said the al-Qaeda men were under the orders of "emirs" whose chain of command went up to Abu Yahya al Libi, a Libyan al-Qaeda commander who took over from al Rahman as the director of al-Qaeda's operations in the tribal areas. Merah claimed it was one of the al-Qaeda figures he met during this time who encouraged him to return home to launch an attack in France.[55]

Investigators later established that a Tunisian Jihadist named Moez Garsallaoui, who had resided for years in Europe and later developed deep connections to al-Qaeda's senior leadership in FATA, mentored Merah and agreed to take responsibility for any attack he launched.[56]

In March 2012 Merah carried out a series of shootings in southwest France that killed seven people, including several French paratroopers (his principal target) and three Jewish schoolchildren. Merah was eventually killed in a hail of gunfire during a siege at his apartment. Before he died, he spoke with negotiators through a walkie-talkie over a 17-hour period in which he outlined how he had plotted the attacks. Some of his claims were at first treated skeptically by French officials, but subsequent investigations corroborated a significant number of details he provided.[57]

Merah told the negotiators he had decided to focus on gun training in the encampment in Pakistan. He told negotiators:

> They proposed me to do the bomb training, but I didn't want to, because I told them the products you need for the bombs are pretty well monitored in France, so I would have risked arrest before managing to do anything. So I told them train me just with guns.[58]

His account was corroborated by a communiqué Garsallaoui posted online just over a week after he was killed: "He did not desire to train in explosives, even though that was available to him within a very narrow circle of no more than three individuals. He preferred fighting with weapons, as he told me."[59]

Back in 2012, before ISIS and other Jihadist groups took control of vast stretches of territory in Iraq and Syria, Merah's experience seemed to represent a paradigm shift in the nature of Jihadist terrorist training. There could not have been a starker contrast between Merah's cursory and self-tailored training and the regimented training courses mandated by al-Qaeda in Afghanistan in the late 1990s. It was a departure even from the sort of training al-Qaeda provided to recruits as late as 2008. As outlined above, that year American recruits Bryant Neal Vinas and Najibullah Zazi had received relatively closely controlled instruction from the group.

Since Merah trained in Pakistan the continued attrition of drone strikes has likely further eroded al-Qaeda's ability to teach terrorist skills in the tribal areas of Pakistan.[60] But the pressure appears to have eased again. Partly because the drone program had been so successful, there were only 48 strikes in Pakistan in 2012 according to the New America Foundation and 26 in 2013, the lowest number since 2007.[61] The reduction of strikes was also reflective of al-Qaeda shifting its center of gravity from the Afghanistan-Pakistan border region to the Arab world to take advantage of political turmoil caused by the Arab Spring, notably the civil war in Syria. A significant number of Arab al-Qaeda operatives are believed to have migrated back to their home countries, including some of the group's explosives instructors.[62] Even so, by November 2014 al-Qaeda's ability to train recruits in the tribal areas of Pakistan had diminished significantly since the 7/7 bombers were trained there ten years previously.

Yemen

Al-Qaeda in the Arabian Peninsula (AQAP) took advantage of political turmoil in Yemen to set up safe houses and small training camps in the tribal areas of the country in the period after 2006. In the fall of 2009 Omar Farouk Abdulmutallab, a Nigerian graduate of University College London, was personally instructed in a training encampment in Yemen in how to detonate an "underwear" device by AQAP chief bomb-maker Ibrahim al Asiri – who himself had built the bomb. According to court documents, al-Asiri had the "defendant practice the manner in which the bomb would be detonated, that is, by pushing the plunger of a syringe, causing two chemicals to mix, and initiating a fire (which would then detonate the explosive)."[63]

In the spring of 2012 a Saudi-British spy run by Saudi counter-terrorism was instructed how to detonate a similar device in another training facility in Yemen, possibly inside a safe house, which the group intended him to smuggle onto a U.S. bound airliner.[64]

But while both Abdulmutallab and the agent were instructed how to operate explosive devices, they were not taught how to build them. As yet there is no conclusive evidence the Yemeni affiliate has provided Western recruits with hands-on training in how to make bombs from everyday chemicals in the West like its counterpart in Pakistan, despite the capability to do so, the sine qua non for any sustained campaign of bombings on Western soil. There have not yet been any plots in the West involving Westerners trained in making bombs in Yemen. This is perhaps surprising given the bomb-making expertise of al Asiri and a team of bomb-makers he has trained. Fears that Western recruits may be double agents may have made the group cautious. Instead the group has sought to spread bomb-making knowledge and other aspects of terrorist tradecraft in the pages of its English-language magazine *Inspire.*

By the summer of 2012 a Yemeni military campaign and drone strikes had driven AQAP away from several population centers in southern Yemen, reducing its footprint and access to resources in the country. But the group, rather than being defeated, was largely able to melt away into desert hideouts, and retained an experienced group of explosive experts and a network of safe houses in the country's urban areas.[65] In the spring of 2014 a surge in attacks by the group prompted a renewed air and ground campaign against it by U.S. drones and Yemeni forces. Around the same time fears emerged that AQAP bomb-makers trained by al Asiri might be sharing bomb-making expertise with Jihadist groups in Syria, prompting new airplane security regulations in the West, including the requirement that all electronic devices brought into the cabin should be able to be powered on.[66]

Somalia

According to U.S. counter-terrorism officials most Al Shabaab training camps have provided only very rudimentary training for Western recruits. However, starting in 2006 al-Qaeda's east Africa cell set up more advanced training camps for a select group of recruits from Somalia, East Africa, Asia, North America, and Europe.[67] Two of the masterminds of al-Qaeda's attack on the U.S. embassies in Nairobi and Dar es Salam in 1998 – Saleh al Nabhan, a Kenyan operative, and Fazul Abdul Mohammed, a Comores native – set up camps in the Mogadishu area.

Fazul wrote in a 2009 memoir that Western recruits should be provided with training rather than instantly dispatched as suicide bombers within Somalia "to build sleeping cells around the world."[68] He described the camps as intensive courses spanning a month, focusing on the art of spying, gathering intelligence, policing, strategic defense and how to form secret cells in wartime situations.

However training programs were limited by cash constraints. Fazul revealed that a lack of funds had prevented the group from being able to put together "large scale" attacks against the U.S and its allies.[69] Fazul wrote that Nabhan focused on "specialized training, for he possessed extensive experience in this domain, including explosive charges."[70]

According to a July 2009 U.S. diplomatic assessment:

> camps associated with Nabhan have come to look increasingly like those run by al-Qaeda. They have been using foreign instructors, making use of similar training aids, physical training formations/activities, explosives and special tactics training undergone by al-Qaeda trainees elsewhere. The regimen followed by recruits at Nabhan's camps differs markedly from that traditionally used in Somali camps, and seems to produce better results.[71]

By then hundreds of militants had traveled to fight in Somalia from the West, most of them ethnic Somalis. The U.S. assessment did not make clear whether the camps offered training specifically tailored to launching attacks in the West, including making bombs out of easily purchasable chemicals.[72]

The State Department cable found that

> although Nabhan alumni have to date confined their fighting to Somalia, there are indications that foreigners trained in his Somalia-based camps could return to their countries of origin in order to conduct attacks. Nabhan is currently associated with efforts to attack Denmark, Nigeria, Kenya, and the UK. Danish trainees assessed to be attending one of the Nabhan-administered camps are reported to intend to return to Denmark to conduct unspecified attacks.[73]

Nabhan was killed in a U.S. Navy seal raid in September 2009 and Fazul was killed in an ambush in Mogadishu in June 2011. Authorities recovered a document on a thumb drive found on his person – probably authored by a British recruit – proposing a Mumbai-style attack on targets including Eton College and five star hotels in the UK. The author proposed two months training in Somalia for Western recruits selected for the attack, including on target reconnaissance, how to take hostages, weapons, and counter-surveillance. Although the plan was aspirational, it illustrated the ambition of Somalia based Jihadists to attack the West.[74]

Since late 2011 Al Shabaab has lost control of a significant amount of territory in Somalia and it is not clear the degree to which Nabhan and Fazul's camps remained in operation. Internal disputes and negative experiences suffered by Western recruits also slowed the flow of recruits from Europe and the United States. But the victory in a power struggle by a hardline faction led by Al Shabaab emir Muktar al Zubayr (also known as Ahmad Abdi Godane) led to a merger between the group and al-Qaeda in February 2012, strengthening the group's commitment to global Jihad.[75] Al Zubayr vowed to attack the U.S homeland and before his death in a U.S. strike in September 2014 set about transforming al Shabaab from a Somali-based insurgent force into a fully-fledged regional terrorist affiliate of al-Qaeda.[76]

By the time of the merger one of Nabhan's protégés – a Kenyan-Somali known as Ikrimah who spent time in Norway and UK between 2004 and 2008 – had emerged as a key planner of external operations for Al Shabaab. As well as attacking Kenya, Ikrimah hoped to send Western recruits back to the West to launch attacks. In 2010 he formulated a plan with AQAP American cleric Anwar al Awlaki to send Shabaab recruits, including Westerners, to Yemen for terrorist training, so that they could then be sent back to Somalia or on to the West.[77] The fact that Ikrimah regarded it as necessary to send recruits to Yemen for this training rather than train them in Somalia suggested that despite Nabhan and Fazul's best efforts the quality of training on offer in Somalia was inferior to that on offer in Yemen.

Kenyan intelligence services believe Ikrimah masterminded the September 2013 attack on the Westgate mall in Nairobi, which would have likely required fire-arms training. One of the gunmen was named as a Norwegian-Somali Al Shabaab recruit Hassan Abdi Dhuhulow. This worried European counter-terrorism officials because the Norwegian could theoretically have been used by the group to launch an attack in the West.[78]

There has been a number of terrorism cases in the West linked to training in Somalia. One of the two Danish-Somali brothers allegedly plotting a terrorist attack in Denmark in May 2012 received instruction in an al Shabaab training camp according to Danish authorities. This was a new phenomenon for a terror arrest in Denmark linked to the Somali diaspora community, according to Hans Jørgen Bonnichsen, a former director of Danish intelligence agency PET. “It makes a difference whether or not they have been trained or are self-active,” Bonnichsen said in an interview. “It is the first time in Denmark that an arrested individual has been trained, and the odds of success increases greatly the moment that one is operationally capable.” The two brothers were arrested after Danish security services heard them discussing potential targets, methods of attack, and use of weaponry.[79]

North and West Africa

Until a military offensive by French troops, a hodgepodge of Jihadist groups, including al-Qaeda in the Islamic Maghreb (AQIM), enjoyed a safe-haven in northern Mali. No evidence, however, emerged that Western militants were receiving training in the region in how to make bombs out of items purchasable in the West.

Al-Qaeda affiliated Jihadists, including a “Katiba” led by Moktar Belmoktar have moved from northern Mali to the “Salvador Pass” border region between Algeria, Libya, and Niger. Westerners have been recruited into Belmoktar’s group, including two Canadian nationals who took part in the attack on the In Amenas gas facility in Algeria in January 2013, creating concern that some might be trained and sent back to launch attacks in the West.[80]

As of November 2014 there were several training camps run by pro-al-Qaeda Jihadists in Libya including in the Jebel Akhdar region (Green Mountains) between Benghazi and Derna and in the southern desert.[81] A number of Libyan militants with European residence papers are believed to have joined militant groups in the region.[82] The growing instability in the country may see it become an even greater safe haven for Jihadists.

Iraq and Syria

The terrorist safe-haven now stretching across Iraq and Syria poses arguably the greatest potential threat to the West since al-Qaeda operated training camps in Taliban-run Afghanistan. ISIS now operates large training camps in the region, including near its stronghold in Raqqa, Syria, where until recently it could train fighters with relative impunity.[83]

The threat from the region is significantly greater than at the height of the Iraqi insurgency. In the years after the U.S invasion, Iraq did not emerge as a center for terrorist training for Western militants. Fewer Western militants travelled there compared to the Afghanistan–Pakistan border region. Though some Western militants travelled to the country and learned urban warfare skills there from insurgent groups, a high proportion of their number was killed in suicide bombings or has been killed in fighting.[84] Al-Qaeda was never able to establish significant training camps in Iraq because of the country's flat terrain, years of U.S. counter-insurgency operations, and a backlash against the group amongst Sunnis. Most skills were learned by recruits inside urban safe houses or in the process of actual fighting. Nor did al-Qaeda's affiliate in Iraq – focused on creating an Islamic state in Iraq – see attacking the West as a high priority.

However, a Jihadist resurgence in Iraq and Syria after the total withdrawal of U.S. troops from Iraq and three years of civil war in Syria has created an unprecedented potential national security threat for Western countries.

During the Iraqi insurgency against U.S. ground forces in Iraq, the most notable Western terrorism case linked to Iraq was that of Swedish-Iraqi terrorist Taimour Abdulwahab al Abdaly, who carried out a suicide bombing in Stockholm in December 2010 killing only himself. According to British court documents, al Abdaly travelled to Iraq in 2009 to receive training. The head of Iraq's counter-terrorism unit, General Dhai Kanani, disclosed in January 2011 that a detained al-Qaeda suspect had confirmed that Al Abdaly received three months' bomb-making training in Mosul, a stronghold of al-Qaeda in Iraq.[85]

European counter-terrorism officials are concerned that some of the growing number of Western militants travelling to fight with Jihadist groups in Syria in the struggle to overthrow the Assad regime may be taught bomb-making skills useful for launching terrorist attacks in the West.[86] By November 2014 around 3,000 Europeans had travelled to fight in Syria, many with Jihadist groups including Jabhat al Nusra and ISIS.[87] Both groups have experienced bomb-makers providing explosives instruction to new recruits. But their focus to date has been on toppling the Assad regime so it remains to be seen the degree to which they will provide training specifically tailored to launching attacks in the West.[88]

The fact that both groups are sympathetic to al-Qaeda's global Jihadist ambitions has created significant concern. If and when hardline Jihadist groups start to lose ground in Syria it is possible they may blame and lash out against the West. Jabhat al Nusra may pose a serious long-term threat to the West because of the fealty sworn by its leaders to al-Qaeda's senior leadership, though it is far from clear how the group will evolve, as a significant proportion of the Syrians fighting with it does not have a deep commitment to a global Jihadist agenda.[89] In February 2014 al-Qaeda's General Command expelled ISIS from al-Qaeda because of its increasingly brutal sectarian agenda and targeting of other Islamist factions, including al-Nusra.

One group of veteran al-Qaeda fighters linked to Jabhat al Nusra has already made attacking the West a priority. In September 2014 U.S. officials said the Khorasan group, based in and around Idblib province, was plotting attacks against U.S. aviation with sophisticated devices concealed in electronics and bombs made out of clothing dipped in explosive solution. They also said it was plotting to train European recruits and send them back home to launch attacks.[90]

As of November 2014 there was little evidence ISIS's leadership had yet initiated a program to launch terrorist attacks in the West. Its focus remained holding onto and expanding territory in Syria and Iraq and administering Islamic rule.

But after the United States launched air strikes against ISIS fighters in Iraq in August 2014, and ISIS executed American and British journalists in retaliation, there was concern the group could pivot towards plotting terrorist attacks in the West. The group's spokesman Abu Muhammad al Adnani promised to eventually bring the war to the West in an audio tape released in September 2014.[91] ISIS had unprecedented capability to unleash carnage in Europe if it chose to because it had more European recruits than any other Jihadist group, training camps on a scale last seen in Taliban-run Afghanistan, and likely tens of millions of dollars of cash reserves.[92]

The fact that several members of a Jordanian terrorist cell who planned to attack Amman in late 2012 spent time fighting in Syria illustrated the potential future threat of urban warfare skills learned in Syria, especially if in any transitional period Jihadists take control of swaths of the country. According to a Western intelligence official the experience provided the Jordanian cell with new skills and a radicalized outlook.[93]

In October 2013 British police broke up an alleged plot to launch an assassination or Mumbai-style attack by two extremists who were recorded by security services discussing the time they had spent in Syria.[94]

In late 2013 and early 2014 officials warned that Western fighters in Syria were undertaking training to prepare them for possibly returning to their home countries to carry out attacks. Though intelligence gathering has been difficult because of the chaos of Syria's civil war one worry was that experienced AQAP bomb-makers were now operating on Syrian soil and sharing their expertise with groups like Khorasan. "We're seeing now the appearance of training complexes in Syria to train people to go back to their countries, and, of course, conduct more terrorist attacks," U.S. Director of National Intelligence James Clapper[95] testified. Richard Walton, the head of Scotland Yard's counter-terrorism command, warned they had started to see signs of fighters being turned around to launch attacks in the UK. "I don't think the public realizes the seriousness of the problem," he said. "The penny hasn't dropped. But Syria is a game-changer."[96]

The urgency of the threat was soon made clear. In February 2014 French security services in the Cannes area thwarted the first alleged terrorist plot on Western soil linked to Syria. Two French nationals were arrested after returning from Syria, where they were suspected of having fought with Jabhat al Nusra and then ISIS. French authorities said they found three soda cans filled with almost a kilogram of TATP in the apartment of one of the suspects – Ibrahim Boudina, and they suspected Boudina learned how to make the explosive while he was in Syria. Both men denied the allegations against them. It is possible they were acting under their own steam. No evidence has yet emerged that the ISIS leadership signed off on the alleged plot.[97]

In May 2014 Mehdi Nemmouche, a French national who intelligence services believed had also fought with ISIS in Syria, shot four people dead in a Jewish museum in Brussels. Investigators believe he was likely acting under his own steam rather than being directed to launch the attack by the ISIS leadership.[98]

In October 2014 British authorities charged five men with an "ISIS-inspired" plot to shoot police in the UK. One of those charged was alleged to have returned to the UK after spending time with Jihadists in Syria.[99]

Despite U.S. and coalition airstrikes in Syria and Iraq as of November 2014, there was little indication that ISIS or Jabhat al Nusra had yet made training operatives for attacks in the West a priority or diverted significant resources to it.

Both ISIS and Jabhat al Nusra operate training camps inside Syria for new recruits, including Westerners. At first the facilities tended to be temporary and located in urban neighborhoods controlled by the groups or in the countryside, but as ISIS has solidified its control over parts of northern Syria it has started running training camps on a scale last seen in Taliban-run Afghanistan.[100]

Videos filmed for propoganda purposes by ISIS and Jabhat al Nusra recall the iconic "monkeybar" training scenes al-Qaeda filmed in Taliban run-Afghanistan. One Jabhat al Nusra video posted in early 2014 claiming to show the "Ayman al Zawahiri training camp" featured two dozen masked fighters negotiating an obstacle course in a ramshackle urban courtyard and then learning to clear buildings.[101]

New recruits to Jabhat al Nusra are typically given two weeks of basic physical and weapons training to ascertain how they can be best be deployed by the group. Some of those who show an interest or ability in bomb-making are provided training and sent to work in bomb-making factories manufacturing everything from grenades to car bombs (VBIEDs). Counter-terrorism officials worry about Western recruits working in these facilities. Although many of the devices these recruits build are based on materials and chemicals not readily available in the West, their skill-set makes them dangerous. Some of the more rudimentary car bomb designs could be adapted on their return to Europe or the United States.[102]

Urban areas of the Arab world

Political turmoil across the Arab world has provided new space for militant groups to operate in urban areas, creating concern that Western extremists could receive bomb-making training inside safe houses in cities. For example in October 2012 Egyptian police arrested several members of a terrorist cell based in an apartment in Nasr City in Cairo.[103]

How training creates capability

Bomb-making

The Breivik case in Norway illustrated what Western counter-terrorism officials had long feared: that a self-recruited entirely self-taught terrorist living in the West could meticulously plan and carry out a devastating attack.

It showed how modern technological tools, especially the availability of vast amounts of information useful for bomb-making and targeting, have made autonomously operating terrorists more dangerous than ever before. Like the Fort Hood shooting in November 2009 by Major Nidal Hasan, as well as other non-terror-related shooting massacres like a gun and grenade attack in Liège, Belgium in December 2011, an attack on a movie theater in Aurora, Colorado in July 2012, and the Newton, Connecticut school shooting in December 2012, the Norway attack illustrated just how deadly a shooting attack could be.

It was, however, Breivik' s ability to assemble a huge car bomb from scratch that was his most significant technical accomplishment given how tricky it is to make bombs from chemicals legally purchasable in the West. According to explosives experts many of the explosive compounds featured in online do-it-yourself recipes, for example the primary "detonating" explosive HMTD which was used in the 7/7 bombings, are highly unstable and dangerous to handle.[104]

Breivik included a detailed log of his bomb-making in a manifesto he wrote before his attacks. In this he described "googling for 200 hours" during the initial phases of his explosives research, and trawling through bomb-making websites, including ones authored by pro-al-Qaeda Jihadists.[105]

> Despite what others might say; you don't need any formal training in order to manufacture explosives. I have studied hundreds of various guides, recipes and instructions and can honestly say that it is a relatively safe venture as long as you take the necessary precautions and avoid the most volatile explosives,

wrote Breivik.[106]

The Breivik manifesto underlined the risk involved in making bombs out of readily purchasable chemicals. In a section containing a recipe for an explosive substance, the manifesto describes a "30% chance of blowing yourself up." Breivik's account demonstrated that his ability to tinker and experiment with explosive chemicals at a farm that he purchased in an isolated location was crucial for him to safely build a functioning device. So too was his ability to test the explosives at a second remote location.[107]

It is not impossible that a self-taught Islamist terrorist could build a device like Breivik's, especially after the Norwegian left detailed instructions for doing so in his manifesto. According to a Western explosives expert who has reviewed the manual, it was more alarming than those released in recent years by pro-al-Qaeda militants. "On a scale of 1 to 10 where most of the manuals released by al-Qaeda and its associates score 1 or 2, this manual scores around a 6," said the explosives expert.[108]

But Breivik had so many exceptional attributes that most Islamist terrorists will find it difficult to follow in his footsteps. His Nordic appearance, integration into society, and deep financial pockets gave him opportunities in this regard that few Jihadists terrorist wannabes enjoyed. His purchase of a farm, for example, allowed him to buy large quantities of fertilizer and test them without arousing suspicion.

Breivik also showed unwavering determination and exceptional discipline over a number of years in planning and carrying out his attack. In cases of Islamist terrorism in recent years such traits have mostly been associated with individuals whose dedication and resolve were fortified by time spent in Jihadist encampments overseas.

Few Islamist terrorists in the West have succeeded in building a viable terrorist bomb without instruction in a Jihadist encampment overseas. As stated above the only one to have successfully built a "high explosive" device was Andrew Ibrahim, a Bristol-based "lone wolf" Islamist extremist, who successfully manufactured the high explosive HMTD by downloading instructions from the internet before being arrested by British police in April 2008. He was later convicted of a plot to bomb a Bristol shopping center with a suicide vest.[109]

Such do-it-yourself terrorism has been championed in particular by AQAP through its English online magazine *Inspire*, which includes step-by-step bomb-making recipes. According to authorities one such formula "How to Make a Bomb in Your Mom's Kitchen," published in the summer 2010 edition of *Inspire* has been downloaded by Islamist militants plotting terrorist attacks in both the U.S. and the UK. It contains instructions for making low explosive pressure cooker devices and pipe bombs. According to the eight-page article in *Inspire*, which includes photographs and sketches, one of the advantages of the recipe was that it used readily available ingredients that would not arouse suspicion and could easily be "disposed of if the enemy searches your home."[110]

The recipe was allegedly used by Boston bombers Tamerlan and Dzhokhar Tsarnaev to build low explosive devices. The latter allegedly downloaded it to his computer at some point before the bombings. Explosives experts say the "Mom's kitchen" recipe is unlikely to work if followed to the letter so they likely tinkered with it to make the devices work.[111] They also adapted the instructions by building a triggering mechanism out of a toy car remote control instead of a timer.[112]

The recipe was also downloaded by a group of British militants who were convicted of a plot to bomb the London Stock Exchange and other targets in December 2010. It was also allegedly downloaded by Naser Jason Abdo, a Muslim U.S. soldier who allegedly plotted to attack the Fort Hood military base before his arrest in July 2011. And the instructions were used by Jose Pimentel, an Islamist extremist from New York who began assembling a pipe bomb before his arrest in November 2011.[113]

Explosives experts say many online recipes are unreliable. Bomb-making manuals produced by al-Qaeda and its sympathizers such as the "The Explosives Course by Abu Khabbab al Misri," a manual detailing how to make a variety of high explosive substances which was compiled by al-Qaeda supporters and released onto Jihadist web forums in early 2011, tend to contain many errors, according to Sidney Alford, a British explosives expert who has reviewed many such manuals, making following their instructions hazardous. "The overall quality of the documents that I've seen so far has been pretty low, but some of the content is certainly technically viable and they do contain, among some pages of rubbish – quite practical means of causing havoc," said Alford, who is widely regarded as one of the world's top authorities on explosives.[114]

"There are quite a lot of inherent dangers in some of those mixtures. Some of them will probably injure the person making them," said Alford. He said that a bomb-making novice with little chemistry background could be fortunate enough to select a recipe that works and succeed in making a viable device. "Once you've got the substances, you need only the most rudimentary equipment and no special facilities, no special place to do it," said Alford.[115]

Alford believes that by the law of averages some Islamist terrorist plotters will succeed in making bombs with no formal training.

> If enough people get hooked on this sort of publication, and practice what it preaches then some of them will succeed in causing the havoc and harm that they set out to do inspired in part by this publication, but some of them won't get that far, some of them will get bored because mix after mix doesn't work or they can't get the chemicals. Some of them will maintain their enthusiasm but will die [from an unintended explosion] before they get to use these devices or mixtures operationally.[116]

The unreliability of online explosive manuals has led to ongoing demand from Western extremists for hands-on bomb-making training in Jihadist encampments overseas.

The convicted ringleader of the 2006 transatlantic airline plot, Abdulla Ahmed Ali, testified at his trial that while his group looked up websites in its bomb-making research, he found them to be "wishy-washy" and stated that

> the whole point of us learning how to do it from someone who's done it before – or someone [who] would know about the thing – is obviously [that] it's quite dangerous dealing with these materials. We don't want to injure ourselves or anything.[117]

In the decade after 9/11 several waves of Western Islamist militants received bomb-making training in the Afghanistan-Pakistan border region and then used their explosives know-how to plot attacks back in their home countries. Trials and interrogation reports have shed light on the type of training they received. Similar information is not yet available for Syria, but the fear is that ISIS and Jabhat al Nusra could one day provide Western recruits there with similar training tailored to launching attacks against the West.

Crevice plot

Several of the British terrorist cell plotting to blow up targets in London in early 2004 with fertilizer bombs received bomb-making training in north-western Pakistan and the border region with Afghanistan in 2003. Their plot was thwarted by the British police investigation "Operation Crevice" in which 650 kg of ammonium nitrate (fertilizer) and a smaller amount of aluminum powder were recovered from a west London lock-up used by the cell.[118]

In June 2003, several of the leading figures in the group, including plot ringleader Omar Khyam, received explosives training from a freelance Afghan bomb-maker, in an apartment they rented in Kohat. The two-day training course included theoretical instruction in how to make bombs out of easily purchasable fertilizer and aluminum powder, including an initial grounding in substances and ratios necessary to make explosions.[119] In Kohat they also received hands-on instruction in making a fertilizer-based device which they subsequently tested in a nearby river. (They were also taught how to manufacture the deadly poison ricin, a potentially useful tool for assassination.)[120]

The following month a larger contingent of Khyam's group, including future 7/7 ringleader Mohammed Siddique Khan, traveled to Malakand, a mountainous region to the north of Kohat in the North-West Frontier Province (NWFP). The training camp lasted four to five weeks in total, though not all the group attended for that long.

According to prosecutors, "ingredients including ammonium nitrate and aluminum powder were taken to the camp." On the last day of training the group successfully tested a device using these ingredients. "Those experiments with explosives at the training camp appear to crystallize the view that the essential ingredients required to manufacture successful bombs were ammonium nitrate, aluminum powder and a detonator," prosecutors stated.[121] Mohammed Junaid Babar, who attended the camp, later testified that as a result of his training, Khyam acquired a significant degree of skill in making explosive devices with ammonium nitrate.[122]

7/7 bombings

British al-Qaeda operative Rashid Rauf revealed that bomb-making training was vital to the success of the July 7, 2005 London bombings. He wrote that he arranged for ringleaders Mohammed Siddique Khan and Shahzad Tanweer to meet with Abu Ubaidah al Masri in the tribal areas of Pakistan, who organized instruction in making a bomb out of hexamine peroxide (HMTD) detonators and hydrogen peroxide by Marwan Suri, an experienced and methodical bomb-making trainer. He described how Siddique Khan and Tanweer test-detonated a 300-gram hydrogen peroxide mixture they had made in the tribal areas. Rauf made clear that the successful test significantly bolstered the duo's confidence that they could carry out a successful terrorist attack in the UK. "Siddique [Khan] was always saying to me I hope these mixtures are as good as you say they are. After he tested the mixtures he was very happy," Rauf wrote.[123]

Rauf revealed that the 7/7 ringleaders continued to receive bomb-making guidance after they returned to the UK. Shortly before the attack, Siddique Khan began to boil down hydrogen peroxide to turn it into explosive in an apartment in Leeds. But he told Rauf he was unsure whether he had reduced it to the correct strength. Rauf wrote that he subsequently provided Siddique Khan with technical guidance to ensure he got the bomb mixture right. Additionally, when the bombers travelled to London on the morning of July 7, 2005 they followed instructions – passed on by Rauf – to leave the tops off the containers containing their explosives to prevent their overheating.[124]

21/7 failed London bombings

Rauf wrote that unlike the 7/7 ringleaders, he lost touch with Muktar Said Ibrahim, the ringleader of the 21/7 terrorist plot after Ibrahim returned to the UK. When this cell attempted to blow up London transport targets on July 21, 2005 their devices only partially worked. The TATP detonators in the devices detonated but failed to detonate the hydrogen peroxide main charge, leaving smoke and a strong smell but no blast.[125]

Rauf subsequently wrote the reason the explosives malfunctioned was because he failed to pass on the same fix he had provided Siddique Khan. Another contributory factor in the failure, according to Rauf, was that Said Ibrahim was taught by a less methodical and experienced instructor than the 7/7 ringleaders, and had to leave Pakistan before he felt confident in the bomb-making knowledge he had been given.[126]

2006 airline plot

Rauf also coordinated the 2006 transatlantic airline plot, a conspiracy which arguably more than any other demonstrated how formal bomb-making training can make terrorist cells immeasurably more dangerous.

Without being instructed in making devices by al-Qaeda technicians who had spent weeks brainstorming how to bring down an aircraft, the cell would very unlikely have been capable of mounting what remains by far the most serious plot since 9/11.

In his report Rauf described how he arranged for Abdulla Ahmed Ali, the UK ringleader in the plot, to receive instruction in explosives under the supervision of Abu Ubaidah al Masri in the tribal areas of Pakistan in late 2004. The Egyptian that winter arranged for Ali to receive weapons and explosives training, according to Rauf's account.[127]

Ali returned to Pakistan in June 2005, and was sent back to the UK late in 2005 with the task of putting together a plot to hit a wide range of targets, according to Rauf. At this stage al-Qaeda had not yet come up with the plan to target airliners, Rauf stated. He revealed that because al-Qaeda believed it would be difficult to purchase hydrogen peroxide in the UK in the wake of the London bombings that July, instructors taught Ali on this trip how to make explosive devices with gas.[128]

Eventually, Rauf and others in al-Qaeda decided the best plan was to smuggle hydrogen peroxide (a liquid) from Pakistan to the UK in rosewater bottles which would be packed to look unopened. Rauf wrote:

> We then analyzed the various machines that were used for checking baggage and persons at airports. We found it was very difficult to detect liquids explosives. After analysis that it would be possible to take concentrated hydrogen peroxide on board, the thought came to our mind: would it be possible to detonate the hydrogen aboard an airplane?[129]

That was the moment when the liquid bomb plot was conceived. "The way they progressed from London bomb plots to airline was very unexpected and brilliant," according to a senior U.S. counter-terrorism official.[130]

Rauf wrote that al Masri and other al-Qaeda operatives then brainstormed on how such a device could be constructed from easily purchasable chemicals that could be smuggled onto aircraft undetected.

In Jihadist encampments in Pakistan the al-Qaeda operatives experimented until they had perfected the design: half-liter soft-drink bottles transformed into hydrogen peroxide-based liquid bombs connected to HMTD detonators inserted in hollowed out batteries.[131]

Al-Qaeda's choice of HMTD appears to have been deliberate. In the failed July 21 attack, Al-Qaeda operatives had built detonators using a different substance – TATP – but it had failed to explode the main charge. Prosecutors claimed that this was the reason the Airline Plotters reverted to HMTD.[132]

"Haji/al Masri decided they would convert AA batteries into detonators," Rauf wrote. "We also practiced how to open a drinks bottle, empty it, and replace it with hydrogen peroxide, to make it seem unopened."[133]

According to Rauf al-Qaeda believed half a kilo of liquid explosive would without a doubt destroy an airplane, having noted a similar amount of plastic explosive Semtex had destroyed Pan Am Flight 103 over Lockerbie, Scotland, in 1988.[134]

Ali was recalled from the UK to Pakistan in the spring of 2006 so he could receive specially tailored training in how to construct the hydrogen peroxide based "liquid" bombs. "We trained (him) quickly, and wanted him to leave ASAP. He was also told to do anti-surveillance measures when he got back and only start work when he was comfortable that everything was clear," wrote Rauf.[135] In this final training Ali was taking advantage of weeks of research by al-Qaeda technicians in the tribal areas of Pakistan: figuratively standing on their shoulders.

Rauf indicated that he arranged terrorist training for several other of Ali's UK circle in FATA in 2005–2006, including Assad Sarwar the future bomb-chemist. British investigators believe that Rauf himself provided bomb-making guidance to Sarwar in Pakistan and in e-mail communications back to Ali and Sarwar in the UK.[136]

At trial Ali claimed he did not need detailed help after he returned to the UK:

> Regarding the electronics, I've got a degree in computer systems engineering and in that we did about three or four modules in electronics or digital electronics so I was familiar with the electronics so I didn't need his help as of yet, but if I would have got stuck I would have probably asked him.[137]

After the plotters returned to the UK, Rauf appears to have continued to pass on technical guidance to the plotters. In court Sarwar testified that when he met with "Jamil" (a codename for Rauf) in Islamabad, Pakistan in June 2006 he exchanged code books with him so that Rauf could "troubleshoot" if he ran into problems manufacturing the HMTD detonators and hydrogen peroxide explosives in the UK.[138]

Rauf described how, as the plotters prepared their attack, he was in extensive contact with Sarwar and Ali and another plotter through emails, text messages, phone calls, and instant messaging. At times Rauf appeared to be almost micro-managing the plotters. For example on July 24, Ali received a coded text message from him asking him: "Have you spoken to NB [Sarwar] about the colors?" referring to Ali's plan to dye the liquid explosives to make them resemble the colors of sports drinks.[139]

At some point during this period al-Qaeda conducted a successful test run with the bomb components minus the explosives on a flight, according to Rauf's account. "The purpose of the exercise was to test that none of the components would be checked manually," Rauf wrote.[140]

The plot, which targeted nine aircraft leaving for North America from Heathrow, was broken up on August 9, 2006. When Ali and the other plotters were arrested they had assembled all the components and chemicals necessary to make what authorities said would have been viable devices with enough explosive power to bring down the aircraft.[141]

2007 Sauerland plot

Four German residents making up the so-called "Sauerland group" which plotted car bomb attacks against American soldiers in Germany before their arrest in September 2007 received bomb-making training in Pakistan. They spent three months during summer 2006 training at an IJU (Islamic Jihad Union) encampment near Mir Ali, where they were also instructed in weapons, martial arts, and poisons.

Just like the 7/7 ringleaders they were taught how to make hydrogen peroxide bombs. They practiced boiling down small amounts of hydrogen peroxide in metal pots on the same stoves they used to cook their strictly rationed food. In the courtyard of their simple lodgings, they experimented to see whether the explosive mixture worked, creating small holes in the soil. They were also taught how to make timing detonators from Casio watches, which were later recovered from their safe house in the Sauerland region by German police.

Even though the facility where they trained appears to have been small, the training was very methodical. One group member recalled an examination in which circuits had to be assembled from memory.[142] Police found several hundred kilograms of hydrogen peroxide and more than two dozen detonators obtained from a source in Turkey in their safe house.[143] Prosecutors claimed they were trying to convert hydrogen peroxide into bombs when they were arrested.[144]

2007 Denmark plot

Another Western-based militant who received bomb-making training in the tribal areas of Pakistan was Hammad Khurshid, an al-Qaeda recruit arrested by Danish authorities in September 2007 for plotting terrorist attacks in Denmark with bombs made out of TATP detonators and hydrogen peroxide.

According to U.S. and Danish intelligence sources, he was trained in Mir Ali in how to make the explosive devices under the tutelage of Abu Ubaidah al-Masri, the Egyptian al-Qaeda operative who masterminded the 2006 airline plot. (Al-Masri died of hepatitis in December 2007.) Khurshid received handwritten instructions from his handlers on how to make TATP. On his return Danish intelligence filmed him in his Copenhagen apartment making 10 grams of TATP – a significant quantity – in only 27 minutes, something experts say would have required considerable training.[145]

2009 Manchester and New York cells

The continuing bomb-making guidance provided by British al-Qaeda plot coordinator Rashid Rauf in the 7/7 bomb plot and the 2006 transatlantic airline plot was also a feature of later plots launched by the operative in Pakistan.

An alleged terrorist cell arrested in Manchester in April 2009 for plotting an attack on a shopping center exchanged coded emails about the bombs they were planning to make with "Ahmad," an al-Qaeda facilitator in Pakistan, after their return to the UK.[146] New York Subway plot ringleader Najibullah Zazi was in email touch from Colorado with the same al-Qaeda facilitator about a missing section in his notes on a bomb-making recipe as his cell made final preparations to launch a terrorist attack in New York in September 2009, but was arrested before the facilitator could reply.[147] Times Square bomber Faisal Shahzad maintained regular contact with members of the Pakistani Taliban (TTP) over the internet in the weeks leading up to his failed attack on May 1, 2010. "Using software programs that were installed on his laptop computer while he was in Pakistan, Shahzad and the TTP were able to exchange information about the bomb he was building," according to court documents.[148]

Zazi, the New York subway plotter, provided the most detailed ever public account of al-Qaeda bomb-making training in the tribal areas of Pakistan in court testimony in April 2012. He recalled that he and his two co-conspirators received basic weapons training in a camp in North Waziristan in September 2008, and were persuaded by Adnan Shukrijumah, a senior American-Saudi member of al-Qaeda's external operations unit to launch a suicide bombing back in the United States. Zazi went on to receive specialized bomb-making instruction from a trio of al-Qaeda instructors in a small compound with high walls in South Waziristan in November 2008, where he was the only trainee.[149]

The training lasted only a week, a relatively short period of time compared to the length of time recruits had to learn in Afghanistan a decade earlier or even the time spent on such training by the 7/7 ringleaders. Zazi described taking notes in a small room as one of the instructors wrote on a white-board about safety procedures in handling explosive chemicals, the procedures for mixing the chemicals, and how to collect the necessary chemicals and other necessary items from department stores in the United States.[150]

The instructor then told Zazi about the different required elements in the explosive devices he was being taught to make. Some of these had been developed by al-Qaeda technicians for the 2006 airline plot: exposed lightbulb filaments as triggering mechanisms, detonating substance compacted into hollowed out batteries, a main explosive charge which could be put into a large plastic bottle, and a plate of ball bearings or nails to act as shrapnel.[151]

Zazi was taught the tricky task of making the unstable compounds HMTD and TATP for his detonators – and how long these explosives would last once made. "[The instructor] taught me mixing it, turning it into a solid, and then filtering and then the last part is to dry it," Zazi testified.

He was shown various techniques to make the task of making such a detonator easier:

> They told me that you could make something similar to the cap of pen and ... you could you use milk carton and roll it up, and then you use the scotch tape to [cover] the top completely. And what you need to do is put the powder in it and press it.

After Zazi made TATP detonators he successfully tested them under the watchful eyes of his instructors in the yard inside the compound.[152]

He was also taught how to boil hydrogen peroxide to turn it into an explosive for the main charge. In total he was taught four different types of main charge explosive to give him different options when he returned to the United States. He tested them in the mountains a short drive away from the compound, while one of his instructors recorded with a video camera for propaganda purposes. Zazi testified:

> We test them by putting them under a rock, and we had a wire, electric wire, long wire that was connected to a big battery, and we were like 30, 40 yards away from it, and all we did was just press the button ... it had an explosion and a big loud sound.[153]

During the training Shukrijumah visited the camp:

> He just give me a few safety tips that, you know, making a bomb is simple. But all you need is to find a place such as your family house and your bedroom or kitchen and, you know, just don't do any extra action, extra, you know, thing to bring attention from your family. Just do a simple what you have been taught and – or if you can't find your apartment, find something, some place that is more secure.

The American al-Qaeda recruit took these lessons on board. Months later in the kitchen of a Denver hotel room by referring to his notes he made a significant quantity of TATP. He took some of it to an outdoor area and successfully tested it to make sure it would work.

Zazi succeeded despite later acknowledging his training in Pakistan had been quite rushed. In court he revealed he had decided to build the device on his own in Denver "because I wasn't trained that good. I would definitely have little bit difficulty with it and I would have to take my time."[154]

On September 10, 2009, just days after U.S. authorities first became aware a plot was in the works, bomb sniffer dogs failed to detect the explosive as he drove into New York City. He was arrested soon afterwards.[155]

Drone strikes intensified further in the years after Zazi trained in the tribal areas of Pakistan, killing a significant number of al-Qaeda operatives, restricting the movement of Western militants, and making it more difficult to organize such training. In 2011 al-Qaeda operatives began migrating back to the Arab world as the terrorist group sought to capitalize on the political turmoil caused by the Arab Spring, further reducing the number of experienced instructors.

But the threat from explosives training courses in Pakistan has not vanished. As long as al-Qaeda and its affiliates retain a number of operatives skilled in making explosives in the tribal areas of Pakistan and access to bomb-making chemicals, it will remain possible for them to arrange the type of training Zazi received.

2011 Birmingham plot

A recent case in point was the training provided to two of the ringleaders of the September 2011 Birmingham terrorist plot.

In the early months of 2011 two Birmingham-based Islamist militants, who were allegedly part of a UK suicide bombing plot thwarted by British police in September 2011, received terrorist training in Waziristan, Pakistan.[156]

Irfan Naseer and Irfan Khalid were instructed how to make explosive devices and then test detonate them.[157] On their return to the UK, in conversations bugged by British police, they tried to dissuade other members of the group from travelling to Pakistan for such training, telling them their handlers had instructed them to "spread the knowledge" back in the UK.

"AQ's main aim is … the knowledge that they give us, [they] want more and more people [to] have [the knowledge] in Europe. So they can start whacking [attacking] there, yeah, do you understand?" Naseer told one of the co-conspirators in the UK.[158]

Khalid echoed this: "that is the command now … the thing is you gotta realise them guys yeah, they've told us that we want to spread this knowledge in Europe that people that they shouldn't be you know coming up." He added: "they'd rather people who are in these countries stay here."[159]

Their comments suggested al-Qaeda and allied groups, under pressure in Pakistan's tribal areas were envisioning a teacher-training approach, in which a select few Western militants schooled in the art of making homemade explosives would be tasked with disseminating this knowledge back in their home countries: a worrying scenario for counter-terrorism officials if it were to more fully materialize as a strategy because it could multiply the numbers skilled in making explosives in the West.

Naseer suggested that nothing that could be taught in Pakistan couldn't also be taught in the UK: "[In the tribal areas] you get bit of an experience in fighting – but you know the rest of the stuff that could be taught – they taught us in a room."[160] The trial made clear Naseer had amassed significant knowledge in bomb-making, including around 40 pages of notes from his training sessions in Pakistan. He showed members of the cell in Birmingham how to make improvised explosive devices. According to expert witnesses consulted in the trial, Naseer had acquired the correct knowledge to teach recruits how to make viable bombs.[161]

At one point, Naseer warned one of the co-conspirators that if he hit or rubbed the explosives in a certain way it could explode and potentially kill him; just one of the many lessons he passed on from his training in Waziristan.[162]

In the days before their arrest, Naseer experimented and tinkered with potential bomb-making chemicals in a Birmingham apartment, including the chemicals inside sports injury cold packs. The recordings revealed that his instructors in Pakistan had taught him that ammonium nitrate – a potentially high powered explosive – could be extracted from such packs, demonstrating al-Qaeda's continued inventiveness despite the increasing pressure placed on them.[163]

The bomb-making training received by the Birmingham plot ringleaders bucked the wider trend of training becoming increasingly low-quality and rushed in the tribal areas of Pakistan.

According to investigations by French authorities, in September 2011 Toulouse shooter Mohammed Merah was given less than a half day of training by the al-Qaeda affiliated group Jund al Khilafah in the North Waziristan area, where he was offered but declined explosives instruction, preferring to concentrate on gun skills.[164]

Weapons training

As the pressure has grown on al-Qaeda, and conditions in the tribal areas have grown more chaotic, the terrorist group has increasingly emphasized gun attacks, which require less specialized training to mount and may have a better chance of success.

Toulouse shooter Mohammed Merah, before he was killed, told negotiators outside his apartment that the fact he had received less than a half day of training on guns in the tribal areas of Pakistan before carrying out his shooting spree "was a proof that brothers without any military experience could do the same thing as me."[165]

Norwegian Islamophobic terrorist Anders Breivik put it this way:

> All Justiciar Knights must ask themselves; should I manufacture explosives or just stick to fire arms? This decision really boils down to what you expect to accomplish. A well planned assault with an assault rifle

may kill 30 category B traitors, while an unsuccessful explosive manufacturing process might result in 1 dead Justiciar Knight and thus 0 traitor executions.[166]

In 2010 senior al-Qaeda operatives Ilyas Kashmiri and Younes al Mauretani orchestrated a plot for recruits to attack European cities with "Mumbai-style" gun and hostage execution attacks, leading to an unprecedented State Department travel warning in October 2010. Just two months later a Swedish terrorist cell with links to Kashmiri's network were arrested after driving down to Copenhagen with a submachine gun and wrist bands to restrain hostages at the *Jyllands-Posten* newspaper, the publication responsible for printing the cartoons that caused controversy in the Muslim world. Western intelligence agencies believe plot ringleader Mounir Dhahri received training in North Waziristan, Pakistan in 2010 and met with senior al-Qaeda operatives Ilyas Kashmiri and Atiyah abd al Rahman.[167]

While it is not clear whether he received specific instruction in how to carry out such an attack, he would have had plenty of opportunity to be taught weapons handling and how to mount armed assaults by any number of militant groups launching cross border raids into Afghanistan.[168]

Further indicating al-Qaeda's new stress on gun attacks, in mid-2011 German police recovered an al-Qaeda document from a suspected al-Qaeda operative in Berlin entitled "Future Works" thought to have been authored in around 2009 calling for Mumbai-style attacks in Europe.[169]

The increasing pressure from drones also saw Al-Qaeda's senior leadership endorse AQAP's calls for followers in the West to take matters into their own hands through "individual terrorism." In June 2011 Zawahiri and other senior leaders appeared in a video entitled "You are Only Responsible for Yourself."[170] In the recording, al-Qaeda spokesman Adam Gadahn (born in Oregon) said it was easy for American al-Qaeda sympathizers to go to a gun show and purchase an automatic assault rifle – without having to submit to a background check.[171]

AQAP's *Inspire* magazine has featured practical advice on how to launch shooting attacks, including an article entitled "Qualities of an Urban Assassin" published in May 2012. The article stressed the need for Western militants choosing this path to educate themselves in the art of assassination and sharpen their skills through practice.[172]

In July 2011 al-Qaeda and its followers could not fail to have noticed the death toll caused by Breivik's shooting rampage on Utoeya Island. In September 2013 Al Shabaab gunmen, including a Norwegian recruit, killed more than 60 in the Westgate shopping mall in Nairobi, Kenya. If shootings begin to replace bombings as al-Qaeda's signature form of attack there will be less premium in the future on formal instruction in training camps.

Several shooting attacks in recent years have been launched by terrorists who never received any formal training. In November 2004

Dutch-Moroccan militant Mohammed Bouyeri shot dead Dutch film-maker Theo Van Gogh without ever having attended a camp. Norway shooter Breivik had been a member of a shooting club in Norway. Both he and Arid Uka, a German lone wolf terrorist who shot two American soldiers dead at Frankfurt airport in March 2011, honed their trigger skills in immersing themselves in shoot-'em-up video games.[173]

Training can, however, enhance the skills of those looking to carry out shootings. Toulouse shooter Merah, who was also a near obsessive player of shoot'-em-up video games, learned how to handle real guns in a Jak encampment in the tribal areas of Pakistan in September 2011. "When he arrived to the main training center, he expressed a great desire to own a weapon and he desired to have the best and the most powerful," claimed one of the group's operatives.[174]

Merah later told negotiators during the siege at his apartment that during his half day of training he was quickly taught how to assemble and dismantle guns and that he fired a variety of different weapons including hand guns and a Kalashnikov. He stated he also bought a weapon at some point after arriving in the tribal areas.[175]

Mehdi Nemmouche, who allegedly shot dead four people at a Jewish museum in Brussels in May 2014 is believed to have learned gun handling skills while fighting with ISIS in Syria.

Most Western Jihadist who spent time in Jihadist encampments overseas were in a better position to carry out gun attacks back in their home countries than militants who never made the trip. This was because they received basic weapons training as a first phase in their terrorist instruction, before graduating to more advance instruction, like bomb-making.

New York subway plot ringleader Najibullah Zazi described being taught how to use a hand gun and other heavier weapons, including machine guns and Ak-47s, before being taken to a firing range during his first training stint in the tribal areas. "We were told about parts of the weapons, opening and putting it back, cleaning it after using it. And we were also told or shown the position of each gun, how to hold military kinds of exercise, and also shown how to fire," he testified.[176]

But Westerners can also learn such skills without training overseas. The Boston bombers went to a gun range in New Hampshire a month before the attack.[177] Videos produced by Jihadist groups indicate that other European militants have trained and participated in armed gun assaults – or so-called "fedayeen" attacks – in the free fire areas in the Afghanistan-Pakistan border region and in Syria and Iraq, which may enhance their capability to launch similar attacks if they return to Europe.

Terrorist tradecraft

Western militants who received hands-on instruction from trainers in Jihadist encampments in target reconnaissance, operational security, counter-surveillance, and secure communications have had better tradecraft than those who autonomously launched plots.

What has been striking about most home-grown plots is their amateurism. One example was a plot to attack the Fort Dix military base in New Jersey 2007 which came onto the FBI's radar screen after the plotters, who were subsequently convicted, dropped off a video of themselves, shouting Jihadist slogans at a firing range, with a video store so that it could be copied onto a DVD.[178]

In another example of amateurism, Maryland militant Antonio Martinez did not help his cause when he posted a message on Facebook "the sword is cummin." And incredibly he did not grow suspicious when an individual, in reality working for the FBI, contacted him over Facebook. He later revealed to him his hope to attack a military recruiting station, leading to his arrest in December 2010, and subsequent guilty plea.[179]

By contrast, Western militants trained in Jihadist encampments tend to be taught at the very least basic operational tradecraft. While the quality of instruction has varied, and appears to have declined in recent years in the Af-Pak region because of the loss of senior al-Qaeda operatives with extensive operational experience, such instruction has paid off in several terrorist plots.

Before 9/11 al-Qaeda had much more time to train recruits in terrorist tradecraft than it did in subsequent years. As the testimony of British shoe bomber Saajid Badat illustrated, courses then tended to last for several weeks, and sometimes even months. In the years after 9/11, al-Qaeda recruits tended to receive only a week or two of training (though some received more), more recently some received only days of instruction.

The 9/11 plot is exhibit A in illustrating the importance of terrorist tradecraft. The hijackers received instruction in terrorist tradecraft in al-Qaeda camps in Afghanistan. Several, including Khalid al Midhar, attended an elite training course at the Mes Aynak camp in the fall of 1999 which focused on physical fitness, firearms, close-quarters combat, shooting from a motorcycle and night operations. In meetings in Karachi and Kandahar Khalid Sheikh Mohammed instructed several of the hijackers on Western culture and using code in communications.[180]

Later all the so-called "muscle hijackers" received special training in the al Matat training complex around late 2000 and early 2001 where they were taught how to conduct hijackings, disarm air marshals, and storm the cockpit at the earliest possible opportunity, and worry about taking control of the rest of the plan later. Their trainers even made them butcher sheep and a camel with a knife to prepare them for the attack.[181]

Mohammed Atta's Hamburg cell, the group at the heart of the plot, also received instruction in terrorist tradecraft in Afghanistan and Pakistan in late 1999 and early 2000. On their way back to Germany, they travelled to Karachi, where Khalid Sheikh Mohammed taught them about operational security and how to live without arousing suspicion in the United

States.[182] The 9/11 hijackers were also "well coached" to make sure their financial communications would not attract attention, according to the report. When Atta provided updates to Ramzi Binalshib on progress on the plot by phone he did so over previously agreed code-words.[183] The plotters demonstrated a high degree of operational tradecraft in the lead up to the attacks. Atta and the other pilots conducted cross-country surveillance flights, and tested to see if they could carry box-cutters onto flights.[184]

Al-Qaeda has never been able to match the level and sheer variety of training it was able to provide recruits in the pre-9/11 years under the protection of the Taliban in Afghanistan, where a recruit could spend months learning new skills at different camps. But once it re-established itself in the tribal area of Pakistan it was able for a period to continue to offer recruits a variety of different training courses.

The importance of good terrorist tradecraft was later emphasized by the British al-Qaeda operative Rashid Rauf, whom U.S. counter-terrorism officials describe as being one of the most skilled and careful operatives in the years after 9/11.[185] He especially stressed the importance of secure communications to the operatives he recruited.

In his internal report he revealed he communicated through coded emails, phone calls, and instant messenger communications with the ringleaders of the 7/7 London bombings and the 2006 Airline plot, in a system he instructed the plotters in while they were in Pakistan. He and Siddique Khan would change email addresses frequently, informing each other of the new addresses in code to prevent detection. "We would also exchange mobile numbers in code," he wrote. In the Airline plot Rauf set up a system of cell-phone communications in which the UK group would regularly change the SIM cards in their phones and provide details of new numbers through coded messages.

Rauf wrote that getting to know the plotters in the tribal areas of Pakistan helped greatly when it came to his communications with them. This underscored the value of face-to-face contact between Western militants embarking on plotting and experienced terrorist operatives overseas. In 2012 AQAP, through *Inspire* magazine, urged individuals who wanted to launch attacks to get in touch with them through email, but those at either end of such communications would have little way of knowing who they were really dealing with.[186]

The 7/7 bombers appear to also have been taught the importance of reconnaissance of targets. After the four suicide bombers travelled to London for this purpose, Siddique Khan was in touch with Rauf about what targets should be hit.

Rauf revealed he instructed a number of British recruits in counter-surveillance methods in Pakistan, including 21/7 plot ringleader Muktar Said Ibrahim.[187] Rauf also instructed the Transatlantic Airline plot ringleader to take counter-surveillance measures when he returned to the UK in the Spring of 2006 and only start preparing the attack "when he was comfortable everything was clear," presumably along lines he had been taught in the tribal areas of Pakistan.[188]

The value of such training was brought home to Rauf when he met 7/7 ringleaders Mohammed Siddique Khan and Shehzad Tanweer in Pakistan in late 2004. They told Rauf that they had expected to be arrested at any moment when British police in February 2004 thwarted the so-called "Crevice" plot – in which they planned to participate as suicide bombers – to blow up fertilizer bombs in London.

They revealed to Rauf that they threw the British security services off their tail by engaging in anti-surveillance measures, including pretending to be un-Islamic and joking loudly in cinemas and restaurants.[189] Rauf believed MI5, who conducted surveillance of the duo during the investigation of the fertilizer plot, were deceived by these tactics. It is not clear how Siddique Khan and Tanweer picked up these methods, but it is possible Siddique Khan was instructed in counter-surveillance when he attended training sessions with the "Crevice" plotters in Pakistan in 2003.[190]

Terrorist tradecraft was also emphasized in later plots launched by Rauf. In the fall of 2008 the alleged Western-based ringleaders of plots to target New York (Najibullah Zazi), Manchester (Abid Naseer) and Copenhagen (Mikael Davud) – are all believed to have received instruction in secure communications from al-Qaeda in the tribal areas of Pakistan. All three were handled by a Peshawar-based facilitator named Ahmad, who prearranged with them a system of communicating through code by email.[191]

Ahmad's decision to use the same email for communications with all three is what led to the thwarting of Zazi's plot to bomb New York in September 2009, according to a senior U.S. counter-terrorism official; the latter said that Rauf, who was killed in a drone strike in November 2008, would never have exhibited such poor operational tradecraft.[192]

Zazi later described two phases of training. The first in an encampment in North Waziristan in September 2008 which lasted ten days involved basic weapons training, and the second in a camp in South Waziristan which lasted a week focused on bomb-making. In court he did not specifically mention getting any detailed lessons on target reconnaissance, counter-surveillance, or operational security. It is possible that the short period of training meant there was not enough time for al-Qaeda to focus on these issues, and that the group instead decided to focus on teaching him the key skill of making a bomb.

But it is also possible these aspects were just not brought out in court. Zazi spent significant time with his handler Ahmad in Peshawar before returning to the United States, who would have had the time to pass on more operational advice.

Saleh al Somali, the al-Qaeda operative in charge of the plot did provide Zazi's co-conspirator Adis Medunjanin with at least some advice in tradecraft when he told him to act in the same way as before when he returned to the United States so as not to attract attention.[193]

Zazi later testified that, when he returned to the United States from Pakistan in early 2009, he moved from New York to Denver on the logic that he would be less likely to be on the radar screen of US counter-terrorism agencies if he moved to a place that was remote from the friends he had travelled to Pakistan with, illustrating that he had developed a sound awareness for operational security.[194]

Coded communications continued to be emphasized by Al-Qaeda and affiliated Jihadist groups in subsequent plots. Times Square bomber Faisal Shahzad was able to communicate with the Pakistani Taliban without being detected after they installed software on his computer while he was in Pakistan. Ahmad Sidiqi, a Hamburg militant who was recruited into al-Qaeda's plans to strike Europe in the fall of 2010, revealed to interrogators that a senior al-Qaeda operative Younes al Mauretani provided his group with instructions on how to communicate secretly on the internet, showing them how to use an encrypted email communication site called "Mujahideen Secrets."[195]

But the attrition of drone strikes meant that the overall trend by the early 2010s was a steepening decline in the breadth and quality of instruction offered to recruits. Forced for security reasons to provide recruits with increasingly hasty training, al-Qaeda and allied terrorist groups in the tribal areas of Pakistan appear to have increasingly emphasized teaching recruits bomb-making and shooting essentials rather than teach them *ancillary* skills such as counter-surveillance.

For example, in September 2012 the future Toulouse shooter Mohammed Merah appears to have received little instruction in terrorist tradecraft. He later told negotiators he received less than a half day of weapons training from his handlers in the tribal areas of Pakistan, but did not describe any training in counter-surveillance or other aspects of terrorist tradecraft.

It remains to be seen what level of tradecraft training ISIS and Jabhat al Nusra provide to Western recruits in Syria and Iraq. Given their control of territory and large-scale training camps, they have every opportunity to match the quality of training al-Qaeda provided prior to 9/11.

How training cultivates commitment

Arguably as significant as the specific skills which Western militants have acquired in Jihadist encampments is their greater sense of commitment to al-Qaeda's cause after receiving training, making them more likely than those who remain at home to see plots through to their conclusion.

In their home countries, al-Qaeda's global Jihad for these Western militants had previously been an abstract concept, filling up their computer screens, but actual contact with Jihadist operatives has given recruits a sense they are part of a greater cause and a sense of mission.

In order to persuade them to return home to launch attacks their handlers have psychologically manipulated them by claiming it is their religious duty to carry out such attacks. They argue that since they have Western passports, only they have a good chance of success in such operations.

This pattern has been seen in several cases. In the second half of 2003 it was only after receiving training in north-western Pakistan and meeting with senior al-Qaeda operatives in the tribal areas of Pakistan that the "Crevice" plotters became more serious in their terrorist enterprise.[196]

British al-Qaeda point-man Rashid Rauf described the importance of having a senior al-Qaeda leader meet with new Western recruits, with the charisma to persuade militants to return home to launch attacks.

When Mohammed Siddique Khan and Shehzad Tanweer met senior al-Qaeda operative Abu Ubaidah al Masri in the tribal areas of Pakistan in late 2004 it had a "profound effect on the brothers" and it took the Egyptian just a few days to persuade the duo to conduct a suicide bombing in the UK, instead of fighting in Afghanistan, he wrote.[197]

Rauf wrote that after Transatlantic Airline plot ringleader Abdulla Ahmed Ali received training in weapons, explosives, and electronics in the tribal areas under the supervision of al Masri "a big change came over his temperament [between] when he first came and when he left." Ali had also initially wanted to fight against US and coalition troops in Afghanistan.

As already noted in this chapter, Rauf also described how important the successful manufacture and test-detonation of bombs in the tribal areas was in persuading Western militants they could be capable of carrying out an attack.

There were also other ways to boost commitment. Rauf described how he helped the 7/7 bombers record a martyrdom video, after Abu Ubaidah al Masri insisted they tape one. While such tapes were primarily for propaganda purposes, they had the additional benefit of binding operatives to following through with their operation.[198]

Birmingham terrorist plotter Irfan Naseer, who recorded a martyrdom video in Pakistan in 2011, described the doubts that had started to creep in on his return to the UK to a co-conspirator:

> Spiritually this is harder here bro it's a bigger test and you know when you go there you don't get attacked that much by Shaytan [Devil] when you're here when you do something here yeah Shaytan [Devil] comes and comes.[199]

When it came to some other groups of Western militants, Jihadist operatives in FATA had to work harder to cajole them to launch attacks back home. In 2006 the German "Sauerland" group was not initially keen when their IJU handlers broached the idea of returning to the West to launch attacks, preferring to fight against US troops in Afghanistan. Initially they indicated they were not "particularly suitable" for such operations, but after a personal request from IJU leaders they agreed.[200]

In the fall of 2008 al-Qaeda went to great lengths to persuade Najibullah Zazi's three-strong travel group to launch attacks back in the United States, because they too were initially dertermined to participate in Jihad in Afghanistan, according to the testimony of Zazi and a second member of the travel group.[201]

Their meetings with al-Qaeda operatives are worth recounting in detail because they shed light on the ploys used by the terrorist group to persuade Western militants to return to launch attacks back home. After being driven into North Waziristan from Peshawar, the New York trio were taken to a guest house near the town of Miranshah, where they were introduced to Rashid Rauf and Saleh al Somali, a senior east African operative who headed up al-Qaeda's external operations unit. Both operatives wanted the men to return to the United States to plan an attack there. The trio explained that their intention was to fight in Afghanistan and that they did not want to return to the United States. "Keep it in your mind. Take time to think about it," al Somali replied.[202]

At the al-Qaeda camp in South Waziristan where they received basic training, the group met American Adnan Shukrijumah, who had joined al-Qaeda in Afghanistan in 2000. Shukrijumah pressed more forcefully for them to return to the United States, saying they could avenge the U.S. drone attacks and pressure the United States to leave Afghanistan. He told them that other al-Qaeda operatives lacked their travel documents or comfort operating in the United States. Shukrijumah even showed them a series of propaganda videos glorifying previous al-Qaeda attacks in the West to try to enthuse them. Eventually, by their own account, the American trio were worn down by these arguments and agreed to return to the United States to launch an attack. Shukrijumah later helped Zazi record a martyrdom video, binding him more tightly to the operation.[203]

In other cases Western militants appear to have felt the need to connect with an established Jihadist group before being prepared to embark on terrorist operations. One example was Underwear bomber Umar Farouk Abdulmutallab who travelled to Yemen in 2009 to join the Jihad there. When he met his idol, American terrorist cleric Awlaki, in Yemen he "expressed his willingness to become involved in any mission chosen for him, including martyrdom," according to court documents. According to the documents, al-Awlaki also instructed Abdulmutallab about his martyr video, arranging for a professional film crew to film it, binding him to the operation.[204]

Another case in point appears to have been Toulouse shooter Mohammed Merah who spent half a day in a Jund al Khilafah encampment in North Waziristan in September 2011. His brother Abdelkader, who has been charged with conspiring in the plot, told French investigators:

> I now realize that he was searching a for a way to get the contacts he needed to join al-Qaeda and meet an emir who could decide what he should do. To commit the acts which he did, you have to get the sanction of a sheikh or emir.[205]

Not all militants have had to travel to a training camp to get such a sanction. Fort Hood Shooter Nidal Hassan appears to have received spiritual sanction for his attack in the United States in email communications with Awlaki in Yemen.[206]

How training camps impact target selection

The large majority of Western militants had little intention of launching attacks back home when they arrived in Jihadist encampments overseas. Most arriving in the tribal areas of Pakistan sought training to participate in fighting U.S. and NATO forces in Afghanistan.

It is al-Qaeda and affiliated groups that have persuaded them to return home to launch attacks. But often their handlers only sketched out vague attack plans, leaving designated cell leaders to decide what specific targets were best to attack when they returned. Often target selection was the object of discussion or negotiation between recruits and senior operatives, and sometimes Western militants rejected their mission orders in favor of their own preferred targets.

One notable exception was the 2006 airline plot in which senior al-Qaeda operatives in Pakistan determined the target would be transatlantic aircraft, informed the British cell's ringleaders of this decision, and micro-managed the preparations for the attack from Pakistan.[207]

If al-Qaeda and its affiliates stressed anything consistently it was that attacks should create both as many casualties and economic damage as possible, without jeopardizing the success of the mission by being too ambitious.

This was the advice given by Najibullah Zazi's handlers in Pakistan, a case which has shed significant light on the process of target selection. After Zazi's group agreed to Shukrijumah's request for them to return to the United States to launch an attack, they discussed possible targets with him, each pitching in with their own ideas. These included New York Stock Exchange, Times Square, Penn Station, Grand Central Station, movie theaters, and the New York subway – because it was "the heart of everything in New York City," Zazi testified. "We discussed about targets that could damage U. S. economy. And so, we were targeting which areas or which public areas that would, you know, have impact – better impact, aftermath," he recalled.[208]

Shukrijumah suggested striking a Wal-Mart because it would damage the U.S. economy and told them that if they could not carry out a large attack, they should launch something smaller. Zarein Ahmedzay, Zazi's co-conspirator testified:

> He said that if you can't make something, the actual bomb, if you can't make a big bomb, then make something smaller. And he also said that in other missions people have failed because they tried to do something big and ended up doing nothing.[209]

"[Shukrijumah] was emphasizing [an] explosive suicide mission," Zazi recalled.[210]

When Zazi received specialized explosives training in a camp in South Waziristan soon after President Obama won the U.S Presidential election in November 2008, Shukrijumah came to the camp and told Zazi he wanted his group to send a message to shake the incoming president by launching an attack during the transition before he took office. Zazi pushed back, telling him he'd need more time, illustrating the pull and push often evident between Western recruits and their terrorist handlers in the hatching of plots.

When Zazi's group returned to the United States the cell had been given no specific target by al-Qaeda. Eventually they settled on a plan to launch a huge car bomb outside a New York building during Ramadan 2009. Cell member Zarein Ahmedzay testified:

> Najibullah, he was confident that he could make a bomb. But he also agreed that if he couldn't make the – big bomb like a car bomb, he would make something smaller. And if not that, we would have to do something else, like doing something with guns ... using guns to shoot up people and places.

In August 2009, however, the group decided they could not build such a large device because of missing pages in Zazi's notes. They changed the plan to blowing up subway cars by building smaller suicide vests. Their plan was to fulfill al-Qaeda's hopes by targeting the subways during morning or afternoon rush hour so they could inflict the heaviest number of casualties, instill fear, and inflict maximum economic damage.

> We were being told that our attack is about ... making America [suffer] financial-wise. So, you know, we said that that will be the better target because people will have a lot of fear [because] and these lines are like the busiest and most useful lines of all,

Zazi testified.[211]

In the 7/7 bombings, al-Qaeda left it up to cell leader Mohammed Siddique Khan to select the exact target, according to Rashid Rauf. Rauf wrote that the three potential targets given to them were the Bank of England, the upcoming G-8 summit in Gleneagles, Scotland, scheduled for July 6–8, 2005, or the London Underground. Siddique Khan decided on attacking the London Underground because of the difficulties in obtaining sufficient quantities of hydrogen peroxide to attack the other targets.[212]

The "Crevice" plotters were also provided significant autonomy by al-Qaeda on target selection. Initially the group wanted to fight in Afghanistan. But in early 2003 Abdul Hadi, their al-Qaeda handler, sent word that al-Qaeda already "had enough people" in Afghanistan and that "if they really wanted to do something they could go back [to Britain] and do something there." After they settled on this course of action, and soon after they organized their training, they had consultations with one of Hadi's deputies in the Afghan-Pakistan border region about the operation. The plot ringleader Omar Khyam came out of that meeting wanting to carry out multiple bombings at the same time, according to co-conspirator turned "super-grass" Mohammed Junaid Babar, who said their group discussed between themselves which targets in the UK to attack.[213]

There was significant back-and-forth between the German "Sauerland" cell and their handlers in Pakistan on what targets to strike. The cell agreed to the IJU's request that they launch attacks in Europe on condition they could target Americans and carry out the attack in Germany where they felt most comfortable operating. They later testified that they themselves selected locations to inflict "targeted retribution" against American interests in Germany. At one point, the IJU requested that the group target German politicians for assassination, but the members, demonstrating their autonomy, resisted because they viewed such attacks as too expensive.[214]

It is worth outlining the conversations Toulouse shooter Mohammed Merah had with his handlers in Pakistan on the question of targets in some detail because it illustrates the ad hoc nature of how terrorists come up with their targets. Merah said that after he joined forces with al-Qaeda in the tribal areas of Pakistan,

> some brothers came to visit me who were tasked with sending brothers to other countries. They proposed an attack in America, in Canada and so on. And I told them, that as I was French, it was easier for me and more simple to attack France.[215]

Merah said various fighters he met pitched in with ideas for targets. One Arab recruit suggested killing "gay" people back in France. But like several other Western al-Qaeda recruits before him it took some time for him to be won over. "To start with I didn't want to return. I was very happy to be with them but later there were things coming from them that didn't suit me," Merah recalled.[216]

Merah said it was an al-Qaeda figure who had once spent time in France who won him over to launching an attack back home. "He told me, 'Return to France, kill them in France' ... and I thought about it, took some time to think seriously about it, and said to myself, 'Go for it, I'll take my chance'."[217]

Merah's account indicated that as of late 2012 al-Qaeda continued to press new recruits from the West to return home to launch attacks, but also revealed their approach to be less systematic than before. Rather than being recruited into a specific mission by the al-Qaeda hierarchy, Merah appears to have warmed to the idea of targeting France during the course of several "camp-fire" chats with al-Qaeda comrades.

Merah spent time with the al-Qaeda subgroup Jund al Khilafah (JaK), during which time he received a half day's weapons training. As outlined previously in the chapter his key handler was Moez Garsallaoui, a Tunisian militant who established close connections to the senior leadership of al-Qaeda, after travelling to the tribal areas from Europe in early 2008 with the blessing of his wife Malika el Aroud, the Belgian widow of the al-Qaeda assassin who killed Northern Alliance leader Ahmed Shah Massoud.[218] Garsallaoui formed JaK as a small start-up Jihadist group in Pakistan's tribal areas in 2011 with around a dozen Kazakh, Tajiik, and Arab fighters.[219]

Merah stated the group at first suggested he attack an embassy in Islamabad, but the plan was dropped because there was too much security around it. "I saw that it couldn't work so I decided to return to France," he recalled.[220]

Garsallaoui, using the Jihadist name Abu al-Qa'qa' al-Andalus, corroborated Merah's account in an April 1, 2012 communiqué:

> An agreement was initially reached with him that he would carry out a martyrdom-seeking operation against one of the important enemy sites using an explosive belt after raiding it with weapons and bombs. He was extremely happy with the idea and insisted, despite the late hour at night, to accompany me to the depot to see with his own eyes the explosive belt that he was supposed to use in his mission. Once he put it on, he showed signs of joy and his face shined with happiness and pride. Martyrdom was his goal and the hope that was always on his mind.... One day after the program was changed for reasons that cannot be explained, Yusuf [Merah] started his return to France, promising to accomplish what he is capable.[221]

During the siege that led to his death, Merah claimed his handlers in Pakistan proposed he assassinate several public figures in France, including the Indian ambassador in Paris, and offered to finance the operation.

"According to his desire and his physical and mental abilities, assassinations were more appropriate for him," Garsallaoui stated in the April 1 statement.[222] Merah however decided against this plan. "I'm not somebody from Paris, and so on, and it would have been difficult for me, so I settled with what was doable. One shouldn't go over one's head," he told negotiators.[223]

I told them, "Look, I'll return to France, and find money myself and I'll take as a target anything that is military, police etc."[224] According to the interogation of a fellow member of JaK, Garsallaoui and Merah agreed that Garsallaoui would claim responsibility for any attack Merah put together.[225] French investigators established that when Merah returned to France he concentrated his energies on tracking down French paratroopers.[226]

Merah said his handlers provided him with an email address to stay in touch, but he decided to cut contacts after returning to France. He insisted he alone planned the attack on the French paratroopers. "I spoke to nobody so that I could be sure it would work," he told negotiators. He said his attack on the Jewish school on March 19, 2012 was improvised after he failed to find one of the paratroopers at home.[227]

Though Merah's back and forth with his handlers about which targets to strike was hardly unprecedented, the degree to which he set his own course, tailored his training according to his own wishes, and decided on his own which targets to attack, set him apart from many of the Western militants recruited into terrorist groups in the preceding years. In part this was likely a function of his short stay in the tribal areas of Pakistan and the drone campaign's progressive diminution of al-Qaeda's command structures.

The Merah case suggests the emergence of a new type of terrorist plotter: the militant who tailors their own training – à la carte – in a Jihadist encampments overseas, selecting targets themeselves, before returning to launch solitary attacks.[228]

Merah may have inspired at least one copy-cat. Alleged Brussels Museum shooter Mehdi Nemmouche was reportedly so enthused with Merah that he asked for a television to be wheeled in to his prison cell during coverage of his attacks and siege. He travelled to Syria after being released from prison and investigators believe he may like Merah have acted under his own steam.[229]

The future threat of terrorist training camps

The ISIS takeover of swaths of territory in Iraq and Syria means that in the coming years Jihadist training encampments overseas will almost certainly continue to pose a serious national security threat to European and Western countries.

While the Boston bombings were horrific they were not nearly as deadly as the London bombings. One of the reasons for this is that the bombs had low explosive power. If the perpetrators had made bombs out of high explosive such as hydrogen peroxide or fertilizer – trickier recipes to perfect – there would likely have been many more casualties, and this means that extremists who want to carry out high casualty attacks will likely continue to seek out bomb-making training overseas.

The threat from training camps in Pakistan is significantly less than it once was. Pressure from drone strikes has disrupted al-Qaeda's ability to train recruits in the tribal areas of Pakistan, with the result that it can no longer offer anything like the quality, range, or duration of training it once was able to offer in Afghanistan. Furthermore the Abbottabad documents revealed that Zawahiri has prioritized launching attacks on Western interests in the Muslim world over plotting attacks on Western soil. Since the beginning of the Arab Spring many operatives, including explosives instructors, have also moved back to the Arab world.[230]

This has seen al-Qaeda embrace the concept of "individual terrorism." Both Zawahiri and the leadership of AQAP in Yemen have actively encouraged followers living in the West to carry out attacks in al-Qaeda's name with no formal training.

This was to encourage a trend already in place. The majority of Islamist terrorist plots in Western countries the decade after 9/11 were launched by autonomous self-starting cells radicalized through exposure to radical study circles or Jihadists websites and radical online forums. More recently, there has been an increased concern over "lone wolf" terrorism – militants with no ties to formal organizations plotting alone – a function partly of the growth of social media, a mode of communication which has acted as an interactive echo chamber for like-minded extremists.

But many of these plots have been amateurish, and easily intercepted in the early stages by security services. By contrast, in a majority of the most serious plots in the West in the last decade – those that became weaponized and were not assisted by undercover agents – cell members trained overseas.

Paramilitary training in Jihadist encampments overseas will likely continue to make Western militants more dangerous terrorists by teaching them bomb-making skills and various aspects of terrorist tradecraft, as well as increasing their commitment to carrying out an attack because of the sense of mission and shared purpose inculcated in Jihadist encampments overseas.

There is a need therefore for further research into the critical ingredients that training encampments overseas provide. Any framework would include "classroom instruction" in explosives, weapons, counter-surveillance, target reconnaissance and secure communications, as well as "field instruction" in building bombs and test-detonating them, and shooting practice.

The menu offered by al-Qaeda in the Afghanistan-Pakistan border region has been constricted in recent years. The pressure of drone strikes by constraining movement in the tribal areas of Pakistan, and killing experienced trainers, has made it more difficult to bring recruits to encampments where they can be trained by instructors. And as illustrated by the "Future Works" document found in Berlin in 2011 it led to calls within the group for Western volunteers to be trained and sent back home to launch attacks more quickly than before.[231]

In 2012 it seemed possible that Toulouse terrorist shooter Mohammed Merah might become the new normal: Western militants on arrival in

Jihadist encampments overseas not so much recruited into terrorist networks as much as persuaded to return home to launch attacks immediately after a minimum of training. But the recent emergence of a new Jihadist training camp infrastructure in Syria and Iraq, and the arrival of veteran "Khorasan" al-Qaeda operatives determined to strike the West, suggests formal training will likely continue to pose a significant threat to the West.

Despite al-Qaeda's new stress on individual terrorism and its increasing stress on the rapid turnaround of recruits in the tribal areas of Pakistan, it has made clear it still wishes to launch large-scale attacks. The "Future Works" al-Qaeda strategy document advocated a twin-track strategy of low-cost, low-tech attacks combined with large-scale attacks.[232] This means that while the group may place more emphasis on shooting attacks in the future, because it requires less training, it will likely continue to try where possible to offer recruits bomb-making training in the hopes of producing a mass casualty, economically disruptive attack.

Terrorist groups in Syria are particularly well placed to provide such instruction due to the significant potential recruiting pool of Western fighters. The Khorasan group has already plotted ambitious attacks against Western aviation. Instructors in Syria may be able to encourage recruits in turn to pass on their expertise to militants in the West – the approach allegedly taken in the September 2011 Birmingham suicide bomb plot.

When it comes to the recruitment opportunities available to al-Qaeda and its allies in the tribal areas of Pakistan much will depend on the future flow of Western militants to the region. In the last decade the large majority of Western militants travelling to the region were hoping to fight U.S. and NATO troops in Afghanistan and it is possible that the withdrawal of these forces after 2014 will see the numbers wishing to travel reduced. But an expanded safe-haven for al-Qaeda in the region could itself prove a draw to recruits.

A significant training camp threat in the next several years may come from Yemen. AQAP remains well placed to offer Western militants bomb-making training, despite being driven away from several population centers in southern Yemen in 2012 by Yemeni forces and U.S. drone strikes. The group still has an extensive network of safe houses in the country's urban areas, large areas in which it can hide out in the southern tribal areas, and an experienced cadre of bomb-makers, including Ibrahim al Asiri.[233]

As in the case of Pakistan, however, much will depend on whether Western militants continue to travel to the country. And much will also depend on the degree to which AQAP prioritizes attacks against the West. While al-Qaeda "Central" in Pakistan prioritized launching attacks against the West because of bin Laden's determination to attack the home soil of the United States and its allies, to date AQAP's emir Nasir al Wuhayshi has only dedicated a fraction of the group's resources to attacking the West.

Thus far AQAP has limited itself to providing recruits, like Umar Farouk Abdulmutallab, and the British spy who recently infiltrated the group with pre-constructed improvised explosive devices in its continued determination to bring down a U.S.-bound passenger jet, but it is possible the group could switch to directly providing Westerners who join the group with bomb-making training, and task them to launch attacks in their home countries. In March 2014 Wuhayshi told fighters at a large gathering that confronting the United States and its allies remained a priority.[234]

There is also concern Western militants may in the future be able to access training in several other countries, including parts of northern Mali which still have a terrorist presence, the Sinai region of Egypt, and even in safe houses inside major cities such as Cairo.[235] The September 11, 2012 terrorist attack on the U.S. Consulate in Benghazi illustrated the dangers posed by Jihadist groups operating in eastern and southern Libya. U.S. and Libyan officials believe Jihadist groups, including some with affiliations to al-Qaeda, are operating several training camps in these regions. A worrying scenario for European counter-terrorism officials would be radicalized Europeans training in camps in Libya.[236]

But nowhere is the future potential threat from training camps greater than in Syria and Iraq because of the sheer amount of territory under the control of terrorist groups like ISIS. In August 2014 British Prime Minister David Cameron warned of a potential terrorist state emerging on the shores of the Mediterrenean. If ISIS decides to pivot towards attacking the West it will be able to take advantage of an unprecedented number of European militants in its ranks, tens of millions of dollars of cash reserves, and training camps on a scale last seen in Taliban-run Afghanistan.

Notes

1 See Nic Robertson, Paul Cruickshank, and Tim Lister, "Documents Give New Details on al-Qaeda's London Bombings," CNN, April 30, 2012.
2 Ibid.
3 Ibid.
4 According to Sageman,

> Formal training matters, both al-Qaeda core and al-Qaeda affiliate formal training resulted in an approximate success rate of 20%, while lack of training led to a success rate of 10%. So, training doubles the probability of success in a terrorist network. However, if the assassination of Theo van Gogh is eliminated from the sample, the resulting rate of success of the untrained networks falls to 5%. In this case, training would quadruple the probability of success in a terrorist network.

Sageman wrote the murder of van Gogh required no training. See Marc Sageman, "Confronting al-Qaeda: Understanding the Threat in Afghanistan and beyond." Testimony to the Senate Foreign Relations Committee, October 7, 2009.

5 The author began began tallying the survey for a New America Foundation paper, and updated it for this report. See Paul Cruickshank, "The Militant Pipeline between the Afghanistan–Pakistan Border Region and the West," 2nd Edition, New America Foundation Counterterrorism Initiative Strategy Document, July 2011 [1st Edition, 2010]. This report updates the July 2011 New America Foundation study by including the serious plots that emerged between its publication and November 1, 2014.

6 Ibid.

7 The Ottawa shooter could have killed significant numbers inside Canada's parliament if he had not been shot dead by Seargent-At-Arms Kevin Vickers.

8 See Paul Cruickshank, "French Bomb-Maker with Khorasan Radicalized over Several Years," CNN, November 1, 2014; U.S Treasury Designation of "Al Qaida Leaders in Syria," May 14, 2014; Paul Cruickshank and Tim Lister, "Foreign Legion in Iraq and Syria may Bring Jihad to the West," CNN, June 17, 2014; United Nations Sanctions, Security Council Resolution 2170 (2014).

9 Paul Cruickshank, "Could ISIS retaliate against the West?" CNN, August 9, 2014; David Cameron Downing Street Press Conference, August 29, 2014; Dina Temple-Raston, "Western Fighters Answer MidEast Extremists' Clarion Call," NPR, June 28, 2014.

10 Paul Cruickshank, "Norway Shooter's Explosive Legacy," CNN, July 27, 2011.

11 An alleged home-grown cell from north-west England, which allegedly planned to target the English Defence League at a Yorkshirre rally before the plot was thwarted in July 2012, built what was described as a "home-made explosive device." It is unclear, however, if it had the potential to inflict serious damage. The plotters also obtained a variety of other weapons. Matthew Taylor, "Three Accused of Plotting Attack on English Defence League," *Guardian*, July 10, 2012.

12 Cruickshank "The Militant Pipeline," 2nd Edition.

13 See Sageman, "Confronting al-Qaeda: Understanding the Threat in Afghanistan and Beyond."

14 Indictment: *USA* v. *Dzokhar Tsarnaev*, June 27, 2013. The brothers allegedly appear to have adapted the *Inspire* magazine recipe which included a timer mechanism rather than a remote-control detonation mechanism.

15 According to authorities the Boston bombers hatched a follow-up plot after the attack to bomb New York City's Times Square with their remaining devices.

16 Tim Lister and Paul Cruickshank, "Dead Boston Bomb Suspect Posted Video of Jihadist, analysis shows," CNN, April 22, 2013.

17 "Boston Bombing Investigation Moves Forward," CNN, April 25, 2013.

18 *United States* v. *Dzhokhar Tsarnaev*, Government's Opposition to Defendant's Motion to Supress Statements, United States District Court, District of Massachusetts, May 21, 2014.

19 Indictment: *USA* v. *Dzokhar Tsarnaev*, June 27, 2013.

20 For an explanation of the difference between low and high explosives, see Paige Williams, "Stories in the Smoke: What a Bomb Expert Sees," *The New Yorker*, April 17, 2013.

21 Cruickshank, "Norway Shooter's Explosive Legacy."

22 Paul Cruickshank, "Investigations Shed New Light on Toulouse Terrorist Shootings," CNN, June 13, 2012.

23 Janet Reitman, "Johar's World," *Rolling Stone*, July 17, 2013.

24 See National Commission on Terrorist Attacks upon the United States, *The 9/11 Commission Report*, New York: W. W. Norton, 2004, p. 67; David Rohde and C. J. Chivers, "A Nation Challenged: Qaeda's Grocery Lists and Manuals of Killing," *New York Times*, March 17, 2002.

25 Author interview with former Jihadist who travelled through Afghanistan in this period.
26 Videotaped Testimony of Saajid Badat, *USA* v. *Adis Medunjanin*. Deposition recorded March 29, 2012.
27 Cruickshank, "The Militant Pipeline," 1st Edition.
28 Ibid.
29 Ibid.
30 Ibid.
31 Testimony of Mohammed Junaid Babar, *Regina* v. *Omar Khyam et al.*, April 3, 2006.
32 See Robertson *et al.*, "Documents Give New Details on al-Qaeda's London Bombings" and "Document Shows Origins of 2006 Plot for Liquid Bombs on Planes," CNN, April 30, 2012.
33 Ibid.
34 Robertson *et al.*, "Documents Give New Details on al-Qaeda's London Bombings"; Information from source briefed on Rauf document, spring 2012.
35 Robertson *et al.*, "Document Shows Origins of 2006 Plot for Liquid Bombs on Planes."
36 Peter Bergen and Katherine Tiedemann, "The Year of the Drone," New America Foundation, February 2010; Peter Bergen and Katherine Tiedemann, "New America Foundation Counterterrorism Strategy Initiative: Drone Statistics," available at http://securitydata.newamerica.net, accessed December 2012.
37 Nic Robertson and Paul Cruickshank, "Recruits Reveal al-Qaeda's Sprawling Web," CNN, July 30, 2009.
38 Paul Cruickshank, "Inside the Plot to Devastate New York," CNN, May 2, 2012.
39 Ibid.
40 Ibid. See also Paul Cruickshank, "Enlisting Terror: Al-Qaeda 's Recruiting Challenges," *Jane's*, November 2009.
41 Cruickshank, "The Militant Pipeline," 1st Edition.
42 Nic Robertson, Paul Cruickshank, and Tim Lister, "Documents Reveal al-Qaeda's Plans for Seizing Cruise Ships, Carnage in Europe," CNN, April 30, 2012.
43 Bergen and Tiedemann, "New America Foundation Counterterrorism Strategy Initiative: Drone Statistics."
44 Cruickshank, "The Militant Pipeline," 2nd Edition.
45 Ibid.
46 West Point Combating Terrorism Center Harmony Documents SOCOM-2012–0000015, released May 2012.
47 Ibid.
48 Prosecution Opening *Regina* v. *Irfan Naseer, Irfan Khalid and Ashik Ali*, October 19, 2012; Paul Cruickshank, "UK Trial Reveals Al-Qaeda Strategy to Hit the West," CNN, February 21, 2013; Nick McCarthy, "Terrorism Plot Wanted to Turn Birmingham into War Zone, Court Hears," *Birmingham Mail*, October 24, 2012.
49 McCarthy, "Terrorism Plot Wanted to Turn Birmingham Into War Zone, Court Hears"; Tom Whitehead and Sam Marsden, "British Al-Qaeda Gang Planned Another '9/11' in the UK," *Daily Telegraph*, October 22, 2012; Paul Cruickshank, "Alleged UK Cell Charged with Terror Training in Pakistan," CNN, September 26, 2011.
50 Prosecution Opening *Regina* v. *Irfan Naseer, Irfan Khalid and Ashik Ali*, October 19, 2012, p. 240.
51 Ibid.
52 Ibid; Cruickshank, "UK Trial Reveals Al-Qaeda Strategy to Hit the West."

53 "Exclusif – Transcription des conversations entre Mohamed Merah et les négociateurs," *Libération*, July 17, 2012; Cruickshank, "Investigations Shed New Light on Toulouse Terrorist Shootings."
54 "Exclusif – Transcription des conversations entre Mohamed Merah et les négociateurs."
55 Ibid.
56 See Paul Cruickshank, "Taking Tea with a Terrorist," CNN, October 17, 2012; Personal communication with Erlan Karin, Kazakhstan terrorism expert, March 2014; Eric Pelletier, "Les coulisses du djihad de Mohamed Merah au Pakistan enfin dévoilées," *L'Express*, April 7, 2014.
57 Cruickshank, "Taking Tea with a Terrorist."
58 "Exclusif – Transcription des conversations entre Mohamed Merah et les négociateurs."
59 Garsallaoui was killed in a drone strike in October 2012. He was the husband of Malika el Aroud and recruited Walid Othmani and other members of the 2008 Belgian-French travel group. See Cruickshank, "Taking Tea with a Terrorist" and "Investigations Shed New Light on Toulouse Terrorist Shootings"; "Jund al-Khilafah Describes Travels, Motivations of Mohammed Merah," SITE Intelligence Group, April 1, 2012.
60 Bergen and Tiedemann, "New America Foundation Counterterrorism Strategy Initiative: Drone Statistics."
61 The exception when it came to the declining skill sets of those implicated in terrorism plots in the West were two alleged Chechen al-Qaeda operatives arrested by Spanish police in August 2012 and accused of plotting terrorist attacks. According to experts, they constituted one of the most skilled and experienced terrorist cells seen in recent times. But both men appear to have developed most of their terrorist skills several years before the onset of the drone campaign in Pakistan. The alleged cell ringleader Eldar Magomedov – who Spanish authorities allege spent time in North Waziristan between 2008 and 2011 – appears to have developed the bulk of his expertise earlier in his life during his time as a Russian Spetsnaz commando, though he did also allegedly subsequently receive training in Lashkar e Taiba camps in Afghanistan and Pakistan. Spanish security sources said his associate Muhammad Adamov was trained in explosives in a training camp in Afghanistan, which would indicate he learned the bulk of his terrorist skills before terrorist groups were forced out of the country after 9/11. See Paul Cruickshank, "Spain al-Qaeda Cell May Have Planned Strike to Coincide with the Olympics," CNN, August 7, 2012.
62 Author interviews with Western intelligence officials, 2011–2012.
63 Paul Cruickshank, Nic Robertson, and Ken Shiffman, "Al-Qaeda's Biggest Threat," CNN, February 16, 2012.
64 Nic Robertson, Paul Cruickshank, and Brian Todd, "Saudi Agent in Bomb Plot Held UK Passport Source Says," CNN, May 11, 2012.
65 Author interview with regional security source, August 2012.
66 Cruickshank, "Could ISIS Retaliate against the West?"
67 "Somalia – The TFG, Al-Shabaab, and Al-Qaeda," July 6, 2009, State Department Cable by U.S. Ambassador Michael E. Ranneberger, available at http://wikileaks.org/cable/2009/07/09NAIROBI1395.html, accessed 2014.
68 Nelly Lahoud, "Beware of Imitators: Al Qa'ida Through the Lens of Its Confidential Secretary," West Point Combating Terrorism Center, June 2012. In mid-2012 a European intelligence official told the author that senior al-Qaeda operative Fazul Abdullah Mohammed stated in his memoir that some Western recruits were providing bomb-making training, but did not elaborate on the nature of that training.

69 Lahoud, "Beware of Imitators."
70 Ibid, p. 88.
71 "Somalia – The TFG, Al-Shabaab, and Al-Qaeda."
72 Ibid.
73 Ibid.
74 Paul Cruickshank, Tim Lister and Nic Robertson, "Evidence Suggests Al Shabaab is Shifting Focus to Soft Targets," CNN, September 27, 2013.
75 Ibid.
76 See Paul Cruickshank and Tim Lister, "Ruthless Leader Aims to Extend Reach of Al Shabaab, Eyes the West," CNN, September 24, 2013.
77 Paul Cruickshank, Tim Lister and Nic Robertson, "Former Spy: Kenya Mall Attack Could Have Been Prevented," CNN, November 4, 2013.
78 "Norwegian Suspected of Being Kenya Mall Attacker Named," BBC, October 17, 2013.
79 Christian Wenande, "PET Foil Concrete Terror Action," *Copenhagen Post*, May 19, 2012.
80 Paul Cruickshank and Nic Robertson, "Analysis: Pro al-Qaeda Militants Seen as Lead Suspects in American Teacher's Death," CNN, December 5, 2013; Alistair Macdonald, "Second Canadian was among Algeria Terrorists," *Wall Street Journal*, March 23, 2013; Paul Cruickshank and Tim Lister, "Algeria Attack Changes Terror Landscape in North Africa," CNN, January 23, 2013; Paul Cruickshank and Tim Lister, "Algeria Attack May Have Links to Libyan Camps," CNN, January 23, 2013.
81 Ibid.
82 Author interviews with Libyan counter-terrorism sources 2012–2013.
83 Temple-Raston, "Western Fighters Answer MidEast Extremists' Clarion Call."
84 Truls Hallberg Tønnessen, "Training on a Battlefield: Iraq as a Training Ground for Global Jihadis," *Terrorism and Political Violence*, 20: 4 (2008), 543–562.
85 Paul Cruickshank, Tim Lister, and Per Nyberg, "The Last Days of a Suicide Bomber," CNN, December 9, 2011.
86 Author interviews with European counter-terrorism officials, autumn 2012.
87 Author interviews with European counter-terrorism officials, 2013; Michael Birnbaum, "E.U. Security Officials Note Increased Flow of Fighters to Syria's Civil War," *Washington Post*, December 5, 2013.
88 See Nic Robertson and Paul Cruickshank, "Sources: Pro al-Qaeda Group Steps up Suicide Bombings in Syria," CNN, October 4, 2012.
89 This analysis is based in part on information from Mohanad Hage Ali, an investigative journalist and regional terrorism expert with sources on the ground in Syria.
90 Paul Cruickshank, "French Bomb-Maker with Khorasan Radicalized over Several Years," CNN, November 1, 2014.
91 Abu Muhammad al Adnani Audio Statement, September 21, 2014.
92 Personal communication with regional security source May 2014; Cruickshank, "Could ISIS Retaliate Against the West?"
93 Joby Warrick, "In Foiled Jordanian Terror Plot, Officials See Hand of Resurgent al-Qaeda in Iraq," *Washington Post*, December 2, 2012.
94 "Terror arrests linked to Syria travel," ITV News, October 19, 2013; Victoria Ward, "Secret Terror Trial hears Erol Incedal used code word 'sausage' to buy a gun," *Daily Telegraph*, October 16, 2014.
95 Barbara Starr, "Officials: al-Qaeda tries to recruit Americans in Syria," CNN, February 21, 2014; Michael Schmidt and Eric Smitt, "Syria Militants Said to Recruit Vising Americans to Attack the U.S.," *New York Times*, January 9, 2014.

96 Paul Cruickshank, "Al-Qaeda's Still a Major League Threat," *Politico,* January 24, 2014.
97 For more on the alleged Cannes plot, see Paul Cruickshank, "Raid on ISIS Suspect in the French Riviera," CNN, August 28, 2014.
98 Ibid; Cruickshank, "Could ISIS Retaliate against the West?"
99 "Terror Plot Suspects Planned to Kill Police, Court Hears," BBC, October 17, 2014; "Arrests may have Foiled Terror Attack Planning in the UK," BBC, October 7, 2014.
100 See, for example, Temple-Raston, "Western Fighters Answer MidEast Extremists' Clarion Call."
101 Bill Roggio, "Al Nusrah Front Names Traing Camps after Top al-Qaeda Leaders," *Long War Journal,* March 17, 2014.
102 This section is in part based on information provided by Laith Alkhouri, a senior analyst at Flashpoint Partners.
103 Thomas Jocelyn, "More al-Qaeda Links to Cairo Cell Reportedly Found," *Long War Journal,* November 9, 2012.
104 Author interview with Western explosives expert, summer 2011.
105 Anders Breivik, "2083: A European Declaration of Independence," released July 2011; Cruickshank, "Norway Shooter's Explosive Legacy."
106 Breivik, "2083: A European Declaration of Independence."
107 Ibid.
108 Cruickshank, "Norway Shooter's Explosive Legacy."
109 Ibid.
110 *Inspire* Magazine, Summer 2010.
111 Paul Cruickshank and Tim Lister, "The Lone Wolf: The Unknowable Terror," CNN, September 7, 2011.
112 Indictment: *USA* v. *Dzokhar Tsarnaev,* June 27, 2013.
113 Paul Cruickshank, "New York Bomb Plotter Monitored for Two Years, Officials Say," CNN, November 21, 2011.
114 Cruickshank and Lister, "The Lone Wolf: The Unknowable Terror."
115 Sidney Alford interview with author, UK, summer 2011.
116 Ibid.
117 *Regina* v. *Ali et al.,* June 3, 2008, p. 29.
118 *Regina* v. *Omar Khyam et al.,* Prosecution Opening Statement, Central Criminal Court, March 21, 2006.
119 Cruickshank, "The Militant Pipeline," 1st Edition; *Regina* v. *Omar Khyam et al.,* Prosecution Opening Statement, Central Criminal, Court, March 21, 2006.
120 Cruickshank, "The Militant Pipeline," 1st Edition; "Accused Help Made Test Bomb," BBC, July 24, 2006.
121 *Regina* v. *Omar Khyam et al.,* Prosecution Opening Statement, Central Criminal, Court, March 21, 2006.
122 Mohammed Junaid Babar Testimony, *Regina* v. *Omar Khyam et al.,* March 30, 2006.
123 See Robertson *et al.,* "Documents Give New Details on al-Qaeda's London Bombings."
124 Ibid.
125 Ibid; Cruickshank, "The Militant Pipeline," 2nd Edition.
126 Robertson *et al.,* "Documents Give New Details on al-Qaeda's London Bombings."
127 Nic Robertson, Paul Cruickshank, and Tim Lister, "Document Shows Origins of 2006 Plot for Liquid Bombs on Planes," CNN, April 30, 2012.
128 Ibid.
129 Ibid.
130 Ibid.

131 Richard Greenberg, Paul Cruickshank, and Chris Hansen, "Inside the Terror Plot that 'Rivaled 9/11'," *Dateline NBC*, September 14, 2009.
132 *Regina* v. *Ali et al.*, June 4, 2008, at 143; Duncan Gardham, "We've Never Seen a Bomb Like 21/7 Devices," *Daily Telegraph*, July 11, 2007.
133 Robertson *et al.*, "Document Shows Origins of 2006 Plot for Liquid Bombs on Planes," CNN, April 30, 2012.
134 Ibid.
135 Ibid.
136 Ibid; Interview by telephone with Metropolitan Police Investigator, London, March 2009. Assad Sarwar, the "bomb chemist" of the Airline Plot testified at trial that he received bomb-making instruction from a certain "Jamil" in Pakistan and had established a code so that further instructions could be transmitted by email and text message. The Investigator said that they had established that Jamil was almost certainly an alias for Rashid Rauf.
137 *Regina* v. *Ali et al.*, June 3, 2008, at 39. Ali testified he was in touch by phone with "Yusuf" during this period, the Pakistan-based trainer who had instructed Sarwar in making HMTD and hydrogen peroxide bomb manufacture in June 2006. According to British counter-terrorism sources this individual was in fact almost certainly Rauf.
138 *Regina* v. *Ali et al.*, June 12, 2008, at 68.
139 *Regina* v. *Ali et al.*, May 6, 2008, at 29; British Press Association Timeline of the Airline Plot.
140 Robertson *et al.*, "Document Shows Origins of 2006 Plot for Liquid Bombs on Planes."
141 Greenberg *et al.*, "Inside the Plot that 'Rivaled 9/11'"; Robertson *et al.*, "Documents Give New Details on al-Qaeda's London Bombings" and "Document Shows Origins of 2006 Plot for Liquid Bombs on Planes."
142 See Cruickshank, "The Militant Pipeline," 1st Edition.
143 "Sauerland Cell Trial Opens in Düsseldorf," *Deutsche Welle*, April 22, 2009; "German Prosecutors Seek Arrest Warrant for Turkish Terror Suspect," *Der Spiegel*, August 17, 2009.
144 Yassin Musharbash, "Sauerland Cell Testifies: Jihadists Describe Hatred of US as Reason for Terror Plot," *Der Spiegel*, August 12, 2009.
145 Sauerland Cell Trial Opens in Düsseldorf"; "German Prosecutors Seek Arrest Warrant for Turkish Terror Suspect."
146 Duncan Gardham, "Manchester 'Terror Plot' Emails," *Daily Telegraph*, July 30, 2009.
147 Cruickshank, "Inside the Plot to Devastate New York."
148 Cruickshank, "The Militant Pipeline," 2nd Edition.
149 Trial Transcript *USA* v. *Adis Medunjanin*, April 18, 2012, pp. 561–583.
150 Ibid.
151 Ibid.
152 Ibid.
153 Ibid.
154 Trial Transcript *USA* v. *Adis Medunjanin*, April 18, 2012, p. 610.
155 Ibid; Cruickshank, "Inside the Plot to Devastate New York."
156 McCarthy, "Terrorism Plot Wanted to Turn Birmingham Into War Zone, Court Hears."
157 Prosecution Opening *Regina* v. *Irfan Naseer, Irfan Khalid and Ashik Ali*, October 19, 2012; Cruickshank, "UK Trial Reveals Al-Qaeda Strategy to Hit the West."
158 Ibid.
159 Ibid.
160 Ibid.
161 Ibid.

162 Ibid.
163 Ibid.
164 Cruickshank, "Investigations Shed New Light on Toulouse Terrorist Shootings."
165 Exclusif – Transcription des conversations entre Mohamed Merah et les négociateurs," *Libération*, July 17, 2012.
166 Breivik, "2083: A European Declaration of Independence."
167 Author interview with Western Intelligence Official, Summer 2012.
168 Cruickshank, "Four Convicted in Scandinavian Mumbai-Style Terror Plot"; Cruickshank, "The Militant Pipeline," 2nd Edition.
169 Nic Robertson, Paul Cruickshank, and Tim Lister, "Documents Reveal al-Qaeda's Plans for Seizing Cruise Ships, Carnage in Europe," CNN, May 1, 2012.
170 Paul Cruickshank and Tim Lister, "The Lone Wolf: The Unknowable Terror."
171 Ibid.
172 *Inspire* Magazine Issue 9 (winter 2012), published May 2012.
173 "Oslo Shooting Club Reveal Behring Breivik's Membership," AFP, July 28, 2011; Cruickshank, "Investigations Shed New Light on Toulouse Terrorist Shootings"; Personal interview with German Intelligence Official, March 2011.
174 Cruickshank, "Investigations Shed New Light on Toulouse Terrorist Shootings"; JaK Operative's Statement on Toulouse Shootings, March 31, 2012.
175 "Exclusif – Transcription des conversations entre Mohamed Merah et les négociateurs."
176 Trial Transcript *USA* v. *Adis Medunjanin*, April 17, 2012, pp. 498–504.
177 Indictment: *USA* v. *Dzokhar Tsarnaev*, June 27, 2013.
178 Russakoff and Eggen, "Six Charged in Plot to Attack Fort Dix."
179 "Baltimore Bomb Suspect Indicted."
180 *9/11 Commission Report*, pp. 157, 526.
181 Ibid., pp. 235–236.
182 Ibid., p. 167; "11 September: The Plot and the Plotters," CIA, June 1, 2003, obtained under FOIA by IntelWire.Com after declassification in April 2012 from "top secret" status.
183 *9/11 Commission Report*, p. 245.
184 Ibid., pp. 242–245.
185 Personal Communication with Senior U.S. Counter-terrorism Official, April 2012.
186 Paul Cruickshank, "Al-Qaeda in Yemen Advertises Online for Western Recruits," CNN, June 11, 2012.
187 Robertson *et al.*, "Documents Give New Details on al-Qaeda's London Bombings."
188 Robertson *et al.*, "Document Shows Origins of 2006 Plot for Liquid Bombs on Planes."
189 Ibid.
190 Ibid.
191 Cruickshank, "The Militant Pipeline," 2nd Edition.
192 Personal Communication with Senior U.S. Counter-terrorism Official, April 2012.
193 Trial Transcript *USA* v. *Adis Medunjanin*, April 17, 2012.
194 Ibid., pp. 597–598.
195 Cruickshank, "The Militant Pipeline," 2nd Edition.
196 Cruickshank, "The Militant Pipeline," 1st Edition.
197 Robertson *et al.*, "Documents Give New Details on al-Qaeda's London Bombings."

198 Robertson *et al.*, "Documents Give New Details on al-Qaeda's London Bombings" and "Document Shows Origins of 2006 Plot for Liquid Bombs on Planes."
199 Prosecution Opening *Regina* v. *Irfan Naseer, Irfan Khalid and Ashik Ali,* October 19, 2012.
200 Cruickshank, "The Militant Pipeline," 1st Edition.
201 Trial Transcript *USA* v. *Adis Medunjanin,* April 17–18, 2012; Cruickshank, "Inside the Plot to Devastate New York."
202 Ibid.
203 Ibid.
204 Kiran Khalid and Paul Cruickshank, "Awlaki Directed Underwear Bomb Plot," CNN, February 10, 2012.
205 Cruickshank, "Investigations Shed New Light on Toulouse Terrorist Shootings."
206 David Morgan, "Justice Memo Authorized Killing of Awlaki," CBS News, October 1, 2011.
207 Robertson *et al.*, "Document Shows Origins of 2006 Plot for Liquid Bombs on Planes."
208 Trial Transcript *USA* v. *Adis Medunjanin,* April 18, 2012, p. 544.
209 Ibid., April 16, 2012, p. 166.
210 Ibid., April 18, 2012, p. 547.
211 Ibid., April 18, 2012, p. 619.
212 See Robertson *et al.*, "Documents Give New Details on al-Qaeda's London Bombings."
213 Testimony of Mohammed Junaid Babar, *Regina* v. *Omar Khyam et al.*, March 29, 2006.
214 Cruickshank, "The Militant Pipeline," 1st Edition.
215 "Exclusif – Transcription des conversations entre Mohamed Merah et les négociateurs."
216 Ibid.
217 Ibid.
218 Cruickshank, "Taking Tea with a Terrorist"; Personal communication with Erlan Karin, Kazakhstan terrorism expert, March 2014; Pelletier, "Les coulisses du djihad de Mohamed Merah au Pakistan enfin dévoilées."
219 Personal communication with Erlan Karin, Kazakhstan terrorism expert, March 2014; Pelletier, "Les coulisses du djihad de Mohamed Merah au Pakistan enfin dévoilées."
220 "Exclusif – Transcription des conversations entre Mohamed Merah et les négociateurs."
221 "Jund al-Khilafah Describes Travels, Motivations of Mohammed Merah," SITE Intelligence Group, April 1, 2012.
222 Ibid.
223 "Exclusif – Transcription des conversations entre Mohamed Merah et les négociateurs."
224 Ibid.
225 Pelletier, "Les coulisses du djihad de Mohamed Merah au Pakistan enfin dévoilées."
226 Cruickshank, "Investigations Shed New Light on Toulouse Terrorist Shootings."
227 Ibid.
228 Ibid.
229 Jacques Duplessy, Emilie Blachère et Laurence Nhek, "Djihad Made in France," *Paris Match,* June 7, 2014; Author interview with European counterterrorism official, June 2014.

230 Paul Cruickshank, "Terror Warning May Be Linked to Al-Qaeda Choice of Chief Deputy," CNN, August 3, 2013.
231 Robertson *et al.*, "Documents Reveal al-Qaeda's Plans for Seizing Cruise Ships, Carnage in Europe."
232 Ibid.
233 Author interview with regional security source, August 2012.
234 Barbara Starr, "Video Shows Large al-Qaeda Meeting in Yemen," CNN, April 16, 2014.
235 Conversations with Western intelligence sources, 2012.
236 See Nic Robertson, Paul Cruickshank and Tim Lister, "Pro Al-Qaeda Group Seen Behind Deadly Benghazi Attack," CNN, September 13, 2012; Interviews with regional security sources January 2013.

8 Innovation in terrorists' counter-surveillance

The case of al-Qaeda and its affiliates

Joshua Sinai

This chapter attempts to assess the effectiveness, in terms of operational and technical innovation, of the tradecraft in pre-incident counter-surveillance by al-Qaeda and its affiliates. In the first section, a conceptual framework is developed based on the components of pre-incident adversarial surveillance that terrorist groups and their operatives employ to evade and counter in order to launch their planned terrorist operations. The second section of the chapter presents a discussion of selected cases of terrorist operations by al-Qaeda and its affiliates where their attempts at counter-surveillance played a role in the success or failure of these plots, based on the surveillance measures that were employed against them by their counter-terrorism adversaries. Since much of the activities of terrorism and counter-terrorism takes place covertly, the information used in this analysis is based on published open sources, so while it is likely to be incomplete and fragmentary, it is still valuable in providing insight on how these operatives employed counter-surveillance in their pre-incident attack planning and preparations.

Defining surveillance and counter-surveillance

Within the context of the constant "cat and mouse games" between governments' counter-terrorism campaigns and their terrorist adversaries, one of the most important instruments used by governments is their continuous surveillance of their terrorist adversaries' operatives, facilities and activities especially during the formative pre-incident phases of a potential attack, because identifying such warning indicators is crucial in thwarting them at the earliest possible phases preceding a potential attack. A government's overall surveillance capability is also crucial in post-attack investigations in order to apprehend the perpetrators of such attacks and their associated networks in case they succeed in escaping from the incident's location.

Within the context of counter-terrorism, surveillance involves covertly monitoring the behaviors and activities of one's terrorist adversaries, whether in physical or cyber environments (including telecommunications), in order to gain "information dominance" over them, which will enable the counter-terrorist services (in a best case scenario) to disrupt their warfare-related pre-incident preparations and activities, including being able to identify their operatives and associates following an incident.

While surveillance, in the context of counter-terrorism, is crucial for governments to pre-emptively protect themselves against potential and actual threats posed by their terrorist adversaries, for terrorist operatives, in turn, their own surveillance activities are a means to enable them to prepare the battle-space for their forthcoming attacks. Since terrorist operatives, however, also confront the added hurdle of having their government adversaries surveilling their activities, they also need to develop the capability to counter-surveil and thereby evade detection by the watchful activities of their government adversaries in order for their covert operations to take their intended targets by surprise and thereby succeed in the attack, and, if possible, to escape from the scene of their attacks.

Counter-surveillance, therefore, is the method for detecting, avoiding and mitigating adversarial surveillance or making surveillance for one's adversary sufficiently difficult so as to disrupt such monitoring measures. A sub-component of counter-surveillance is inverse surveillance, in which the intended target of a surveillance activity uncovers it and, if possible, reverses the surveillance on the unsuspecting adversaries, who will be unaware that they are being surveiled.

Various methods and techniques of surveillance and counter-surveillance have been conducted by government security agencies and their terrorist adversaries throughout history (especially since the advent of modern terrorism in the nineteenth century), but in recent years the introduction of relatively easily obtainable highly sophisticated and wide reaching surveillance technologies has empowered the respective adversarial parties involved in surveillance and counter-surveillance with new tools of high scope and complexity in which to conduct such activities against each other, although traditional methods used over the years, such as human surveillance, continue to be employed.

Types of counter-surveillance methods required by terrorist groups to evade adversarial detection

To effectively counter the surveillance activities being waged against them, terrorist groups and their operatives need to develop the capabilities and technologies to overcome them. The following is a comprehensive listing of the types of physical and electronic surveillance activities and technologies that terrorist groups need to evade.[1]

Physical surveillance

Human surveillance

The physical surveillance of the movements of a terrorist adversary has been one of the first techniques employed by governments since the advent of modern terrorism in the nineteenth century. Terrorist operatives, in turn, need to evade such surveillance of their activities, whether through deploying operatives to surreptitiously follow their movements from a close distance or through infiltration (e.g., planting government agents in a terrorist organization or influencing members of an organization to act as informants by disclosing information they hold on the organization and its members). In response, terrorist operatives need to develop a counter-surveillance capability in these three areas that will enable them to function unimpeded in their intended areas of pre-incident operation.

Electronic surveillance

Identification and credentialing

With identification and credentialing documents used by governments to verify the identity of their citizens and those holding other types of residency permits, documents such as identity cards, passports, driver's licenses, social security numbers, banking and credit cards, are highly sought after by terrorist operatives and their associates to enable them to operate in these regulated environments, whether in their own regions or foreign countries. Like their criminal counterparts, terrorists have to overcome the hurdles presented by machine-readable identity documents that use encoded magnetic stripes or identification numbers to corroborate their identifying data, since these are used to create electronic trails when checked and scanned, which can be used by their adversaries to track their activities, movements and transactions.

Personal computers

Safeguarding personal computers is another important concern for terrorists' counter-surveillance. They must ensure these do not become a surveillance target because of operational-related data that may be stored in them. To do so, they must ensure that no unauthorized entity is able to physically or remotely penetrate them by surreptitiously installing bugging software on their computer systems or use other means to copy and mirror-image their data.

Internet communications

With broadband internet traffic (such as emails, web traffic, instant messaging) being monitored by governments, terrorist operatives need to develop appropriate countermeasures in response. In their email communications, for example, they need to come up with coded words or phrases that would not serve as "triggers" that alert those monitoring them that they may be planning an operation, avoid visiting certain types of websites of concern to those monitoring them that might trigger "red flags," or communicate via email or chat rooms with associates or militant groups that might link them to other networks that are being monitored and mapped (particularly via a government's social network analysis tools, as discussed below).

Telephone communications

Terrorist operatives need to adopt counter-surveillance methods to address telephone and VoIP (voice over internet protocol) communications (including calling histories) that are likely being monitored by their government adversaries, whether by human agents or speech-to-text software that is processed by automated call-analysis programs. Moreover, with cell phones monitored to collect geographical location data about the persons carrying them (even when a phone is not being used), mobile phones need to be obtained that are as untraceable as possible. As with emails, operatives need to avoid using certain language such as (for illustrative purposes) "the wedding is on," that might trigger red flags about potentially suspicious activities.

Surveillance devices

Terrorist operatives need to evade detection by surveillance devices, or "bugs." These include electronic devices that are surreptitiously planted to capture, record, and transmit data about an operative's locations, movements, and activities. Such devices can be planted in places such as an operative's residence, car, telephone or personal computer.

Social network analysis tools and databases

Although these are indirect surveillance methods, terrorist operatives need to evade surveillance by social networking analysis (SNA) tools and databases that create maps of their organizational-related networks and activities based on the collected data about them. In addition to intelligence collection-generated data, additional data about terrorists can be obtained from social networking sites such as Facebook, Twitter and YouTube. To evade detection as nodes, hubs or clusters in such social networks, terrorist organizations employ counter-surveillance techniques that have transformed their organizations from highly hierarchical to ones that are decentralized, leaderless, loosely affiliated and geographically dispersed as possible.

Biometric surveillance

Biometric surveillance technologies that measure and analyze an individual's physical and behavioral characteristics for authentication, identification, or screening purposes also require countermeasures by terrorist operatives. Examples of biometric surveillance indicators that terrorist operatives need to evade include fingerprints, DNA and facial patterns, while behavioral indicators they need to disguise might include gait (a person's manner of walking) and voice or speech.

Surveillance cameras

Surveillance cameras, including video recording devices, whether as single devices or centralized networks of CCTVs, are employed to observe and record an area for potential suspicious activity. Their capabilities range from being watched in real-time by security personnel to facial recognition software that scans a crowd at an event for potential terrorists who might be exhibiting pre-incident type nervousness, with such digital footage fed into searchable biometric-based databases that can be used to further track people's movements throughout a region, including identifying who they associate with. Even cameras used for traffic control can be used for general surveillance, for example, in indicating a suspicious car's travel movements and destination. The surveillance capabilities of such cameras need to be countered by terrorist operatives and their associates in order to enable them to maneuver unimpeded in their intended areas of pre-incident operation.

Aerial surveillance

Terrorist operatives also have to evade adversarial aerial surveillance in the form of visual imagery or video from an airborne vehicle—such as an unmanned aerial vehicle (UAV), helicopter or other types of aircraft. This is especially crucial in regions such as Pakistan and Yemen that are frequently targeted by a government's surveillance drones.

Satellite imagery

In addition to surveillance by UAVs and other types of imagery-capable aircraft, terrorist operatives in their areas of operations, whether in safe havens or other conflict zones, need to evade surveillance by high-altitude intelligence satellites that can penetrate cloud cover to detect their movements on the ground, the location of their bomb-making "factories" or laboratories (especially those that might leave chemical traces that can be picked up by satellite sensors), as well as identify "underground bunkers," that would make terrorists, their associates and their facilities vulnerable to attack.

Data mining and profiling

Terrorist operatives also need to evade data mining surveillance technologies that employ statistical techniques and programmatic algorithms to discover what might previously have been hidden relationships within the data about their transactions (whether financial, such as credit card purchases, or social, such as telephone calls and emails), with such information correlated in order to generate a profile of their pre-incident suspicious patterns and behaviors. Such data profiling about them can then be incorporated into social network maps about their networks and activities. The end result is an "electronic trail" that terrorists and their associates need to evade, whether by using stolen credit cards or not using any transaction-related documentation that can be digitized and traced electronically.

Assessing Al-Qaeda's counter-surveillance tradecraft

Throughout its existence, al-Qaeda top leaders and operational managers have been cognizant of the need to overcome their government adversaries' surveillance measures against them. This has taken several forms, beginning with the formulation of their early doctrine on counter-surveillance, which was outlined in their training manual, published in the 1990s.[2] In this manual, al-Qaeda operatives and their combat teams are instructed in countering surveillance methods such as "#1 Human Surveillance" by avoiding unnecessary attention and suspicion in their operations by obeying traffic laws, parking in inconspicuous locations, using false license plates and filling their cars' fuel tanks at small, inconspicuous gasoline stations. They are also instructed, particularly in countering surveillance measures "#10 Aerial Surveillance" and "#11 Satellite Imagery" to hide evidence of any physical training activity, with their trainees being informed of as few of their fellow trainees as possible to prevent them, in case of capture, from possibly exposing a training network.

According to the al-Qaeda training manual, within their combat cells, three types of members are given instructions in observing security maintenance that apply to their different roles within the organization. The first type, "overt" members, are instructed to be discreet in their conversations with others, avoid carrying the names and addresses of fellow members and limit their visits to areas that might arouse what they term "controversy." When using a telephone, for example, they are instructed to avoid mentioning any information that might "be of use to the enemy," with the same safeguards used in sending letters in the mail. The second type, "undercover members," are told to take even greater security precautions, including in their "general appearance … not [to indicate] Islamic orientation (beard, toothpick, book, [long] shirt, small Koran)." They are also instructed to "Avoid visiting famous Islamic places (mosques, libraries, Islamic fairs, etc.)." Most importantly, they must "Have complete and accurate knowledge of the security status related to those around him in his place of work and residence, so that no danger or harm would catch him unaware." In addition, they must avoid attracting attention to themselves by showing any changes in their daily habits, avoid getting into theological arguments, use coded words in their telephone conversations, attempt to detect any surveillance that might be conducted against them and avoid meeting in places where informers or government agents might be present. Finally, in the third type, a cell's "commander" must increase for himself all the security precautions being adopted by his cell members. In a final note, cell members who are married are reminded to avoid "talking with their wives about Jihad work," and not to "travel with their wives. A wife with an Islamic appearance (veil) attracts attention."

Cases of effective counter-surveillance

Illustrative examples of effective innovation in counter-surveillance techniques used by operatives belonging to al-Qaeda and its affiliates include the following case studies.

9/11 attacks against America

It is generally recognized that the effectiveness of al-Qaeda's counter-surveillance measures greatly contributed to the overall success of their team's 9/11 attacks. For example, in countering surveillance measure "#2 Identification and Credentialing" most of the operatives selected for these operations succeeded in obtaining visas to enter the United States and did not arouse suspicion during their flight school training (or even at their physical training at fitness clubs). In fact, Zacarias Moussaoui, allegedly the twentieth hijacker, who was the only operative to be arrested (on August 17, 2001), came to the attention of the American authorities for immigration violations, but he nevertheless succeeded in maintaining operational security by not revealing any details about the impending attacks or the existence and identity of his co-conspirators.

The 9/11 cell also succeeded in obtaining the necessary identity documentation that avoided leaving a digital trail. Mohamed Atta, the operation's leader, varied his name on such documents, using aliases and different spellings of his name such as "Mehan Atta," "Mohammad El Amir," "Muhammad Atta," "Mohamed El Sayed," "Mohamed Elsayed," "Muhammad al-Amir," "Awag Al Sayyid Atta," "Muhammad al-Amir" and

"Awad Al Sayad." While in Germany, Atta registered his name as "Mohamed el-Amir Awad el-Sayed Atta," and went by the name "Mohamed el-Amir" at the Technical University of Hamburg. Moreover, when he arrived in the United States, Atta used the spelling "Mohamed Atta" to evade detection. Finally, in his conversations with Americans, Atta also claimed affiliation with different nationalities, ranging from Egyptian to the United Arab Emirates, in order to disguise his origins.

In another successful 9/11-related measure to counter surveillance measure "#2 Identification and Credentialing," after using his regular passport to enter the United States, Khalid Muhammad Abdallah al-Mihdhar then used a forged passport to return to the United States on July 4, 2001 (via New York's John F. Kennedy International Airport). Later, on August 1, Mihdhar and Hani Hanjour drove to Virginia in order to obtain driver's licenses, for which they paid two Salvadoran immigrants to vouch for them as Virginia residents. With the notarized residency forms they were able to obtain their driver's licenses at a Virginia motor vehicle office. Three additional hijackers, Salem al-Hazmi, Majed Moqed, and Ziad Jarrah then used the same addresses gained from the Salvadorans to obtain their own Virginia State driver's licenses.

In countering surveillance measures "#4 Internet Communications" and "#5 Telephone Communications," the 9/11 conspirators also succeeded in using their own assigned code words in their electronic communications, whether via the internet or land telephones with one another in order to evade any possible detection. The U.S. Capitol, for example, was called "The Faculty of Law," the Pentagon was dubbed "The Faculty of Fine Arts" and the World Trade Center was termed "The Faculty of Town Planning." Moreover, Abu Abdul Rahman, one of the conspirators, sent a "love message" on an internet chat room to his "German girlfriend," who was actually Ramzi bin Shibh. The message said:

> The first semester commences in three weeks [how long it would be before the attacks]. Two high schools and two universities [two high schools refer to the two towers of the World Trade Center, and the universities refer to the Pentagon and the United States Capitol].... This summer will surely be hot ... 19 [the eventual number of hijackers] certificates for private education and four exams [the number of planes used]. Regards to the professor. Goodbye.

Finally, on September 10, Muhammad Atta telephoned Khaled Sheikh Muhammad, the operation's mastermind, who gave him a coded message to proceed with the operation on the following day.

In their effort to counter surveillance measure's "#2 Identification and Credentialing" of leaving a possible digital trail that they were part of a single plot, the 9/11 conspirators paid for their airline tickets separately, using different payment methods. One group, for example, paid for their tickets in cash, while another group made their reservations over the internet on sites such as Travelocity.

Despite these examples of al-Qaeda's effective counter-surveillance measures in 9/11, their brazen operation also succeeded because of a series of errors and inefficiencies by their government adversaries. In fact, prior to the crucial meeting of al-Qaeda's operational planners in Kuala Lumpur, Malaysia, from January 5 to January 8, 2000, that prepared the ground for the USS *Cole* bombing (which took place on October 12, 2000), and the September 11, 2001 attacks, the United States had reportedly intercepted a telephone call to Yemen by al-Mihdhar concerning arrangements for the trip. By request of the CIA, the meeting was videotaped by Malaysian authorities, but without any sound recordings. Thus, even though several of the operatives who attended the meeting were photographed, the "dots were not connected" and the meeting ended up raising no warning flags that major terrorist attacks were imminent.

London airliners plot

In what was known as the Summer 2006 London airliners plot, an al-Qaeda-affiliated cell in England had planned to detonate liquid explosives on board some ten airliners travelling from the United Kingdom to the United States and Canada. The plot was eventually discovered and foiled by British security services during its pre-incident phases. Nevertheless, the plotters were still able to conceal some of their operational activities, with one of the cell members, Dhiren Barot, reportedly trained in anti-surveillance techniques.[3] On one occasion, for example, to avoid possible surveillance by measure "#1 Human Surveillance,", Barot and a co-conspirator travelled 200 miles from London to Swansea to use an internet café to send emails before driving another 200 miles on their return trip.[4]

To avoid possible detection by surveillance measure "#4 Internet Communications," the cell members also created Yahoo email accounts in which they sent coded messages to each other, according to their trial's prosecutor.[5] He added: "These were written in the style of teenagers discussing music, television, etcetera, and using language and employing sexual references which would not normally be considered appropriate to devout Muslims."[6]

In another example of effective counter-surveillance, although all the plotters had mobile telephones, they rarely used them over a lengthy period, thereby countering surveillance measure "#5 Telephone Communications."[7] Moreover, when the members of the cell met, their meeting sites, in order to counter surveillance measure "#1 Human Surveillance" were "often deliberately chosen to make observation difficult, if not impossible," according to the prosecutor.[8] As a result, according to a witness, "They would meet in parks or reservoirs and walk around in the open air to have their discussions."[9]

Usama bin Laden and counter-surveillance

In another example of counter-surveillance awareness, some of the documents found in bin Laden's computer during the course of the raid against his compound in Abbottabad, Pakistan in the overnight hours of May 1–May 2, 2011, focused on the importance of maintaining vigilant counter-surveillance and security (particularly against surveillance measures "#4 Internet Communications" and "#5 Telephone Communications"), with those letters hand delivered by couriers to wherever it was safe to print or email them as attachments.[10] With regard to aerial surveillance of al-Qaeda operatives (measure "#10 Aerial Surveillance"), in his letters bin Laden advised his operatives "to stick to tree cover or to only venture out when there was dense cloud."[11] He even went so far as to urge al-Qaeda operatives in North Africa "to plant more trees to conceal them from all-seeing satellites and pilotless drones."[12]

Bin Laden was also concerned about the vulnerability of financial transactions by his couriers and money launderers, warning them, in countering surveillance measure "#6 Surveillance Devices," that they "should always be careful about [the possibility of] tiny electronic 'chips' concealed in the bags they used to move cash."[13]

In countering surveillance measures "#10 Aerial Surveillance" and "#11 Satellite Imagery," bin Laden also took precautions to safeguard and hide his own presence in his compound, with his residential quarters having virtually no windows, with a tarpaulin covering his movements around a vegetable patch, to protect against potential surveillance drones and satellites.[14]

Attesting to the difficulty of tracking bin Laden's whereabouts, it took several years of intensive and comprehensive multi-disciplinary surveillance for American counter-terrorism agencies to identify one of bin Laden's couriers, which paved the way for ultimately locating his hideaway in Abbottabad.

Cases of ineffective counter-surveillance

Despite such examples of effective and innovative counter-surveillance techniques used by al-Qaeda and its affiliates, however, there are also numerous examples of inefficiencies and failures, which have led to the prevention on several significant occasions of what could have turned out to be catastrophic bombing attacks by their operatives against their intended targets.

London airliners plot

Although, as noted earlier, it was effective on one level, with the plotters of the 2006 transatlantic aircraft operation maintaining effective operational security, their plot was eventually uncovered and foiled by British police before it could be carried out due to operational sloppiness by one of its operatives. It is reported that when Ahmed Ali, one the conspirators (who had already been under police surveillance), had returned from a visit to Pakistan (reportedly to consult with al-Qaeda operational planners), investigators, employing "#1 Human Surveillance," had secretly opened his baggage, where they found a powdered soft drink and a large number of batteries. This raised suspicions and led in the following weeks to what became one of the United Kingdom's largest surveillance operations, and ultimately resulted in the successful apprehension of the entire cell.

Najibullah Zazi's plot to bomb the NYC subway

Although Najibullah Zazi had succeeded in maintaining tight operational security from the start of his plot in late 2008 to bomb the New York City subway system around September 11, 2009, it is reported that on September 6, three days prior to the date of the intended attack, he had sent an email to an IP address that was flagged as an al-Qaeda mail drop associated with Rashid Rauf, a senior al-Qaeda bomb-maker, based in Pakistan (who was allegedly also behind the plot to blow up all the London airplanes in 2006)[15] – thereby failing to counter surveillance measure "#4 Internet Communications." When U.S. authorities discovered that the other IP address was located in Aurora, Colorado, outside Denver, they were able to connect it to Zazi, against whom they immediately employed surveillance measure "#1 Human Surveillance." Interestingly (and, again, failing to counter surveillance measure "#1 Human Surveillance"), Zazi was surreptitiously followed on his drive to Queens, New York. While staying with friends, Zazi had carelessly left his laptop in his parked car, which was immediately towed away by the FBI agents who were tracking him; they were able to copy the contents of his laptop, without Zazi aware of this when he retrieved his car from the pound. Such operational sloppiness by Zazi ended up revealing details about his plot to blow up the New York subways, resulting in the eventual apprehension and arrest of his cell.

The targeted assassinations of Al-Qaeda leaders

Inefficiencies in the way couriers were employed to transmit information and other operational services between al-Qaeda's top leaders and operational managers also led to uncovering vital intelligence about the whereabouts of the organization's top leaders, resulting in the killings of top leaders such as bin Laden, in Pakistan, on May 2, 2011 and Anwar al-Awlaki (together with Samir Khan), in Yemen, on September 30 of that year. In these cases, although both leaders had maintained effective counter-surveillance techniques, such as avoiding using mobile phones (surveillance measure "#5 Telephone Communications") that could reveal their locations, their reliance on couriers to deliver their messages and provide for other needs eventually led to their demise since they were left vulnerable to surveillance measure "#1 Human Surveillance." Most importantly for those attempting to identify their whereabouts, upon their potential detection, apprehension and interrogation, although "such 'gophers' would not necessarily be privy to inner circle discussions, [they] would nonetheless know the all-important details of where their masters were hiding out."[16]

In the case of al-Awlaki, despite his maintenance of tight internal security, his reliance on couriers ended up being a vulnerability, when, in a key breakthrough by Yemini security forces, working on behalf of the CIA, and employing surveillance measure "#1 Human Surveillance," they succeeded in apprehending a junior courier in al-Awlaki's inner circle, who proceeded to reveal details about Awlaki's whereabouts.[17] This information led to the successful drone strike killing against al-Awlaki's convoy (which included Samir Khan, his associate and the publisher of *Inspire* magazine) as it drove through a remote province in Yemen's mountainous region.

The impact on Al-Qaeda of Snowden's revelations on surveillance

In mid-2013, the revelations by Edward Snowden, a former contractor for the American National Security Agency (NSA), of the classified operational and technical surveillance measures employed by Western intelligence agencies against terrorist groups such as al-Qaeda, were expected to provide it with additional means to overcome them by changing the way they communicate with one another. As explained by Sir John Sawers, the head of British MI6, al-Qaeda was likely "rubbing its hands with glee" over such revelations detailing American and British surveillance programs around the world.[18] This was further elaborated by Matthew Olsen, the director of the American National Counterterrorism Center (NCTC), who stated:

> We have seen, in response to the Snowden leaks, al-Qaeda and affiliated groups seeking to change their tactics, looking to see what they can learn from what is in the press and seek to change how they communicate to avoid detection.[19]

Specifically, in the wake of such leaks, to make it more difficult and complex for its online and cellphone communications to be intercepted, al-Qaeda and its affiliates were reported to have developed new and stronger encryption software tools, following several years of employing "Mujahideen Secrets," a previous generation platform to encrypt emails, according to a report by Recorded Futures, an American firm that tracked such developments.[20]

Conclusions

At their best, the operations by al-Qaeda and its affiliates have been characterized by meticulous and long-term planning and preparation, including the crucial pre-operational surveillance of their potential targets. During its heyday, especially prior to 9/11, when al-Qaeda had established training camps in Afghanistan under Taliban patronage, its training programs, including counter-surveillance, had the luxury of uninterrupted and extensive periods in which to conduct such training unhindered by its government adversaries. As a result, their operatives were considered to be highly proficient, at least by terrorist training standards (which were still not comparable to that of soldiers in regular militaries), in basic and sophisticated surveillance techniques that might be used against them and those they needed to employ against their adversaries. Due to a number of developments that were implemented after 9/11, however, particularly their dispersal from their training camps in Afghanistan following the overthrow of the Taliban regime which had provided them safe haven, as well as the upgraded capability by its adversary governments' counterterrorism measures and special forces, including the escalation in targeted killing of many of its operatives, their training programs have become greatly curtailed and reduced, resulting in the degrading of their operational capabilities. Moreover, although al-Qaeda's operatives (whether as AQ Central in Pakistan or their affiliates around the world) have still managed to continue their attempts to conduct terrorist attacks against their adversary's homelands, such attacks have largely been characterized by substantially degraded operational efficiency and a higher rate of disruptions during the formative pre-incident phases.

In one of the latest developments affecting the counter-surveillance capability of al-Qaeda and its affiliates, their operational planners had become greatly concerned about the impact of measure "#10 Aerial Surveillance," particularly the need to evade surveillance (and targeted killing) by their government adversaries' unmanned aerial vehicles (UAVs). The drone campaigns' airstrikes, in particular, had forced al-Qaeda's operatives to take additional precautionary measures to limit their movements in countries where the drones operate, such as Pakistan, the Yemen and Somalia, with the organization reportedly distributing a "strategy guide" to its operatives "advising them how 'to anticipate and defeat' unmanned aircraft."[21] In response, al-Qaeda was reportedly conducting "research projects to develop jammers to interfere with GPS signals and infrared tags that drone operators rely to pinpoint missile targets."[22] It was reported that al-Qaeda was also attempting to develop "observation balloons and small radio-controlled aircraft, or hobby planes, which [they] saw as having potential for monitoring the flight patterns of U.S. drones."[23] In the absence of a technological solution, al-Qaeda operatives began sharing lessons about the importance of adopting basic defensive measures to avoid being identified by drones, including camouflage and "using radio or microwave transmitters to 'confuse the frequencies used to control the drone'."[24]

Despite such research projects by al-Qaeda and its affiliates to counter or neutralize the surveillance and killer drones, however, their "research and development" capability was considered as lacking "the technical knowledge to successfully integrate and deploy a counter-drone strike system."[25] However, with al-Qaeda-affiliated insurgents in Iraq able to use off-the-shelf software to hack into video feeds from some Predator and Shadow drones in 2009, seizing actual control of drones from their mission control ground stations began looming as a potential future possibility.[26] Attaining such a capability, although considered highly unlikely, would dramatically alter the capabilities of al-Qaeda and its affiliates' counter-surveillance tradecraft.

Finally, although it was always possible that the encryption upgrades introduced by al-Qaeda and its affiliates following Snowden's revelations might have been introduced in any case because al-Qaeda, as well as other terrorist groups, have always known about the widespread electronic surveillance capabilities of their government adversaries, in the "cat and mouse" game between terrorist groups and their government adversaries, terrorist groups are always aware of the need to innovate in their counter-surveillance tradecraft to make it more difficult for their stronger government adversaries to track and monitor their activities.

Notes

1 This listing is based on "Surveillance," http://en.wikipedia.org/wiki/Surveillance.
2 An English-language translation of the *Al-Qaeda Training Manual* is located at www.fas.org/irp/world/para/manualpart1.html,
3 "Al-Qaeda Briton used anti-surveillance techniques," November 6, 2006, http://metro.co.uk/2006/11/06/al-qaeda-briton-used-anti-surveillance-techniques-348359.
4 Ibid.
5 Ibid.
6 Ibid.
7 Ibid.
8 Ibid.
9 Ibid.
10 Michael Burleigh, "What the fruits of 'Neptune's Spear' reveal about Osama bin Laden and his management of al Qaeda," *MailOnline*, May 4, 2012, www.dailymail.co.uk/debate/article-2139492/Osama-bin-Laden-letters-The-fruits-Neptunes-Spear-revealed.html.
11 Ibid.
12 Ibid.
13 Ibid.
14 Ibid.
15 The information about Zazi's terrorist plot is based on Joshua Sinai, "Najibullah Zazi's Plot to Bomb the New York City Subway System: A Case Study of How U.S. Domestic Counterterrorism Operates," in James Forest, Russell Howard

and Joanne Moore, editors, *Homeland Security and Terrorism: Readings and Interpretations*, 2nd edition (New York: McGraw-Hill, March 2013), pp. 116–131.

16 Adam Baron, Majid al-Kibsi, Colin Freeman and Sean Rayment, "How America Finally Caught up with Anwar al-Awlaki," *The Telegraph*, October 2, 2011.

17 Ibid.

18 Lee Ferran and Marcus Wilford, "Al Qaeda 'Gleeful' over Snowden Leaks, MI6 Head Says," *ABC News*, November 7, 2013.

19 Thomas Joscelyn, "House Intel Chair: Snowden Leaks Tipped off Al Qaeda," *The Weekly Standard*, November 4, 2013.

20 Rachael King, "Report: Al Qaeda Tries New Encryption Post-Snowden Leaks," *The Wall Street Journal*, May 9, 2014.

21 Craig Whitlock and Barton Gellman, "U.S. Documents Detail al-Qaeda's Efforts to Fight Back against Drones," *The Washington Post*, September 3, 2013.

22 Ibid.

23 Ibid.

24 Ibid.

25 Ibid.

26 Ibid.

Part II

Case studies

9 Innovation and learning in the British jihad

Raffaello Pantucci

The purpose of this chapter is to explore learning in British jihadist plots since September 11, 2001 to see what evidence there is of innovation, information transfer, or learning in the conduct of operations, both failed and successful. By looking in detail at 22 plots or attempted plots between 2001 and 2013, the chapter will look at different stages in the terrorist plot cycle to try to understand what changes are discernable over time and whether these can be attributed to innovation or learning in some way. In concluding, the article will draw some thoughts across the spectrum of plots to try to track some broader trends and touch upon the direction the current terrorist planning cycle may be going.

Dataset

The first element to clarify is the choice of dataset. The author has identified some 22 plots since September 11, 2001 in which either a plot was carried out in some fashion or a well-developed outline for a plot was publicized as the reason why individuals were arrested or disrupted for carrying out an operation. This is by no means the entire spectrum of plots and attempted jihadist operations to have taken place in the United Kingdom since September 11, 2001 or before. In his testimony before the Parliamentary Intelligence and Security Committee, the head of the Secret Intelligence Service (SIS or MI6) Sir John Sawers stated that the security services had disrupted some 34 plots since the July 7, 2005 plot:

> if I go back to 2005, rather than ten years … 7/7 is that there have been 34 plots towards terrorism that have been disrupted in this country, at all sizes and stages. I have referred publicly and previously, and my predecessors have, to the fact that one or two of those were major plots aimed at mass casualty that have been attempted each year. Of that 34, most of them, the vast majority, have been disrupted by active detection and intervention by the Agencies and the police. One or two of them, a small number, have failed because they just failed. The plans did not come together. But the vast majority by intervention.[1]

Clearly the dataset identified for this chapter is smaller than this, but the focus here is on specific plots that have been publicly identified and about which the author was able to gather enough public information (mostly through court documents) to be able to identify a specific plot, methodology and intended (or actual) target in the United Kingdom. So, for example, the case of Rajib Karim, the British Airways IT engineer who was in contact with Anwar al Awlaki, is not included, as it is unclear that Karim was planning an attack himself rather than simply providing information and support for Awlaki's global ambitions. In an email exchange, Awlaki specifically says 'our highest priority is the US. Anything there, even on a smaller scale compared to what we may do in the UK, would be our choice.'[2] From what the author has seen, there was no evidence that Karim was planning an actual attack in the UK at time of arrest, though it was clear he was highly supportive of Awlaki's desire to use Karim from his position within British Airways to launch an attack, most likely towards the United States. Nor is the case of Rangzieb Ahmed and Habib Ahmed (not related) included – in that case the two men were part of a cell with connections to the highest echelons of al-Qaeda that had done some possible reconnaissance online about targets, but there was no publicly available evidence that they had developed a particularly mature plot at time of arrest. Rather, they stood accused of being involved in a network connected to al-Qaeda seeking to advance its goals more generally.[3]

Prior to September 11, 2001 there were also a number of plots that might now be included within a dataset of jihadist plots in the UK, due to the individuals involved being part of a broader community of radicals that made connections with al-Qaeda or other groups abroad or because of their targeting choice and reason given. For example, Faisal Mostafa was convicted for possessing a gun and cleared of planning to build a bomb even though explosive materials, timers and information relating to a visit to Manchester of the Israeli Ambassador to the UK. Mostafa was also in possession of material produced by Hizb ut Tahrir, an organization that at the time was still run by Omar Bakri Mohammed, who went on to found the more radical al-Muhajiroun.[4] Similarly, on December 17, 1999, Amer Mirza, an accountancy student at Humberside University who was an active member in al Muhajiroun[5] was arrested and jailed for six months after throwing a petrol bomb at a Territorial Army barracks in Southall, West London. In sentencing, Judge David Paget stated 'your actions were clearly prompted by the resumption of the bombing on Iraq' and the act was largely dismissed as a juvenile act of protest.[6] And finally, in November 2000 police in Birmingham arrested two men: the aforementioned Faisal Mostafa and Moinul Abedin. Ultimately, only Abedin was convicted of planning to cause an explosion with a variety of chemicals and enough ingredients to make some 40 grams of HMTD (hexamethylene triperoxide diamine), though found in his possession were a variety of al-Qaeda related manuals, and a note which said (in Abedin's hand): 'If I become a martyr, you should marry a 'MOZAHID' [*mujahid*] God willing.'[7] Abedin had in his possession bomb-making instructions that he had learned while at a camp in Afghanistan or Pakistan. Ultimately, Mostafa was cleared, while Abedin was sentenced to 20 years incarceration.

This final case was clearly seen in a threatening light and resulted in heavy sentencing, unlike the previous two which only seen from today have a far more menacing aspect to them. However, for the purposes of this chapter, these are not included as they took place prior to September 11, 2001 before the threat of al-Qaeda to the West had come into clear public focus, something that becomes significant in particular when considering learning and institutional transfer in the later years. As time went on and the idea of jihad and violent Islamist ideologies were more publicly accessible, the direct connection in knowledge and information transfer becomes more complicated, something more relevant within the British context given the rise of the phenomenon of lone actor terrorist plots. Furthermore, prior to September 11, 2001, the whole lexicon and understanding of jihadist terrorism was something that was relatively rare outside a specialized community, meaning that to analyze plots prior to 9/11 using current terms and analysis is imposing the present on an earlier and different time.

Finally, any plots included within the dataset needed to have an element of targeting within the UK, or at the very least an aspect of the active element of the plot cycle needed to take place within the United Kingdom. Consequently, plots like that undertaken by Taimour Abdalwahab al Abdaly in Stockholm is not included – even though he had a support member (or more) based in the United Kingdom.[8] And neither is Umar Farouk Abdulmutallab's AQAP (Al-Qaeda in the Arabian Peninsula) directed 'Underpants' bomb plot included as while his radicalization seems to have taken place in the United Kingdom, his operational cycle did not touch British shores. Finally, AQAP's printer bomb plot from 2012 that was detected in the UK is not included as not only was it not targeting the UK, but there was no clear British element publicly associated with the plot cycle within the UK.

Consequently, for the purposes of this chapter, the author uses the 22 plots listed in Table 9.1[9] as the basic dataset. Drawing on information from court documents, details reported in the media, gathered during the author's research, and other sources, the chapter will then look at each of these plots focusing on six specific phases or aspects of the plot cycle to look at understanding what trends and patterns are discernable over time. The six phases are as follows.

Recruitment

The recruitment phase of the plot is intended as focus on recruitment into al-Qaeda or an affiliate as well as in-cell recruitment (of course two different things, but grouped together here for simplicity to observe

Table 9.1 Plots used as dataset

Year	*Plot*	*Target*
2001	Shoe Bombers	Plot to detonate shoe bombs on transatlantic flights
2003	Ricin Plot	Plot to leave Ricin in public places in London
2004	Operation Crevice	Plot to detonate a large fertilizer-based bomb at a shopping mall outside London
2004	Operation Rhyme	Plot orchestrated by Hindu convert Dhiren Barot to undertake one of a number of plots using explosives in vehicles
2005	July 7 attack, Operation Theseus	Attack by four suicide bombers on London's public transport system
2005	July 21 attack, Operation Vivace	Attempted attack by five suicide bombers on London's public transport system
2005	Kazi Rahman	Purchased RPGs and Uzis from undercover officer to launch a spree attack in the UK
2006	Operation Overt	Plot to bring down up to eight airlines on transatlantic routes using liquid explosives
2006	Operation Gamble	Plot to kidnap and behead an off-duty British soldier and release the video on YouTube
2007	Operation Seagram	Attempted double VBIED attack in downtown London and next day attempted VBIED at Glasgow International Airport
2008	Nicky Reilly	Attempt to detonate an explosive at a Giraffe restaurant in Exeter
2008	Andrew Ibrahim	Attempt to detonate a suicide vest at Bristol shopping mall
2008	Jewel of Medina publisher's attack	Firebombing of publisher of Jewel of Medina's house
2009	Operation Pathway	Plot to undertake explosive attack in northern England
2010	Roshonara Choudhry	Stabbing of MP Stephen Timms
2010	Operation Guava	Plot to detonate explosive at the London Stock Exchange
2011	Mohammed and Shasta Khan	Plot to target Manchester and Oldham's Jewish community
2012	Operation Pitsford	Plot to launch unspecified suicide bombing
2012	Operation Alva	Plot to attack an English Defense League (EDL) using guns, knives and explosives
2012	Operation Movie	Plot to target march at Wootton Bassett
2012	Operation Nimrod	Plot to launch an attack using a remote control vehicle to bomb a Territorial Army barracks
2013	Woolwich	Michael Adebolajo and Michael Adebowale's murder of Lee Rigby with a car and knives

trends). In what way did the cell come together and what were the in-cell dynamics? The purpose of exploring this aspect is to see if patterns in recruitment change in specific active cells over time in response to security services or other exogenous factors. It also offers a look at the connection to al-Qaeda (if present) to see how this evolves over time.

Training

The training phase is intended to focus on what training took place – whether in terms of directly for the plot or more generally. What elements of the cell attended training camps at home or abroad, what was learned and how was the training undertaken? The purpose of exploring this aspect is to see the degree to which training impacts cell effectiveness as well as exploring how the patterns of getting training or undertaking training shift. Where possible it aims also to explain how these patterns change over time.

Communications

How did the group communicate with their al-Qaeda paymasters (if they had any)? How did they communicate with each other? What methods and tools did they use? Did they undertake clandestine methods of communication that they had learned or they choose to deploy in reaction to disruption of other cells?

Funding

How did the group raise money for their operation? How did they raise money in general? Were they paid by al-Qaeda? Did they deploy fraud or petty criminality and why did they make this choice if at all discernable?

Target choice and planning

What target did they choose and why did they choose it? What preparations and reconnaissance did they do in pursuit of the plot? Seen over time, is there a shift in targeting choices and is it possible to understand why this shift took place?

Weapons choice

What weapons did the group choose to carry out their plot? Did they use explosives, guns, knives, etc and how or why did they reach this decision? To what degree did they innovate themselves in their weapons choice versus an externally driven decision making or direction.

Analysis

Recruitment

Overall, there is very little variety in terms of recruitment across the 22 plots. What is seen is that those involved in plots tend to draw from broadly speaking the same pool of individuals, groups and preachers.

While after 2007 there is something of a shift in terms of people emerging from beyond this community, we still see elements in plots subsequent to this point emerge from what has been termed the 'jihadi soup' in the UK. This term is a shorthand for the body of individuals who tend to come from the UK's radical community, often associated with the radical group al-Muhajiroun (or whichever name it is currently operating under) or from the study circles around a select group of prominent radical preachers who have repeatedly surfaced in the background of terrorist plots in the United Kingdom. This is a community that has demonstrated strong and substantial connections to al-Qaeda, with repeated members of the community connecting with the group and training with the group. The question of recruitment is more a case of a connection that was established over time that provided a steady supply of willing individuals who were eager to train and ultimately take orders to try to launch attacks within the United Kingdom.

In fact, looking prior to the dataset assembled, it is easy to see that this same community is at least tenuously linked to the three plots highlighted prior to September 11, 2001. Faisal Mostafa was found in possession of Hizb ut Tahrir material in 1996,[10] Moinul Abedin was part of the radical community hovering around Abu Hamza and the Finsbury Park mosque,[11] and Amer Mirza was a prominent member of al-Muhajiroun who would regularly attend protests outside the US Embassy.[12] However, it is an interesting development that after 2007, there is the emergence of cells and plots that seem to emanate from beyond this British jihadi soup – in some cases, this is the lone actor plots (like Roshonara Choudhry, Andrew Ibrahim or the Khans), but it is also in the case of two larger plots. Specifically, Operation Seagram – Bilal Abdulla and Kafeel Ahmed's two-man campaign to bring Iraqi-style car bombs to UK city streets – and Operation Pathway in which a cell connected to al-Qaeda core was believed to be planning a campaign of terror in northern England. Neither demonstrated substantial links to the 'jihadi soup' though Bilal Abdullah did spend some time in Cambridge in the company of some locally based Hizb ut Tahrir members. There is no evidence that they provided him with links or contacts to the terrorist network he ultimately connected with.

Another notable aspect is that over time, the connection to al-Qaeda or an affiliate becomes more tenuous. In fact, by the time we reach 2010 only one of the plots demonstrates an operationally useful connection to al-Qaeda. In Operation Pitsford, Irfan Naseer was able to lead his cell to receive training from al-Qaeda-linked groups in Pakistan, as well as help others go to training camps there. Crucially, he was able to learn about how to make explosive devices in these camps and by his account was able to receive some direction about how to use this information. The Operation Movie and Nimrod plots were able to reach out to extremists abroad and had deep connections to radical elements in the UK, but they had difficulty operationalizing these in any useful way.

The explanation for these shifts is complicated and of course marred by numerous other exogenous factors that may impact the decision-making processes of individual cells. But broadly speaking, it could be interpreted that over time al-Qaeda and its affiliates became more wary of British plotters eager to connect with them in Pakistan, something that was no doubt exacerbated by the intense scrutiny the 'jihadi soup' community had come under. Evidence for this difficulty can be found outside the United Kingdom in the case of Mohammed Merah and documents found in Germany that showed al-Qaeda sought to train people quickly and send them back to undertake attacks undirected.[13]

In later cells, it is possible to see that individuals sought connections to al-Qaeda or affiliates, but had a great deal of difficulty actually achieving it. For example, Michael Adebolajo tried to go and join al-Shabaab, but was intercepted by Kenyan authorities. Richard Dart's co-conspirator Imran Mahmood had reportedly received some training at camps in Pakistan and tried to help Richard Dart make similar connections, but the white convert Dart encountered great difficulty doing it and made at least one failed trip to the country. Similarly, in Operation Nimrod, Zahid Iqbal had a contact in Pakistan who was believed to be in contact with terrorist networks and was referred to in court by his Security Service codename 'Modern Sleeve.' However, contact with this individual seems to have been sporadic, leading to the group at one point contemplating trying to perhaps explore opportunities of joining jihadist groups in North Africa.[14] In further evidence of the group's eagerness to grow its network, Mohammed Sharfaraz Ahmed was also accused of trying to recruit two cousins of his in Bradford who were later arrested and accused of plotting to go to a terrorist training camp.[15] There is little learning happening here, but we can see how cells in the United Kingdom find it harder to locate and operationalize contacts with al Qaeda.

Similarly, it is possible that one could interpret the emergence of the Operation Pathway cell as evidence of a plot that was specifically concocted from beyond the British 'jihadi soup.' Initially directed by British al-Qaeda planner Rashid Rauf, the plot was one of three disrupted that appeared to be part of a concerted effort by al-Qaeda to launch a wave of attacks in Europe and the United States.[16] The Pathway group demonstrated no known connections in the United Kingdom to the broader community of radicals, with the exception of their choice of staying with Hamza Shinwari, an individual who was already under the scrutiny of the security services.[17] It is possible that the construction of this cell was a conscious effort by al-Qaeda to reach beyond the traditional community of individuals they had used to launch attacks in the United Kingdom.

In many ways, the most interesting development or innovation across the cells is the emergence over time of lone actor plots – cells with no clear directional or other links to al-Qaeda or other radical elements in the United Kingdom. As time goes on, these plots become more common, suggesting that a direct link to al Qaeda (be it direction or simply contact) is not necessary to undertake a terrorist plot. In parallel to this trend, there is also something of a shrinking in the significance of the in-cell recruitment aspect. Looking at earlier cells involving multiple individuals, it is clear that there is an individual at the core of the group who is able to drive the plot forwards on the ground. He is not always the absolute leader, but he is clearly the magnetic center of the network.

For example, the July 7 group is clearly a case of Mohammed Siddique Khan (the older member of the cell) drawing in the three younger boys. Similarly, in Operation Overt, Abdulla Ali is the core who is able to go around and recruit a number of his former school friends into the cell. Reflecting the importance of these figures, in Rashid Rauf's post-facto analysis of a series of plots he claims to have directed he states: 'Zubair [presumed to be a pseudonym for Ali] was a very clever patient brother, and a natural leader.' This trend can be seen in some of the later cells – for example, Irfan Naseer is the beating heart of his cell, able to not only go and train himself, learning the crucial elements of making an effective detonator and seeking out the appropriate chemicals, but he is also able to recruit and facilitate others' travel to training camps.[18]

But his plot is the exception, and looking at the plots post-Operation Pathway, it is harder to distinguish this charismatic core to the cell. Instead, the cells appear to be groups of individuals with certain elements having more prominent aspects, but no one individual as the clear magnetic core.

In conclusion, it increasingly seems as though within the context of recruitment, there has been little shift over time that could be described as active innovation or learning by al-Qaeda or British cells. Cells are drawn to al-Qaeda in the same ways as before and seek out connections and recruit in similar ways. From al-Qaeda's perspective, there has been a lowered willingness to connect with cells linked to the known pool. Another trend is the lowered presence of core magnetic individuals who could be described as the drivers of the cell on the ground. This could correlate with a broader lowering in the caliber of individuals drawn to terrorist cells, something that has been anecdotally been referred to in the media.

But these trends have been matched by a growing tendency towards lone actor plots. The question of whether this is an innovation or demonstration of learning is a complicated one, as it suggests that the trend towards such plots is something that has a causal link to something al-Qaeda has done. It may be true that al-Qaeda and its affiliates have increasingly pushed groups towards the phenomenon of lone actor terrorism through their media propaganda, but it is not clear that one can draw a link between this and the acts of the identified lone actor cells. This difficult question is one that will run through much of the analysis in this chapter.

Training

Over time, there is a noticeable shift in training. All of the cells prior to Operation Overt demonstrate a strong and decisive training component to them involving al-Qaeda core. Admittedly, in some plots, like the Operation Crevice group, it seems as though the group sought the training from networks parallel to al-Qaeda, especially in preparing for the specific plot they were planning to undertake.[19] But the group nevertheless had a strong connection to al-Qaeda core, even if it is not entirely clear the degree to which they may have facilitated the training element of the cell. Since Overt, only the Operation Pathway and Operation Pitsford plots suggest a similar level of al-Qaeda core training. In the case of Operation Seagram, it seems likely that Bilal Abdulla connected with the insurgency in Iraq, but it is not clear that he received training there that was specifically intended to provide him with instruction to launch his intended attack back in London and Glasgow.

But while the level of al-Qaeda training might drop off substantially, the desire by cells to connect with al-Qaeda to receive training remains high. Consistently across the time period covered, individuals seek out training or undertake it themselves. It is a well-known fact that most plotters seek to go to the gym, undertake sporting exercises or other forms of physical training, and this remains consistent through the dataset. Another phenomenon that is notable across the dataset is the attempt to establish their own training camps and regimens. For example, in the Crevice cell, the group decided to establish its own training camp in Malakand where they would learn how to make explosives.[20] In 2010, the Operation Guava group planned to establish a training camp in Kashmir on the grounds of a piece of property owned by one of the cells members and to turn this into a location where British jihadists could go and train.[21] Given this is a habit to have recurred over time, it is not something that can be characterized as innovation, but rather is something that might be described as a reaction by the groups to an inability to connect with al-Qaeda at the time.

The exceptions to these are again the lone actor cells – specifically, Roshonara Choudhry and the Khans. In the cases of Andrew Ibrahim and Nicky Reilly, both attempted to make contact with other radicals and it is possible that this was part of an attempt to develop contacts to receive training. They were unsuccessful. In the other two cases (Choudhry and the Khans), there is no evidence that they tried to connect with specific groups or sought out training to help them in their intended plots (though it worth noting that it is believed that Mohammed Khan may have initially encountered the germ of radical ideas through a period in prison, and they may have appeared on the fringes of prominent and known radical communities). Again, in some ways, the lone actor cells could be characterized as an intense form of the self-established training camps. In these cases, the individuals sought out and obtained all the training and experience they needed by themselves, though it is not always clear that this was done in response to an inability to obtain proper training from al-Qaeda or other networks.

A complicated aspect that bears touching upon at this point is the question of *Inspire* magazine as a source of training issued by al-Qaeda (through its affiliate AQAP). Geared at providing aspirant jihadis with a one-stop shop to find ideology, training and direction, the magazine's intention is in many ways to create an online training manual that people can use in place of undertaking the perilous journey to connect with al-Qaeda or an affiliate. Taken in this way, it is possible to see that as the magazine emerges, growing numbers of cells appear who are both in possession of the publication, but also use it as a guide to build their devices. This is most clear in the Operation Guava and Khans' cells where individuals are arrested with components to build a bomb as indicated by the magazine, but it is also hinted at in the Operation Nimrod group where one of them proudly boasts of being in possession of all of the copies of the magazine. In reading and following the instructions, the groups are demonstrating some sort of remote learning or training suggesting that AQAP's intention in providing a surrogate training camp is somewhat successful.

Looking across the dataset, it is difficult to discern any clear evidence of learning or innovation in the cells covered. We can see as time goes on that individuals find it harder to make the connection with al-Qaeda and receive direct training from them, but what specific innovations in training that do occur – the emergence of *Inspire* magazine – is something that largely happens as a product of external al-Qaeda efforts rather than any learning within the British jihadi community.[22]

Communications

The communications aspect is one that appears more dynamic over time. What is difficult to ascertain is the degree to which this is due to learning from other cells experience or simply a reflection of shifts in technology usage publicly.

Similar to earlier trends in *Training*, there is a noticeable and sharp drop off in regular back and forth communication between al-Qaeda and the cells on the ground after the Operation Overt cell. The Operation Pathway group seem the exception to this, but the other cells that demonstrate a strong connection to al-Qaeda or an affiliate (Operation Seagram, Operation Pitsford and Operation Gamble) demonstrate no communication. Seagram and Gamble will be looked at in greater detail later, but the Pitsford group seems to be a model whereby al-Qaeda deployed a 'fire and forget' approach. In that case, the group received training and then seems to have been sent back with broad instructions, but no particular need to communicate back with the group in Pakistan. In this case, communications seem to have specifically been made unnecessary.

While in the cells that were able to connect with al-Qaeda it is possible to discern specific tactics and choices for methods of in-cell communication that were likely the product of instruction by the group – for example Dhiren Barot's propensity to drive to a different city to send an email, or to go to have conversations with colleagues in isolated places in the rain making surveillance all but impossible, or the Overt cells tendency to go into the middle of parks and lie down facing each other to have conversations. A consistent habit that is noticeable right up to the Pathway group is the use of email full of coded references – for example, Assad Sarwar's conversations with 'Paps' about the concentration of aftershave as a reference for hydrogen peroxide, or Abid Naseer's conversations about the women he was courting, which was also seen as a coded reference to chemical concentrations. These seem to be in-cell communication methods that they had learned or been taught by the core group.

In contrast, we can see how some of the later cells that were unable to make contact use their own innovative ways of communicating within the cell. The most interesting case study in this regard is the Operation Movie group – where Richard Dart and his colleagues used a number of methods to communicate that they thought were safe. For example, at one point the group was observed typing messages into smart phones and then holding them up to each other to talk, without ever sending the message. A similar method was that two individuals would sit in front of a computer with a word-processing document open and then just type into it a conversation that they would immediately delete. Unfortunately for them, this method was ultimately penetrable as a police forensic computer specialist was able to reconstruct the conversations, but the approach was an innovative one. Similarly, but in a more obvious way, one of the Nimrod men, Syed Hussain, was jailed for failing to disclose the password on a USB stick which he had encrypted using a complicated password to hide the radical material he had on it.[23]

In this way it seems fair to say that from a communications perspective British jihadi cells do show some evidence of learning and innovation. The Movie group is particularly instructive in this regard. The question, however, is the degree to which this learning is something that is based on learning from experience of others or whether this is simply a reflection of the increasing pervasiveness of technology. British jihadi networks and cells have repeatedly demonstrated themselves as being at the cutting edge of technological adoption – from the UK being home of the first prominent English-language jihadi websites and seeing their potential as ways of fundraising and recruiting, to the international network of cells disrupted in 2005 that stretched from Canada, to Georgia in the US, to Bosnia, Denmark and across the UK, all seemingly directed out of Bradford by Aabid Khan and his co-conspirator in London Younis Tsouli, also known as Irhabi007.[24] The question is not so much whether learning is taking place, but rather whether this is a reflection of the world that they live in and the tools that they have access to. The interesting thing about the community around Aabid Khan and Younis Tsouli was that they were predominantly a young group for whom an online life is second nature – a generational gap that was brought into sharp focus when the judge in the case asked for some detailed explanation about the internet.

The point is to what degree can this be considered innovation through learning, rather than innovation through broader technological advancement and easier access to technologies. Historically, it is clear that cells use methods that they have heard about elsewhere or that appear tried and tested in their security. For example, using 'dead letter' email accounts, using different phone boxes to make phone calls, or basic codes that appear to mask intent. In some cases, it seems likely that the cell was taught this basic tradecraft in Pakistan, but in others it is possible that they are simply picking up things that they have heard about.

Funding

Much like the majority of the trends identified so far, there has been little evidence of innovation or learning in fundraising. Plots and cells have traditionaly been self-funding in a variety of ways, something that has taken place consistently through the dataset. What has changed over time is al-Qaeda's support of cells. For example, it seems likely that al-Qaeda provided Abdullah Ali with some funding to help him procure the bomb factory the cell was using, and it seems similarly likely that Dhiren Barot was receiving funding from al-Qaeda to support his fairly expensive itinerant life style with no evidence of any particular employment. Whilst by no means a regular feature of the cases examined, over time this phenomenon almost completely disappears, with every plot after Overt clearly and seemingly solely self-funded – with in fact it appearing in many cases like they are raising money to send to the core group. There is some question around the Pathway group, but given the limited information available, there is little more that can be done.

Over time the willingness of communities around them to provide the groups with money easily becomes less evident. Most of the cells prior to Overt came from within a community that was willing to support them. This would include monies that were often given in the knowledge they were likely supporting a radical cause. Over time, this community fundraising becomes more discrete.

One response to this over time is the reliance of cells on fundraising for charitable causes and then redirecting the money for their own ends. We see this most clearly in Operation Gamble and Operation Pitsford groups, where both were involved in public fundraising and this was subsequently proved to be simply a front for raising funds for their own purposes. It is worth noting that both cells were operational in the same part of the United Kingdom – Birmingham – possibly suggestive of a community that was more likely to be willing to provide money towards charitable causes linked to the Muslim community. A possible aspect of learning visible here is that the groups in Birmingham realize that the community is generous to charitable causes and take advantage of them, something that offers a focus for security counter-efforts to heighten local awareness of such duplicitous charities.

A further trend that remains largely consistent over time is the use by the groups of creative ways of conducting fraud to raise funds, whether through fraud in obtaining delinquent loans for which there is no intention of paying back, charity fraud as highlighted above or through more complicated credit card frauds. The later Pitsford cell is further ambitious in trying to multiply their funding through the use of online forex trading, but this goes wrong and the group loses most of its money through bad trades.[25]

It is also notable over time how the amounts of money required for cells to conduct their attacks decreases. In earlier plots, money is required to obtain material, safe houses and more, but over time we see a trend towards less ambitious plots. Later plots rely on knives (Gamble, Roshonara Choudhry, Alva and the two Michaels in Woolwich), or homemade explosives produced in a relatively crude manner (Nicky Reilly, Andrew Ibrahim, the Jewel of Medina, Guava and the Khans, while Movie and Nimrod both seemed to have not yet established how they might assemble devices). All of these are much cheaper to assemble, meaning less fundraising is necessarily required.

Target choice and planning

One of the more obvious shifts over time is the move from directed and allocated plots to self-chosen ones. And as the plots become more self-chosen, the targeting choice becomes much more domestic in its expression. From cells like Rhyme, Overt or the July 7 group who all chose to launch incidents using a targeting package that would reflect an international terrorist networks preoccupations – transport networks, aviation or large buildings. We see over time a shift towards target choices like individual soldiers, less prominent politicians, book publishers and other communities in the UK, whether the Jewish community in Manchester or an English Defence League (EDL) march in Dewsbury.

The trend is a shift towards localized targets that is likely reflective of less command and control from abroad. In the plots directed from abroad, the key feature in targeting terms is the choice to launch an attack on a traditional prominent location where many people will be located and will cause lots of attention. Over time, with less command and control from abroad, the cells instead elect to choose their own targets and in doing this they aim for targets that in some cases are more reflective of issues they are aware of in the world around them rather than ones that would be immediately obvious to someone based abroad. A plotter sitting in Waziristan might be less able to see the significance of attacking EDL, for example, or would be less inclined to command a cell to launch an attack on a publisher.

But while there is shift in what the groups are choosing to target, there is little change in the choice of reconnaissance of potential targets. In most of the cells, there is some evidence of the group scoping out their target in some way prior to launching an attack, whether through going to see the target, seeking to do dry-runs of operations, or using online tools like Google maps or others to scope out an intended target. The methods deployed over time evolve, but these are more reflective of evolving technology rather than any innovation or institutional learning. In much the same way that we see later cells deploying new and creative ways of communicating, these are reflective of the cell simply using technological tools that are more easily available than to earlier cells.

Weapons choice

It is interesting to observe that the groups remain consistent in their desire to use explosives. Across the dataset, the cells remain fixated on building bombs and using them. Of the identified plots, only four do not use explosives in their targeting and in one case, the Alva cell who were planning on attacking an EDL march, the group was actually going to use a mix of knives, explosives and guns. Of the four that did not use explosives, three chose knives (Gamble, Roshonara Choudhry and the Two Michaels in Woolwich), while one wanted to obtain heavy weaponry like guns and rocket propelled grenades.

There are two interesting aspects to this. In the first instance the choice of explosives demonstrates a persistent trend that is not only reflective of al-Qaeda core's direction, but also the groups themselves. There is clearly a fixation with trying to make things go bang, likely a reflection of the larger impact an explosion is going to have, but at the same time this complicates matters for the cell as making a viable explosive without training can be difficult.

None of this is to say that there is no shift or noticeable innovation across the dataset, and in particular the choice of construction of the explosives. For example, Richard Reid and Saajid Badat's use of explosives carefully crafted into their shoes, the Overt group's use of liquid explosives made of concentrated hydrogen peroxide carefully injected into bottles to not break the seals, or the Pitsford cell's use of sports ice packs as sources of chemicals to build bombs. In all of these cases, however, these innovations are not ones that can be described as indigenous to the cell, rather they are practical innovations developed by al-Qaeda core that are taught to the group. In other words, it is not clear that these innovations are generated by individuals from Britain's jihadist community as opposed to foreign al-Qaeda bomb-makers.

This trend towards external innovation is in some ways also reflected in the other cells bomb-making efforts. Looking at the cells to be detained in the wake of the first emergence of *Inspire* magazine, it seems clear that the group was influential in their planning. The Guava cell were arrested as they downloaded *Inspire* seeking to use it to learn how to make a device, the Khans had assembled a number of the elements necessary to build a bomb according to the magazine's instructions, the Alva group had built a pipe bomb similar to those indicated in the magazine, the Pitsford group were recorded talking about emulating some of the tactics discussed in the magazine, and the Nimrod group boasted about having copies of the magazine and their plotting seemed to reflect the publication.

In institutional learning and innovation terms, it is clear that the biggest tactical innovations came from outside, with al-Qaeda either directly teaching the cells or using its publication *Inspire* to pass along training and knowledge. But in what could be described as a feedback loop, it is noticeable that in *Inspire* magazine as well as other media issued by the group and its affiliates they try to harness the main innovation that we see emerge from within the British jihad at home rather than abroad. Specifically, Roshonara Choudhry's attempted killing of Stephen Timms and the Two Michaels' murder of Lee Rigby. Both of these incidents are venerated and highlighted by, respectively, *Inspire* magazine and al-Shabaab who use them as examples that others should emulate.

Conclusions

There are some broader trends that appear observable but are hard to categorize. For example, there seems to be a shift in terms of cell's professionalism. In early plots it is possible to see individuals using quite sophisticated tradecraft, either in obtaining false documentation or in masking their true identities, or (as shown in the communications section) using quite effective methods to mask conversations from observers. In later plots, however, there is a lack of such professionalism – in some cases, like the Khans', there is some evidence of them trying to use professional tools by buying things at Spy Shops, but there is little evidence of them being actually used.

A public trope that has been often advanced in discussions about the British jihad is a tendency towards incompetence and low quality individuals being drawn into cells. In some ways, it could be said that this lack of professionalism is a reflection of this. But the reality is that incompetence is something that can be found across the dataset, as well as individuals of questionable caliber. For example, Dhiren Barot may have been a smart and dedicated individual who was able to write a rather dull book, but it is not always clear what he was doing and whether it was going to lead to anything. When he was captured quite detailed plots were attributed to him, but there was no evidence that he had done anything in his almost decade-long career as an international jihadist. He had travelled the world, met with some of the most senior figures in al-Qaeda, but at the same time the plots he had concocted seem of limited feasibility. Similarly, Saajid Badat was a thoughtful and intelligent individual, while Richard Reid was so shambolic that when Khalid Sheikh Mohammed first came across him he was uncertain what to do with him. In later cells, similar blends can be found: Irfan Naseer was clearly quite intelligent and able to retain the designs for sophisticated detonators, but at the same time it was not clear that some of the other members of his cell were as adept as him.

Overall, there is little evidence of institutional learning within British cells in the UK. Where there is innovation in the groups, for example in terms of types of explosives they use, it is clearly something that they have been taught from outside rather than something indigenous. The one interesting element that could be characterized as an innovation to have emerged domestically is the growing incidence of lone actor terrorists and of individuals using knives to attack less prominent, but symbolic individual targets, an innovation that a number of al-Qaeda affiliates have seen take place and highlighted as something they both approve of and think others should emulate. This can be seen in *Inspire* magazine's praising of Roshonara Choudhry's action or in al-Shabaab's video 'It's an Eye for an Eye: The Woolwich Attack' that praises Michael Adebolajo and Michael Adebowale's brutal murder of Lee Rigby. In both outlets, the groups praise the acts and call on others to copy this approach.

Looking at recruitment in particular, there is little evidence of changes over time, though there is a growing instance of plots emerging beyond the known networks of individuals in the 'jihadi soup' that exists in the UK. There is further a lowering in cells' capacity to connect with the al-Qaeda core, something reflective of the groups inability to make itself available to prospective recruits. For example, in the Movie case, Richard Dart's co-conspirator Imran Mahmood makes a connection in Pakistan, but cannot help Dart get to any camps. In the case of the Nimrod cell, Zahid Iqbal has a connection with al-Qaeda, but loses contact at a certain point and the group starts to consider alternative places to connect with al-Qaeda affiliates. Even in the cell that is able to connect with al-Qaeda, the Pitsford group, it is not clear that they are actually able to maintain this contact as they carry out their plot in the same way that earlier cells like the July 7 group or the Overt cell.

It is also noticeable that the quality of training goes down over time. Fewer demonstrate sophisticated operational security or evidence of having learned this abroad. They are also unable to get training in foreign training camps in the same manner any more – earlier cells are able to connect with al-Qaeda core and receive training tailored to their specific plots. Rather groups instead seek to develop their own training regimes, either through development of their own camps or finding ways of training in the countryside in the UK. None of this, however, lessens the eagerness of cells to try to receive proper training and they continue to seek it throughout the dataset though success rates clearly decrease.

Cell methods of communication change over time, but it is uncertain the degree to which this is due to outside direction or training. While the early cells that get to training camps are taught methods of communication there, the later ones innovate their methods of communication and contact by themselves. However, it if difficult to know whether this can be considered innovation, rather than simply the groups adapting to the world around them and using technology that has become more common since previous plots in their planning.

There is a shift in fundraising in terms of cells no longer receiving funds from abroad to support their activities. In early cells, we can see al-Qaeda investing money into plotters and cells, something that ceases over time. Fundraising through fake charities or fraud remains consistent throughout, and in at least two linked cells it is possible to see cell members talking to each other about their methods, something that could be described as institutional learning between the cells.

In many ways the most interesting innovation is that of targeting choices. As the cells becomes more detached from al-Qaeda, their targeting becomes much more local. Rather than targeting choices that reflect a global al-Qaeda preference of attacking transport, airlines or large economically significant buildings, the cells choose local targets like other communities or specific locally prominent or significant individuals. Again, it is difficult to know whether to characterize this as learning, an absence of direction or more simply an adaption to the more local context that the entire cell is operating within.

Ultimately, there is very little evidence of idea transfer, innovation or learning from cell to cell. Even amongst connected cells where they seem to pursue different targeting choices, weapons choices even though their information and ideology comes from the same place. Inasmuch as there is some evidence of institutional learning and innovation, it seems to be retained within al-Qaeda rather than the cells, whether reflected in the types of devices cells try to assemble or *Inspire* magazine trying to highlight specific effective tactics as something others should emulate.

However, within this, there are in some ways the most significant innovation to have emerged indigenously within Britain's jihadi community, that is the phenomenon of lone actor terrorism. While the UK is by no means the only country to suffer from the phenomenon, nor home to the most successful ones, some of the earliest and most prominent ones occurred there. And they have inspired emulation from others, at home and abroad, something that demonstrates a certain amount of innovation and learning.

A final note to touch upon is the future. Many of the trends highlighted here suggest entropy in Britain's jihad. Specifically, the degradation in ability to connect with al-Qaeda or an affiliate effectively, the ability to access training camps, the ability to maintain communication with leaders abroad, the lowering caliber of individuals involved. All of these aspects may seem on a downward trajectory at the moment, but there is a noticeable potential that they might abruptly turn around in the wake of Syria – a live jihadist battlefield that provides willing British recruits with a place in which they can remedy all of these (the caliber of individuals is of course impacted by other reasons, but given the numbers going to Syria, it is impossible that at least one or two effective operators do not emerge). How this battlefield might impact future trends in innovation and learning is unclear at this early stage, but given the preliminary echoes that people might already be returning home with direction the concern is that many of these entropic trends might be on the cusp of being overturned. Should this happen, the wave of plots that are outlined in this chapter might end up repeating themselves, though now with the further fuel of the viability and permissibility of the lone actor attack existing in the common consciousness as well as *Inspire* magazine and other al-Qaeda affiliates individuals might like to connect with.

Notes

1 Sir John Sawers, uncorrected transcript before the Parliamentary Intelligence Committee, November 7, 2013: https://b1cba9b3-a-5e6631fd-s-sites.googlegroups.com/a/independent.gov.uk/isc/files/20131107_ISC_uncorrected_transcript.pdf.

2 Raffaello Pantucci, 'Al Awlaki recruits Bangladeshi militants for strike on the United States,' *Terrorism Monitor*, vol. 7, no. 7, February 17, 2011.

3 *Regina* v. *Ragzieb Ahmed and Habib Ahmed*, judgment handed down February 25, 2011, [2011] EWCA Crim 184 case no: 200900336 B5 200900496 B5.

4 Steve Panter, 'Muslim cleared of bombing plot jailed for having gun,' *Manchester Evening News*, March 28, 1996; *Regina* v. *Faisal Mostafa*, judgment handed down November 7, 1996, [1996] EWCA Crim 1335, no. 96/2762/X2.

5 Al-Muhajiroun (translated as 'the emigrants') is a prominent British extremist organization originally founded by Omar Bakri Mohammed. It is a proscribed organization in the UK, though its members continue to be quite active.

6 'Iraq protest student jailed for petrol bomb attack,' *Press Association*, March 5, 1999.

7 'Bomb maker jailed for 20 years,' *BBC News*, February 27, 2002; *Regina* v. *Moinul Abedin*, judgment handed down May 14, 2004, [2004] EWCA Crim 2232, no. 02/1794/B4.

8 High Court of Justiciary at Glasgow, indictment of Nasserine Menni by Her Majesty's Advocate, November 16, 2011.

9 In the interests of brevity they are listed here by Operational names or distinguishing features with a brief summary of their target.

10 Panter, 'Muslim cleared of bombing plot jailed for having gun.'

11 'Jobless waiter who conspired to unleash hell on families,' *Birmingham Post*, August 14, 2006.

12 'Iraq protest student jailed for petrol bomb attack,' *Press Association*, March 5, 1999.

13 Paul Cruickshank and Tim Lister, 'How did Mohammed Merah become a jihadist?' *CNN*, March 26, 2012, http://edition.cnn.com/2012/03/26/world/europe/france-shooting-suspect.
14 *Regina* v. *Zahid Iqbal et al.*, April 15, 2013.
15 'Bradford Brothers on terrorism camp charges to face retrial,' *Asian Image*, January 8, 2014.
16 The other two plots were Najibullah Zazi's plot to attack the New York subway and Mikael Davud's half-baked effort to undertake some sort of an incident in Oslo. For more on this network, see Raffaello Pantucci, 'Manchester, New York and Oslo: Three Centrally Directed al-Qaeda plots,' *CTC Sentinel*, August 1, 2010.
17 Ibid.
18 *Regina* v. *Irfan Naseer, Irfan Khalid and Ashik Ali*, Opening Statement, September 19, 2012.
19 *Regina* v. *Omar Khyam et al.*, March 23, 2006.
20 *Regina* v. *Omar Khyam et al.*, March 23, 2006.
21 *Regina* v. *Usman Khan et al.*, Court of Appeals, April 16, 2013.
22 There is a wrinkle that bears mentioning. At least one British individual (Minh Quang Pham), connected to more than one plot identified here, has also been connected to the production of *Inspire* magazine. He might have been involved in authoring the specific training aspects of the magazine. But even if this were the case, the magazine was still being produced under the direction of AQAP meaning that it remains primarily an external innovation rather than one internal to the British jihad at home.
23 Sophie Curtis, 'Man jailed for refusing to divulge USB password,' *Telegraph*, January 15, 2014.
24 Raffaello Pantucci, 'Operation Praline: The Realization of al-Suri's Nizam, la Tanzim?' *Perspectives on Terrorism*, vol. 2, no. 12, 2008.
25 The question was a complicated one for the group who went on to demand the individual who had lost the money sell his car to pay them back. But at the same time, the group realized they could not demand this given that they needed the vehicle to get around.

10 Terrorists lack imagination when choosing targets in Denmark

Morten Skjoldager

When a team of police officers raided a house in a suburb of the Bosnian capital Sarajevo on October 19, 2005, they stumbled upon a terrorist safe house with all the ingredients for a brutal terrorist attack; the white-painted villa near the airport of Sarajevo was packed with explosives, weapons and a suicide belt. But the policemen didn't just find weapons and explosives; they also caught two young men red-handed inside the house. One of the suspects was Abdulkadir Cesur, a 21-year-old Turkish citizen who was born and raised in Copenhagen, Denmark. When he was arrested in Sarajevo he immediately became known as the first Danish-born Muslim who had turned into a terrorist and in Denmark the news of the crackdown on a Scandinavian terrorist cell was received with shock and surprise.

As traditionally Denmark is characterized as a peaceful country with a long democratic tradition there were very few Danes who considered their country a target for Islamic militants in 2005. This sense of immunity has changed dramatically since then; today most Danes have become used to the fact that Islamic militants are constantly planning to attack Denmark.

During the last eight years the Danish public has witnessed an unusually high number of foiled terrorist plots that far exceeds the modest size of the kingdom. Because of Denmark's involvement in the two wars in Iraq and Afghanistan and – probably even more so – the *Cartoon Crisis*, which started when the Danish newspaper *Jyllands-Posten* published 12 cartoons of the prophet Mohammed – Denmark has become a priority target for Islamic terrorists. Since 2005 several terrorist cases have appeared resulting in court cases and both Danish and foreign terrorists have been convicted of planning terrorist attacks against Danish targets.

This chapter describes a selection of Danish terrorism cases and argues that Denmark in many ways can be seen as a microcosm of broader trends in terrorism: among the terrorists convicted of planning attacks against Denmark have been home-grown terrorists, well trained terrorists with international connections, foreign extremists traveling to Denmark and solo terrorists plotting simple attacks. The chapter examines whether the methods and skills used by terrorists have developed over time and shows that the terrorists have actually been innovative when it comes to changing strategies and methods – but that they at the same time have demonstrated a profound lack of imagination when choosing targets.

Home-grown amateurs

As mentioned previously the first Danish terrorism case from 2005 started far from Denmark, in a depressing suburban house near the airport in the Bosnian capital of Sarajevo.

Bosnian authorities had received information that a young Swedish man by the name of Mirsad Bektasevic – better known by his online handle "Maximus" – had come to Bosnia to receive military training in the mountains outside Sarajevo. But when police officers arrived at the house in the afternoon to arrest the Swede, to their surprise they also found another young man inside the house. This was Abdulkadir Cesur from Denmark, who was sitting in a couch in the living room holding a gun in his left hand and who had to be overpowered by a policeman.

When the police team started searching the house they quickly realized that the two young Scandinavians were not just in Sarajevo to get military training in the mountains. In the house the men had been hiding 19.8 kilograms of explosives that was cut into smaller chunks, pressed flat and then stored in a freezer. The Bosnian police also found weapons, ammunition, balaclava hats with holes for the eyes, handheld walkie-talkies and a kitchen timer that could be used as a detonator for a bomb. In the house was also found a home-made suicide belt with designated holes for explosives and parts of the explosives already prepared for the belt.

The investigation revealed that the two Scandinavians had bought the explosives with help from local Bosnian extremists who were also arrested. Bosnian authorities, however, never managed to find out what targets the terrorist cell had planned to attack, but a video film found in Maximus' pocket clearly indicated that the group had been in the final stages of an attack.

In the central part of the film you could see two men dressed in camouflage gear with their faces covered. One of them was showing an AK-47 assault rifle and the other was holding a gun in his hand and had a suicide belt around his waist. The floor of the small room, where they were sitting, was literally covered with explosives, cartridges, grenades, guns and knives. The video showed far more weapons than the Bosnian authorities ever found.

"The brothers are ready to attack. They are showing us the things they will use in the attack", said a voice belonging to Maximus, who was behind the camera.

In Denmark the security service, PET (Politiets Efterretningstjeneste), already knew both Maximus and Abdulkadir Cesur before they were arrested in Sarajevo with all the ingredients for a terrorist attack. PET had been watching Abdulkadir Cesur for more than a year, since the Turk had travelled to London in the summer of 2004 together with 11 other Danish Muslims. The group was going on a study trip to attend a conference where the keynote speaker was the notorious "preacher of hate" Omar Bakri. Abdulkadir Cesur was denied entry to the UK because, as a Turkish citizen, he needed a visa. But the other Danish Muslims spent a week in London where they attended the conference and met Omar Bakri in person.

In Denmark PET noticed that a group of young Muslims had traveled to London to form bonds with Omar Bakri and his protégés. PET subsequently decided to monitor the group – and more than half of the men involved in the journey to London in 2004 have since been subject to police investigation in relation to terrorism. The man who organized the trip was killed in September 2013 in Syria where he was the leader of a group of Danish Muslims fighting against the Assad regime.[1]

Back in 2004 and 2005 surveillance was concentrated on a group of very young Muslims. Most of them were born and raised in Denmark and their families originated from various different countries. They came from different backgrounds; some were brought up in very religious homes while others came from families where religion played no role. Some of them could be described as well integrated while others were unemployed and uneducated. The only thing that the young men all shared was an intense interest in radical Islam and conflicts involving Muslims around the world.

When Abdulkadir Cesur and Maximus were arrested in Sarajevo it came as a surprise to PET that they had come so far in the planning stages. PET suspected that other members of the group in Denmark had been involved in the preparations in Sarajevo and the Secret Service was also afraid that the group was planning something similar in Denmark. A week after the arrests in Sarajevo PET therefore decided to arrest four young men who were charged in the so called Glostrup case: the first case in which anybody was convicted of terrorism in Denmark.

The four men in custody were between 15 and 20 years of age and the trial against them left little doubt that they all shared a radical perception of Islam and that they were inspired by al-Qaeda. They frequently exchanged radical texts, jihadi videos and other extremist material online, and on their computers police found downloaded manuals with recipes for explosives and poisons. There was, however, not much solid evidence that they had been engaged in illegal activities and when the court presented its final verdict only one out of four defendants was found guilty.

That was a boy by the name of Abdul Basit Abu-Lifa who was only 15 years old when he was arrested. One of the reasons that he, in contrast to the others, was convicted was that PET had recorded a phone call where he was talking with his older brother and asked to borrow his backpack. Apparently the original plan of the group had been that the 15-year-old boy was supposed to travel to Sarajevo – that was why he needed a backpack. But Abdul Basit Abu-Lifa's older brother got suspicious when he heard about the younger brother's plans to travel abroad. He went to their father and told him that the young kid was planning to carry out a martyr attack. The father then confiscated his son's passport so that he couldn't travel and Abdulkadir Cesur went to Sarajevo as a replacement. Ironically, it was partly the attempts from a father trying to stop his son from getting involved in terrorism that ended up with the son being convicted of terrorism.

There was, however, other evidence against Abdul Basit Abu-Lifa. The prosecutor managed to connect him to one very important piece of evidence: a document which was published on September 11, 2005 – the fourth anniversary of the attacks against the US. In a manifesto published on the extremist Ansaar forum a small Scandinavian terrorist cell announced the foundation of "al-Qaeda in Northern Europe" and declared that the newly formed group was ready to carry out attacks in revenge for humiliation of Muslims around the world. Referring to the 7/7 terrorist attack in London earlier the same year, the manifesto said:

> We in the organization al-Qaeda in Northern Europe declare that Muslim blood that has been spilled and continues to be spilled should not be wasted, and we swear by Allah, who has raised the sky without support, to give our souls and bodies like soldiers of the religion and raise the flags of the holy warriors in protecting the souls of Muslims, like our brothers, the holy warriors in the UK.

The investigation showed that it was Abdul Basit Abu-Lifa who had written the manifest of "al-Qaeda in Northern Europe" on his computer and he was found guilty of aiding the terrorist cell in Sarajevo. In Bosnia Maximus, Abdulkadir Cesur and local extremists also received long prison sentences.

That meant that several terrorists were convicted in different countries in relation to the first Danish terrorism case. And although it was never revealed where and when the terrorist cell had planned to carry out the attack – whether it was supposed to have taken place in Denmark, Bosnia or a third place – there was no doubt that authorities had prevented an attack. The young men had managed to get hold of weapons, explosives and a suicide belt and they were apprehended very late in the planning stages.

On the other hand there were many things in the investigation that indicated that the members of the terrorist cell were definitely not well trained terrorists and that the fact that they had got so far before the authorities stepped in was more a question of coincidence and luck, than a result of skills. There was nothing in the investigation that indicated that the first convicted Danish terrorists had any connection to foreign terrorist networks or that they had received any kind of training. This was also obvious in the group's behavior.

In the Muslim community in Copenhagen the young men attracted some attention because of their not very subtle behavior which made it almost impossible for PET not to notice them. Evidence presented in court also showed that the members of the cell had been well aware that the Secret Service was watching them. They tried to hide what they were planning but their attempts were limited to talking in primitive coded language on their phones. At the same time they wrote emails and communicated quite openly on chat forums, apparently unaware that they were also being monitored online, thus revealing information about the plans. They were typical home-grown terrorists.

The Cartoon Crises

To understand why the threat against Denmark has increased dramatically since the first terrorism case, it is necessary to mention the Cartoon Crises; these initially started on September 30, 2005 when the Danish newspaper *Jyllands-Posten* published 12 drawings of the prophet Mohammed and thereby offended Muslims around the world.

At that time Denmark was most likely already considered a legitimate target among terrorists because of Danish involvement in the wars in Iraq and Afghanistan, but there is no doubt that the Cartoon Crises played – and still play – a very important role when trying to understand the desire among Islamic militants to attack Denmark. It was the Cartoon Crises that once and for all put Denmark on the terrorist world map.

In the investigation of the Glostrup case PET had picked up a few phone calls where the suspects were talking about the cartoons, but they were arrested before the issue had developed into an international crisis. In the first months after the publication the reaction in the Muslim world was calm – but in the beginning of 2006 the situation spiraled out of control and became one of the worst foreign policy crises in modern Danish history. Danish embassies were attacked by angry crowds and Muslims around the world suddenly associated Denmark with offensive drawings of their Prophet. This firestorm of protests enveloped Danish foreign policy and mass media attention.

In Denmark there were of course also Muslims who were angry about the cartoons, and among them were extremists who were ready to retaliate by committing a terrorist attack. The second Danish terrorist group was caught in the summer of 2006 in Vollsmose, a suburb of the third-largest Danish city of Odense, where the group was hiding fertilizer and a small portion of homemade TATP (acetone peroxide). The leading member of the terrorist cell was clearly motivated by a desire for revenge and the Vollsmose case was the first example of Danish Muslims who were willing to use terrorism in order to retaliate against the cartoons.

The Vollsmose case, however, added litttle new to the profile of a typical Danish terrorist. The surveillance of the group showed that the three cell members had neither been to training camps nor had any guidance from foreign terrorist networks. It was yet another example of home-grown terrorists trying to hide their digital trails in amateurish ways while struggling hard to find a way to get access to explosives (the Vollsmose group at one point got so desperate that they even tried to use goat excrement from a halal slaughterhouse as bomb-making material). The group showed no sign of learning or innovation compared to the first Danish terrorism case: it was another example of young extremists trying to get hold of a bomb to blow up something – but lacking the skills and connections to make it happen.

It was however far from the last terrorism plot motivated by the Cartoon Crises and the next Danish case would show that it was not only local Islamic militants who were determined to do something to revenge the *Jyllands-Posten* cartoons.

Training camp

On March 31, 2007 Hammad Khürshid landed in Copenhagen Airport after a long vacation in Pakistan. The 21-year-old Danish citizen had Pakistani roots and had spent two and a half months in the country. Officially he had been visiting family but PET had received information from a foreign intelligence service that the young man had also spent time in an al-Qaeda training camp in Waziristan. PET therefore made sure that Hammad Khürshid was delayed in passport control in Copenhagen Airport so that agents had time to go through Khürshid's luggage without his knowledge.

In a suitcase PET discovered an envelope with the title "Marriage Contract" written on the front – but Hammad Khürshid had not been married during his stay in Pakistan. Inside the envelope he had tried to hide a 21-page-long manual with handwritten notes. The manual contained accurate and useful recipes for making various kinds of explosives and detonators, passages where it was explained how to produce poison and a short paragraph explaining how to make a suicide belt. One of the bomb recipes in the manual showed how to use a three-volt light bulb as a detonator for a bomb, and in his suitcase Hammad Khürshid had tried to hide wires and a light bulb exactly like the one in the manual. At several places in the manual Hammad Khürshid had added his personal thoughts about the recipes. For example, he had written "1 to 3,000" next to the recipe for a type of explosive. Khürshid later explained in court that by that he had meant that one man could be just as powerful as 3,000 soldiers if he has a bomb like the one mentioned in the recipe. Another place in the manual Khürshid had written "AD Butt". That was the name of the owner of a dry cleaning shop in Copenhagen where Hammad Khürshid planned to buy some of the chemicals used in the recipes.

After these unsettling discoveries PET decided to initiate an intensive surveillance of Hammad Khürshid. At first the surveillance was focused on a larger group but PET quickly realized that the central figures in the network were Hammad Khürshid and his close friend Abdoulghani Tokhi. Tokhi was a 21-year-old refugee from Afghanistan who came from a family with social problems and who had already attracted attention from PET in 2005 when he got into trouble in high school by handing out extremist material stating that Muslim students were not allowed to say "merry Christmas" to the non-Muslims at school.

Hammad Khürshid and Abdoulghani Tokhi spent most days together in Khürshid's apartment on Glasvej in the northern part of Copenhagen. PET was listening to the young men's phone conversations and the communication clearly showed that Khürshid and Tokhi were trying to get the ingredients to produce a bomb. Hammad Khürshid was not satisfied with the three-volt light bulb that he had brought home from Pakistan, and tried to get new light bulbs through a friend who worked as an electrician. Meanwhile Abdoulghani Tokhi was sent to different shops to buy the chemicals used to produce the highly explosive material TATP.

Hammad Khürshid and Abdoulghani Tokhi suspected that they – especially with their foreign backgrounds – risked attracting unwanted attention from authorities when they were buying chemicals. To avoid being reported they therefore made sure not to buy the chemicals in the same shops but to purchase smaller portions in different outlets. Hammad Khürshid also did some research and found out what the chemicals could be used for by people who were not terrorists. On a piece of paper he wrote down peaceful purposes which the chemicals could be used for and gave the paper to Abdoulghani Tokhi so that he would be ready with an explanation if any shop-assistant should start asking questions. Tokhi, however, had no problems buying the household chemicals used to produce TATP, and the two young men also managed to buy scales, bottles and other kinds of technical equipment which they needed to produce explosives.

It was obvious that Hammad Khürshid knew what he was doing. PET believed that he had been recruited and trained in person by a well-known senior al-Qaeda member by the name of Abu Ubaidah al-Masri, who at the time held a high rank in the terrorist network. The connection to Abu Ubaidah al-Masri was based on intelligence that was never disclosed during the open trial, but the intelligence information said that Hammad Khürshid had been to a training camp in Waziristan – something which was backed up by pictures found on his phone – and sent back to Denmark with direct orders from al-Qaeda core members to carry out an attack in revenge for the Danish cartoons.

From Copenhagen Hammad Khürshid kept contact with his connections in Pakistan and other countries. PET intercepted numerous phone calls where Khürshid spoke in cryptic coded language with people in Pakistan.

"All of this is a company. The company has grown into something big. The company does not need any investment. On the contrary, the company has started to invest," Hammad Khürshid said in one of the recorded phone calls. He later explained in court what the different words actually meant: when he spoke with his contacts about "the company" it was code for a group of holy warriors. When they talked about "a manager" he was referring to the leader of the group. And when Khürshid was talking about "investments" he was actually referring to the fact that he had people with him that were prepared to go into holy war.

Hammad Khürshid's communication with his contacts abroad, however, not only involved coded language on the phone. Khürshid and his handlers also communicated by different email accounts which they used as a kind of notice board. Khürshid and his contacts shared passwords to the same email accounts which they would log onto and only use to write draft emails – thereby communicating without emails ever leaving the account. The draft-email method might not seem new today but it was definitely innovative in comparison with the communication used in previous Danish terrorism cases.

When PET realized that Hammad Khürshid and Abdoulghani Tokhi had purchased all the ingredients they needed to make explosives, agents went into the apartment on Glasvej when nobody was there and placed miniature hidden cameras in several places.

That was how PET could follow the events live when Hammad Khürshid, on August 31, 2007, called in sick for work. In reality he was not ill, and on the live video-recordings you could see how Khürshid went into the living room of the apartment carrying several bottles with chemicals. He placed the bottles on the floor and collected the 21-page bomb-making manual which he had hidden inside a pillowcase on the living room couch.

Khürshid then started mixing the different ingredients. He worked patiently with the dangerous chemicals and paused several times to check the temperature of the mixture with an old-fashioned thermometer. In the manual it said that the temperature was important when making TATP and Khürshid had filled a bowl with water and ice cubes so that he was ready to cool down the mixture if necessary. When white crystals began to appear in the cup, Hammad Khürshid got up and left the room. A few seconds later he returned with a dark scarf. He then turned on the camera on his mobile phone and started filming himself as he poured the liquid through the scarf so that the white crystals – the actual explosives – were the only thing left. Hammad Khürshid started to sing a song from a jihad movie that he had watched again and again in recent weeks: "Accept my last greeting, because I am on my way to make a martyr attack. If I become a martyr, then pray for me, because I am going to see my god".

Judging by the images caught by PET's hidden cameras it was obvious that Hammad Khürshid had experience in producing TATP. He was calm and going though the process patiently – looking like a man who had been taught well in the dangerous art of making explosives.

The next day Abdoulghani Tokhi stopped by the apartment at Glasvej, and Hammad Khürshid proudly showed his friend the TATP he had made. The two men decided to test the explosives on the staircases of the apartment building at Glasvej. They emptied a magic marker, filled it with TATP and then placed the small homemade bomb on a windowsill. Abdoulghani Tokhi used his mobile phone to film while Hammad Khürshid ignited the home-made bomb. It exploded with a blast.

When Abdoulghani Tokhi and Hammad Khürshid came back into the apartment – and back in front of PET's hidden cameras – they were ecstatic about the successful test of the explosives and Hammad Khürshid was telling Tokhi about how to make a larger bomb. "The one in twenty kilograms ... you have to hold it like this!", he said while holding his hands in front of his chest to demonstrate where a large bomb would be placed in a suicide attack.

After seeing the disturbing footage from the apartment PET had enough evidence and decided to arrest Abdoulghani Tokhi and Hammad Khürshid. They both pleaded not guilty and PET never found out exactly what they had planned to attack. A drawing and some calculations about quantities needed to produce a large bomb which was found on a piece of paper in the apartment at Glasvej aroused police suspicions that they had planned on a big target with many people, and Khürshid and Tokhi received long sentences.

From a Danish perspective the Glasvej case provided new knowledge about the terrorists wanting to attack Denmark. It made clear that high-ranking members of al-Qaeda were investing time and energy in attacking Denmark in revenge for the Cartoons and it showed that young Muslims go to training camps abroad represents a new threat to Denmark.

Hammad Khürshid and Abdoulghani Tokhi were of course arrested in the planning stages, but that had a lot to do with the fact that PET had been tipped off very early about Khürshid's visit to an al-Qaeda training camp. Compared to the home-grown terrorists from the previous cases Khürshid appeared much more capable of planning a successful terrorist attack. He had learned methods of hiding his communication and most importantly he had learned how to produce explosives. That made him a far more dangerous enemy.

According to PET there is an increasing number of Danish Muslims who travel abroad to take part in fighting or get military training. Danish Muslims have received training in a number of different countries but most go to Somalia – where up to 40 Danish Muslims have been involved in such activities[2] – or Syria where at least 100 Danish Muslims have joined the civil war. The Danish Security Service considers the threat from returning foreign fighters as one of the most serious threats against Denmark.[3]

Reprinting of the cartoons

Six months after the arrests in the Glasvej case, on February 12, 2008, two Tunisians were arrested in a suburb of Denmark's second largest city of Aarhus. PET had for months been watching a network of Islamic radicals in the city and according to the Secret Service the two Tunisians had made plans to kill Kurt Westergaard, the cartoonist behind the most controversial drawing of the Prophet Mohammed in *Jyllands-Posten* – the one depicting the prophet with a bomb in his turban – and to attack Flemming Rose, the cultural editor at *Jyllands-Posten* when the cartoons were published in 2005.

PET however decided not to bring the two Tunisians to court, presumably because the Security Service wanted to protect an informant who had been providing vital information. Instead PET used a special paragraph in Danish law which allows Danish authorities to expel foreign nationals administratively if they are considered a threat to national security.

Consequently it was never revealed in detail what the Tunisians had planned. Small chunks of information did however come out: in the apartment of one of the Tunisians police found several weapons, including axes, a sword, an illegal knife and a malfunctioning gas pistol. PET also found an old calendar in which someone had written the name and address of Kurt Westergaard together with notes in Arabic that indicated that the Tunisians had watched the cartoonist's house and taken notes about his sleeping habits and neighbors.[4] The Tunisians pleaded not guilty to all charges; the case is mentioned in this chapter not because the case itself is very interesting but rather because it ended up indirectly having an enormous influence on the threat level in Denmark.

The police crackdown on the two Tunisians was the first publicly known example of a plan to assassinate employees at *Jyllands-Posten*. In Denmark politically motivated attacks were unprecedented at the time and news of the arrests was obviously a huge story. As a result all major Danish media outlets chose to reprint the most controversial of the cartoons – the one showing the Prophet with a bomb in his turban. The arguments for reprinting the cartoon differed from media to media; some newspapers reprinted the cartoon to show solidarity with *Jyllands-Posten* while others argued that showing the cartoon was basic news documentation when reporting about alleged plans to kill staff at the newspaper that had originally published it.

None of the arguments changed the fact that the reprinting of the most controversial of the 12 cartoons was perceived amongst radicals as a synchronized event joined by virtually all Danish media. In early 2008 most people had forgotten about the Cartoon Crisis, but the reprinting brought attention back to Denmark. Islamic militants viewed the reprinting of the cartoon as even worse than the original publication – the reasoning being that Denmark had already been warned the first time but nonetheless continued to insult Muslims – and according to PET the reprinting actually had a bigger impact on the threat level in Denmark than the original publication in *Jyllands-Posten* in 2005.[5]

Since the reprinting of the cartoons in 2008 Islamic militants have had their eyes focused on targeting Denmark. Among the publicly known examples of terrorists trying to attack *Jyllands-Posten* or its representatives since the reprinting is the case against David Headley, several cases of solo terrorists, a case where three Norwegians were convicted, the arrests of two men in North Africa and a case where a terrorist group from Sweden were arrested on their way to do a Mumbai-style attack in Copenhagen against *Jyllands-Posten*'s offices. In fact there have been so many cases there is no space to write about all of them in this chapter and the following therefore focuses on two different categories of plots.

Solo terrorists

Two terrorists who were equally determined to attack *Jyllands-Posten* or employees from the newspaper in revenge for the cartoons were apprehended in different cases in 2010. From the outside, the main characters in the two cases did not share many common features. One was a Somali father of three, the other a Chechen refugee living with his mother in Belgium. But their choice of target and the strategy they each used was similar.

The "Axe man"

The first of the two cases started on the very first day of 2010. Kurt Westergaard was at home in his house near Aarhus, looking after his five-year-old granddaughter because her parents had been at a New Year's party the night before. Kurt Westergaard and the granddaughter were eating candy and watching "The Wizard of Oz" when they suddenly heard loud noises coming from a door facing the garden. Someone was pounding hard on a glass window in the door and Kurt Westergaard knew immediately that it was probably a terrorist trying to get into the house.

The granddaughter had broken a leg in a skiing accident and had her leg in a cast. Kurt Westergaard, an elderly man who always uses a cane when walking, he knew that he would not be able to carry his grandchild who couldn't walk by herself. Security advisors from PET had also told Kurt Westergaard to think of himself first in a case of an emergency like this, because the assessment was that terrorists wouldn't go after his family members. When Kurt Westergaard heard the banging on the door he reacted promptly and ran to a bathroom which was designed as emergency shelter. Inside the secured bathroom he locked the door and activated an alarm.

The cartoonist therefore did not see the 28-year-old Somali man, Muhudiin Mohamed Geele, who a few seconds later managed to break through the door and came running into the house, armed with an axe and a knife. At first Geele couldn't find Westergaard anywhere in the house but he soon realized that the cartoonist was hiding in the bathroom. He started pounding on the door while shouting curses and threats against Kurt Westergaard on the other side.

Muhudiin Mohamed Geele quickly realized that he could not break through the secured door to the bathroom. The 28-year-old father was also surprised and uncomfortable with the fact that Kurt Westergaard's granddaughter was sitting in front of the TV in the living room. Geele decided to leave the house and on his way out actually tried to comfort the crying girl by saying that she should not be afraid of him.

He then left the house and went to the street where police had already arrived to arrest him. But instead of surrendering Geele attacked the police officers. He threw his axe at one of the officers who then pulled his gun and shot Geele twice. Even after being overpowered and wounded the Somali was shouting insults at the policemen.

Muhudiin Mohamed Geele had links to the Somalian terrorist organization al-Shabaab and to al-Qaeda leaders in Eastern Africa. Geele had even been arrested in July 2009 in Nairobi because the Kenyan authorities suspected him of being part of a network that was planning attacks during an official visit to the country by US Secretary of State, Hillary Clinton. Kenyan authorities believed that Geele had had personal contact with al-Qaeda's former leader in East Africa, Saleh Ali Saleh Nabhan, and that it had been intended that Geele himself should hold a prominent role in the network before he was arrested.[6]

That the 28-year old terrorist made it all the way into the living room of Kurt Westergaard's house gave rise to harsh criticism of PET. Everybody knew that the cartoonist was the most endangered person in Denmark. How then could an armed extremist get access to his house so easily? PET also admitted that the Security Service knew Muhudiin Mohamed Geele well before he was arrested and that he was a part of a network in Denmark that had been under surveillance for months. In other words PET knew that the Somali was a potential terrorist but still had not discovered that he had made plans to attack Kurt Westergaard until the cartoonist pushed the alarm button in his bathroom.

The explanation for the seemingly poor intelligence work was that Muhudiin Mohamed Geele had apparently acted completely alone in this specific plot. The Somali was not a "lone wolf" per se. He was part of an established network and monitored by PET, but no secret service in the world has the capability to watch all suspected terrorists 24 hours a day and there was nothing that indicated that he had communicated with other members of the network about the plans to attack Kurt Westergaard.

The investigation showed that in the days before the attack Geele had spent some time online reading about *Jyllands-Posten* and the cartoons.

He had been watching videos with famous al-Qaeda members who condemned the cartoons, and had tried to find justification for an attack by searching on Google for "Usamah Bin Laden danish cartoons" to find out what the leader of al-Qaeda had said about the issue.

On December 31, 2009 – the day before the attack – Geele went to the hardware store Bauhaus in Glostrup where he bought an axe which he would use to gain access to Kurt Westergaard's house. At home he also sharpened a knife. On the actual day of the attack he went online to find Kurt Westergaard's address – which was still on Yellow Pages even though the cartoonist was under police protection – and bought a train ticket from Copenhagen to Aarhus the same evening. When Geele arrived at the railway station in Aarhus he simply took a taxi and had the driver drop him off close to Kurt Westergaard's address. He then walked the last couple of hundred meters and went into the house. He had shaved his body and used a lot of perfume which could indicate that he was prepared to be martyred in the attack.

Geele did not succeed in killing the biggest symbol of the Cartoon Crises – but he got very close. The case was the first Danish example of the trend towards carrying out very simple terrorist attacks and it was probably the simplicity of the plan and the lack of previous communication which were the main reasons that Geele slipped through the security net.

The one-legged bomber

Later the same year Denmark witnessed a similar example of a terrorist who came close to achieving his goal because he acted on his own.

On September 10, 2010 an explosion suddenly rocked the cheap backpacker hostel, Hotel Jørgensen, in the heart of Copenhagen. The blast came from a bathroom in the basement and a man was seen running away from the hotel – but at first it wasn't clear whether it was caused by a gas explosion or a criminal act. Police launched an extensive search in the center of Copenhagen and after a few hours the man who had been seen running from the hotel was found in a park a few hundred meters away.

By that time police knew for a fact that it was indeed a bomb that had destroyed a small bathroom in the basement of Hotel Jørgensen, and that it was quite a miracle that the unknown perpetrator had survived the explosion. The man now lying on the ground in the park had obvious wounds from the explosion but the police officers present were afraid to approach him because of fear that he was carrying a suicide belt around his waist. It turned out that it was just a money belt and the man was arrested.

The mysterious bomber, however, refused to reveal who he was to Danish police – and he had obviously tried hard to conceal his identity. The investigators could tell that he had bought a return ticket to Belgium, but while he had been in Copenhagen he had been using three different sets of identification papers – either fake or stolen – and moved around Copenhagen as Raoul Foltz, David Francois J. de Vicq the Cumptich or Hans Veller.

The mysterious terrorist did however have some very distinctive physical characteristics. His nose had been broken several times which could indicate that he was a boxer, and he was missing a part of his right leg and was using an artificial prosthetic leg.

Prosthetic legs are always equipped with special identification numbers that reveal when and where they were produced. But when police had the prosthetic leg examined by an expert they discovered that it was not so easy to establish the man's identity from the prosthetic leg. The technical studies showed, astoundingly, that the unknown terrorist had removed the unique identification numbers inside the core of the prosthetic leg. Around two years prior to his visit in Denmark someone had scratched away the serial number. There was no doubt that the man had gone through a lot to hide his identity.

It was the Danish tabloid newspaper *B.T.* that finally managed to find out who the mysterious one-legged bomber was. A reporter at the newspaper simply called different boxing clubs in Belgium until he found a coach who could remember training a one-legged boxer in his gym. That's how it was discovered that the man under arrest was Lors Doukaev. He was born in Chechnya, where he lost his leg on a landmine at the age of ten. Later he and his family moved to Belgium where he was living in Liège before the arrest in Copenhagen.

The investigation showed that Lors Doukaev had arrived by bus to Copenhagen and that he had brought with him a homemade bomb, a loaded gun and multiple sets of identity papers. His plan was to attack *Jyllands-Posten* in revenge for the cartoons but during his stay in Copenhagen, Doukaev apparently realized that security measures at the headquarters of *Jyllands-Posten* had become so strict that it would be practically impossible for him to place the bomb inside the building.

Inside the destroyed bathroom at Hotel Jørgensen police found a tourist map of Copenhagen and on the map Doukaev had written the address of *Jyllands-Posten*'s office in Jutland. He had apparently decided to reconstruct the bomb and make it into a postal bomb, which he planned to send to the newspaper instead of trying to get it into the building himself.

Inside the bathroom police also found a plastic container from a Nintendo videogame that Doukaev had bought in Copenhagen. He had used the plastic box as a container for explosives and hundreds of small metal balls that would make the postal bomb more lethal. But something went wrong while he was rebuilding the bomb and the explosives suddenly went off by accident.

On one hand you can argue that Lors Doukaev was an amateur terrorist who nearly got himself killed by literally blowing up a toilet in a cheap hostel in Copenhagen. Not exactly a proud moment for a terrorist. On the other hand you could find many explosives experts who would be reluctant to build a home-made postal bomb in a small and cramped toilet. It is simply a complicated and dangerous process. And one can also argue that Doukaev actually came very close to being able to carry out a terrorist attack. His idea of sending a postal bomb to *Jyllands-Posten* was not very good, because the newspaper does not receive mail directly to the office anymore, which means that Doukaev, if he had sent the bomb, would most likely have killed a postal worker but no employees at *Jyllands-Posten*.

Nevertheless Doukaev was staying at a hostel in the center of Copenhagen – only a few hundred meters from Nørreport Station, which is the biggest traffic hub in Denmark and an obvious target for terrorists – the day before September 11 and neither PET nor any other intelligence services had any idea that he was there. He could probably have used the bomb on a train without being caught – but like so many other terrorists he was to determined to attack *Jyllands-Posten*.

There are also other aspects of the Doukaev case which are unusual and suggest that the one-legged bomber was not a complete amateur. Doukaev was actually known by several intelligence services in European countries. He was connected to radical circles in Belgium and was captured on surveillance photos when he was visiting extremists in a radical mosque in Bremen. But as in the case of Muhudiin Mohamed Geele he apparently operated on his own when he decided to go to Denmark and attack *Jyllands-Posten*. By following a fairly simple plan and refraining from communicating with anybody, he managed to get to Denmark without anyone in the intelligence community finding out.

Once Lors Doukaev was in Copenhagen he moved around the city acting like a professional spy and changing his appearance like a chameleon. Surveillance photos from different places in Copenhagen showed how Doukaev constantly changed his clothing and his physical appearance. In some pictures he had a hat on his head and a beard around his mouth, but in other pictures he had removed the hat and shaved off his beard. In videos from the shops where he bought items which he needed to construct the postal bomb you could also see that Doukaev always wore gloves or was holding a piece of paper when touching anything so that he did not leave any fingerprints.

Mumbai-style attacks

Three months after Lors Doukaev was arrested, four men from Stockholm in Sweden rented a car at a petrol station and drove towards Denmark. Just like the one-legged bomber the group was planning to attack *Jyllands-Posten* in revenge for the cartoons but the men from Sweden had chosen a totally different strategy. Their plan was to carry out a Mumbai-style attack by going to the newspaper's headquarters to kill as many journalists as possible. The plan was to run away when police arrived, but the group members had already discussed the possibility that it could end up as a martyrdom operation in which they were killed.

Before going into further details about the Swedish cell it is necessary to briefly summarize another case. Just over a year before the Swedish cell headed off towards Copenhagen an American Muslim by the name of David Headley was arrested in Chicago. Headley is a famous character in the history of Islamic terrorism and arguably the most dangerous terrorist who has ever tried to attack Denmark. The Headley case, despite its seriousness, is only briefly mentioned in this chapter simply because it is covered extensively by so many others.[7]

It is, however, worth summarizing parts of the Headley case in relation to the case of the Swedish cell. David Headley was in 2008 the brain behind the terrible terrorist attack in Mumbai in India where more than 160 people were killed by a group of heavily armed terrorists that attacked different locations in Mumbai. After the successful and in many ways innovative attack David Headley and his accomplices in Pakistan started planning a similar attack against *Jyllands-Posten* in Denmark. Headley called the operation the "Mickey Mouse project" – referring to the cartoons – and made two trips to Denmark in 2009. In Denmark he made reconnaissance video recordings in several places and even made it inside the building of *Jyllands-Posten* pretending to be a business man who wanted to place an ad in the paper. Headley made sure not to have contact with extremists in Denmark – most likely to avoid attracting the attention of PET – but he still needed somebody to carry out his plan which involved storming the building, shooting the journalists and throwing their heads out the windows afterwards.

The high-ranking al-Qaeda leader Ilyas Kashmiri – whom Headley knew from his many visits to Pakistan – told Headley to contact some of his friends around Europe for assistance. Headley travelled to the UK to ask for help from two men in Derby, then on to Stockholm where he contacted a shop owner by the name of Farid Lamrabet, a Moroccan man in his late forties who was a well known figure in the Swedish intelligence service SÄPO (Säkerhetspolisen) because of his many connections in the radical circles in Sweden. Farid Lamrabet, however, refused to help David Headley because he – rightly – assumed that he was under surveillance and David Headley was arrested a few months later, before he had carried out the Mickey Mouse project.

The question, then, is if the Swedish cell were on their way to carry out the unfulfilled plan when they drove towards Copenhagen on December 28, 2010. The members of the Swedish cell were Mounir Dhahri, a 44-year-old Tunisian, Sahbi Zalouti, a 37-year-old Tunisian with Swedish citizenship, Munir Awad, a 29-year-old Swedish citizen born in Lebanon, and 30-year-old Omar Abdalla Aboelazm, who was born and raised in Sweden.

Between them they had former convictions for drug selling, violence, threats and sexual offences. Now they had turned to terrorism.

In Stockholm the four men moved in the same radical circles as Farid Lamrabet – the shop owner whom Headley had contacted in Stockholm – and the investigation showed that Mounir Dhahri had had contact with Farid Lamrabet. On December 6, 2010 Mounir Dhahri called Farid Lamrabet three times and in the end of December he called him again.

It is not known what was said in the phone calls but it is certain that Mounir Dhahri a few days later was sitting in a rental car on his way to Copenhagen together with three other members of the terrorist cell, and it is also a fact that the men in the car had a plan remarkably similar to Headley's. Like the American terrorist they wanted to kill and behead employees at *Jyllands-Posten.* At least three of the four terrorists from Sweden had also spent time in Pakistan, but if they had been told by people in Pakistan or by Farid Lamrabet in Stockholm to continue with the plans that Headley didn't finish – or if they were just copy-cats inspired by the American – is uncertain.

Either way the Swedish Security Service already knew everything about the plans to attack *Jyllands-Posten* when the group started the six-hour drive from Stockholm to Copenhagen, and agents from SÄPO were shadowing the rented Toyota Avensis. Inside the trunk of the car the men had hidden a 9mm pistol, a machine gun, ammunition and 200 plastic strips which would be used for handcuffing employees at *Jyllands-Posten.*

On the way through Sweden there was a disagreement in the car and Sahbi Zalouti got out while the other three continued towards Copenhagen. When the group reached the bridge which connects Denmark and Sweden, the SÄPO-cars pulled away and allowed the terrorists to continue across the bridge and when the car entered Danish territory several cars from PET were waiting and picked up the surveillance.

PET knew that the Swedish cell had arranged to borrow an apartment in the suburb of Herlev from a connection in Denmark and agents had already been inside the apartment to install small cameras that made it possible for PET to follow every move the Swedes made. On the way to the apartment there was a short period of panic among the agents when the terrorists suddenly pulled off the highway and drove towards the center of Copenhagen. PET was afraid that the group had decided to go directly to *Jyllands-Posten* to do the attack and the agents feared for a moment that it would be necessary to pull over a car with armed terrorists in the middle of Copenhagen. It turned out that the terrorists just needed to ask for directions at a petrol station and they eventually arrived at the apartment in Herlev in the middle of the night and went straight to bed.

When the three terrorists woke up the next morning they started the day by praying in a corner of the apartment. On the hidden cameras you could see the men kneeling down and you could hear them asking for blessing to "kill the infidels." Just a few seconds after they had finished praying the door to the apartment was knocked open and heavily armed police forces stormed in and arrested the group. At the same time Sahbi Zalouti – who had left the car on the way to Copenhagen – was arrested in his home in Järfälla, Sweden. The four members of the Swedish cell were all sentenced to 12 years in prison, which is a long penalty in Denmark.

Once again it was proven that Denmark has become a priority target for Islamic terrorists because of the cartoons in *Jyllands-Posten* in 2005 and especially because of the reprinting in all Danish media in 2008. But although there has been sign of innovation and learning among the terrorists who have tried to attack Danish targets since 2005, the case of the Swedish cell once again demonstrated that there has been little innovation when it comes to picking out targets to attack. Several different terrorist groups and individuals have repeatedly tried to attack *Jyllands-Posten* – and thereby made it easier for authorities to prevent an attack because the newspaper is obviously heavily protected and because any intelligence information involving extremists who are talking about the Danish cartoons instantly sets off alarms in the international intelligence community.

Documents

The chapter is based on a large number of interviews, police reports and court documents from the Danish terrorist cases which the author has also followed in court. The court documents used are:

Indictment in the case against Mirsad Bektasevic, Abdulkadir Cesur, Bajro Ikanovic, Amir Bajric and Senad Hasanovic, Court of Sarajevo, April 6, 2006.

Court documents from the Glostrup case, Østre Landsret, December 6, 2006.

Court documents from the Glostrup case, Østre Landsret, January 8, 2008.

Court documents in the Vollsmose case, Østre Landsret, September 5, 2007.

Court documents from the Glasvej case, Retten på Frederiksberg, August 11, 2008.

Court documents from the Glasvej case, Østre Landsret, May 15, 2009.

Court documents in the case against Lors Magomedovitch Doukaev, May 31, 2011.

Court documents from the case against Muhudiin Mohamed Geele, Vestre Landsret, June 11, 2011.

Court documents from the case against Sahbi Ben Mohamed Zalouti, Mounir Ben Mohamed Dhahri, Munir Awad and Omar Abdalla Aboelazm, Københavns Byret, June 4, 2012.

Court documents from the case against Munir Awad and Østre Landsret, January 31, 2013.

Notes

1 http://politiken.dk/indland/ECE2086669/danske-salafisters-edderkop-er-draebt-i-syrien.
2 www.pet.dk/Nyheder/2013/~/media/CTA/AlShabaabpdf.ashx.
3 www.pet.dk/CTA/~/media/CTA/20121119FinalFFUKLASS2pdf.ashx.
4 Report from PET presented during a court hearing about the legitimacy of the decision to expel the Tunisians.
5 www.pet.dk/CTA/~/media/VTD%20NTV/NTVjanuar201231012012pdf.ashx.
6 www.pet.dk/Nyheder/2010/KW%20drabsfors%C3%B8g.aspx and http://politiken.dk/indland/ECE877195/al-qaeda-var-ved-at-give-terrorjob-til-oeksemanden.
7 Especially by Sebastian Rotella in countless articles and in his documentary *The Perfect Terrorist.*

11 "Please inform Europe"

Al-Qaeda's Europlot of 2009/2010 in the light of new documents and testimonies

Yassin Musharbash and Guido Steinberg

The backdrop

On September 18, 2009, just over a week before Germany's general elections of that year, al-Qaeda's media operations department al-Sahab published a video message warning of imminent terror attacks in the country.[1] The message was delivered by Bekkay Harrach, a young German of Moroccan origin from the former West German capital Bonn in North Rhine-Westphalia. It was his third message in the name of al-Qaeda, but the first in which he directly warned of attacks within two weeks after the elections if the results did not reflect a clear wish by the German population to withdraw its troops from Afghanistan. Germany's security services took the warning seriously and implemented a number of measures, including massive police presence close to potential targets like railway stations and crucial infrastructure.

The elections, of course, took place as scheduled and no al-Qaeda attack occurred in Germany. And yet, the video message was more than mere propaganda. The threats against Germany were supported by senior al-Qaeda operative Atiyatallah al-Libi exactly one week later in a prominent Arabic-language video entitled "The Dark Tunnel."[2] Just how close Harrach had been to the al-Qaeda leadership became clear much later, when the US government in May 2012 published a set of 17 documents found at Osama bin Laden's residence in Abbottabad, Pakistan. In a letter from the al-Qaeda chief, the Saudi asked that Harrach send him ideas for any "external work" he may be planning, in other words, for a terror attack in the West.[3] Although Harrach died in an insurgent attack near Bagram airbase in Afghanistan in 2010, his video not only set the tone for a propaganda offensive by jihadist groups against Germany, but was also conclusive evidence that al-Qaeda indeed aimed at attacking Germany.[4]

Harrach, who had taken on the nom de guerre "Abu Talha al-Almani,"[5] was only the most prominent of an increasing number of Germans and other European recruits to make his way to the jihadist network of guesthouses and training facilities in North Waziristan, where al-Qaeda had established its headquarters after 2001. Between 2007 and 2011 several dozen young men and (fewer) women from Germany travelled to the Afghan-Pakistani border region with the aim of joining jihadist groups there. Most started out as members of the Islamic Jihad Union (IJU), the Islamic Movement of Uzbekistan (IMU) or the German Taliban Mujahideen, a small group of Germans which only existed between September 2009 and summer 2010 and can best be described as a splinter of the IJU. Some have since returned to Germany or been killed. But quite a few, perhaps as many as a dozen, ended up with al-Qaeda. Of these, we know of several who in the process became part of an al-Qaeda endeavor to plot one or more major attacks in Europe, combined with a long-term infiltration and sleeper cell strategy – the so-called "Europlot".

The idea of the Europlot seems to already have existed when Bekkay Harrach, alias Abu Talha al-Almani, threatened Germany with attacks in September 2009, and it is likely that Harrach at some point became part of those deliberations. Looking back, Harrach's speech was perhaps the earliest indication that something big was afoot. In the following two years, it became increasingly apparent that al-Qaeda was indeed planning attacks in Europe. A number of different threads had come together: There was a definite increase of jihadist "chatter" and there were unconfirmed pieces of information by several Western intelligence services about alleged terror cells having been dispatched to Europe from Waziristan, including unspecified warnings of "Mumbai style" attacks. Even though Europe wasn't struck, new information trickling in over the following months only seemed to confirm already existing concerns.

- First, there was the testimony of Ahmad Wali Sidiqi, an ethnic Afghan al-Qaeda recruit from Germany who had spent time in South and North Waziristan and had been apprehended by US Forces in Kabul in July 2010. By September, some of what Sidiqi had told his American interrogators had filtered through to the public: al-Qaeda was sending attackers to Europe, he had told them, and the al-Qaeda leader behind the plot was a certain Abu Yunis al-Muritani (the Mauretanian). Osama bin Laden himself, Sidiqi informed his interrogators, had signed off on the plan.
- Second, in June 2010, another German jihadist (of Syrian origin), Rami Makanesi, was apprehended in Pakistan and later extradited to Germany. He seconded a lot of what Sidiqi had already shared. Most importantly, he stated that Yunis al-Muritani was al-Qaeda's point man for an attack strategy against targets in Europe. According to Makanesi, the terror network had quite obviously made an effort to bring together as many Western recruits as possible in order to train them and send them back to Europe where they were supposed to recruit even more willing radicals to raise money, to plan attacks right away or to prepare for a later moment in time. Al-Muritani, according to this testimony, put a clear emphasis on training his recruits in

covert communication and the use of the respective software – just one indication that he was pursuing a long-term plan.

- Third, in October 2010, another piece of information surfaced when a British citizen by the name of Abdul Jabbar was killed by a US drone in Pakistan. The BBC and other media reported that Abdul Jabbar had previously been tasked by al-Qaeda with setting up a branch of the network in Great Britain with a view to attacking targets there. Doubts remain, but it is not unreasonable to assume he may have been part of a larger Europlot.
- Last, one month later, in November 2010, a heightened state of alert was declared in Germany when the minister of the interior, Thomas de Maizière, stated in a press conference that there was reason for concern over possibly imminent attacks. He went as far as asking the public to be watchful, a very rare request in Germany's political context. The reason for this request was that only days before a Turkish-Kurdish jihadist from Germany, Emrah Erdogan, had contacted the German Federal Criminal Police (BKA) from his hideout in Waziristan and warned of a group of attackers that according to him had already been set in motion; one of the targets they had in mind, he divulged, was the Reichstag, the landmark building in Berlin's historical city centre that hosts the federal parliament. Immediately, the building was closed to the public.[6]

After these initial hints, two cells connected with al-Qaeda's plans were arrested by German authorities, most likely on the basis of American information, adding to a picture which however still remained somewhat incomplete. The first of these two groups had been operating in the western German city of Düsseldorf in April 2011. The Moroccan ringleader Abdeladhim El-Kebir had also been trained by al-Muritani and he and two of his associates were apprehended after they had started preparing explosives in order to train for an actual attack. For some months after his return, El-Kebir had stayed in touch with Atiyatallah al-Libi, telling him in an e-mail, among other things, that

> we will keep our promise: victory or martyrdom.... O Sheikh, I am training some youths from Europe.... After the training I will, with God's help, start with the slaughtering of the dogs.... Please pass on my regards to all the dear ones, especially Sheikh Yunis.[7]

In internet searches El-Kebir inquired about bullet-proof vests and hunting bows. He contemplated what would happen if he just went for "a small bomb" and "then go back to sleeping afterwards" as that would "break the enemy ... – more pressure." In a notebook investigators found what is probably a list of potential targets. Among them was a note saying "NATO" as well as the name of French satirical magazine "Charlie Hebdo" which had previously published Mohammed cartoons, and also a name of a former German special forces commander in Afghanistan and the names of two German TV personalities, probably because they had make jokes about Islamists.

A few weeks after the breaking up of the Düsseldorf cell two more Western recruits who had been dispatched back to Europe were arrested: Maqsood Lodin from Vienna and Yusuf Ocak from Berlin. Interestingly, these two are reported not to have been trained by al-Muritani, but instead probably by Adnan al-Shukrijumah, a US-Saudi member of al-Qaeda, and by Atiyatallah al-Libi, a senior al-Qaeda operative who commanded its headquarters in North Waziristan and worked together with al-Muritani on the Europlot until he was killed by a US drone in August 2011.

As we have seen, pieces of the puzzle that is the Europlot had already come together before the arrest of Ocak and Lodin. Al-Qaeda was clearly at it, and Europe was on the menu. But the picture was still sketchy. That changed after a set of documents that had been found on the body of Maqsood Lodin was decrypted by the authorities. The documents in question do not contain a detailed plan by al-Qaeda for hitting Europe, but they do provide very informative background information previously unknown about how the terror network perceives itself, what it hopes to achieve and what it thinks it has learnt from past experience. In addition, they contain pieces of evidence as to what kind of attack Lodin and Ocak were planning to perpetrate. But most importantly, they convey a sense of al-Qaeda's strategic deliberations.

In the following section, we will discuss these documents, starting with the "future work," which has turned out to be the main strategy paper allowing the deepest insight into al-Qaeda's ways of thinking, and the information they contain about a single attack most likely foiled by the arrest of Ocak and Lodin. This is followed by an assessment of more strategic documents that inform us about the discussion process leading to the new strategy. This in turn will be followed by an assessment of another document of great value that surfaced in the trial of El-Kebir, namely a hitherto unknown letter by Yunis al-Muritani to Osama bin Laden found at Abottabad and made available to the German court by the US government.

The Ocak/Lodin papers

In the course of the trial against Lodin and Ocak in Berlin in 2011/2012, information emerged about strategic debates among al-Qaeda leaders which had led to the sending of these two recruits back to their countries of origin. When arrested in Berlin, Maqsood Lodin carried a memory device on his body that, among dozens of worthless papers, contained a few much more interesting documents, most importantly an internal al-Qaeda document called "Future work," which analyzed the position of the organization in early 2009 and laid out its future strategy.

The argument in that document was based on a very pessimistic view of the organization's development since 2005, when al-Qaeda had perpetrated its last successful terrorist attack in the Western world. In fact, Lodin's memory device contained two other documents, namely "Report on operations" and "Lessons learnt from operations," which dealt with the organization's attacks in London on July 7 and July 21, 2005, as well as with the transatlantic airliner plot in August 2006. In them, the author, quite possibly the British-Pakistani al-Qaeda-member Rashid Rauf, describes the preparations of these attacks in great detail and tries to draw lessons from a comparison focusing on the reasons for the success of the first and the failure of the other two plots.

"Future work"

In "Future work," an anonymous author depicted the situation of al-Qaeda in early 2009, shortly after Lashkar-e Tayyiba's attacks in Mumbai, lamenting that his organization had not successfully perpetrated an attack in the Western world since the London bombings on July 7, 2005.[8] According to the author, this failure led to a demoralization of al-Qaeda cadres, with the result that even the number of attempted attacks had diminished. But in his view neither this nor the high losses of al-Qaeda due to the US drone war in Pakistan should lead to the organization refraining from planning large, continent-spanning attacks like those of September 11, 2001. Rather, the organization should adopt a two-track strategy and divide its aims into short-term and long-term ones. It should first order its recruits to perpetrate relatively unsophisticated and simple attacks worldwide, which would not need any major assistance of a large organization. By adopting such a strategy, the author argues, al-Qaeda would keep Western security services busy so that they would relax the pressure exerted on al-Qaeda Central in Pakistan and Afghanistan. This in turn would allow the organization to regain the capability to perpetrate more sophisticated attacks in the West.

According to the author of "Future work," this new dual strategy would have to have consequences for the training of al-Qaeda recruits in Pakistan. Every volunteer unknown to the security services of his home country should be trained in a specific activity and sent back as soon as possible – thereby minimizing the danger of losing him to the frequent US airstrikes. After his return to his country of origin, the organization would only contact him in case of urgent need and let the recruit perpetrate the attack on his own authority. Furthermore, al-Qaeda could call upon its followers to execute operations, which did not even require the recruit to come to Pakistan in order to undergo instruction. The organization would only need a recorded will (*wasiya*) to make clear that the perpetrator acted on behalf of the organization, which would then use the material for propaganda purposes. At the same time, al-Qaeda should spread information regarding legitimate aims and tactics via jihadist forums on the internet. In the case of recruits who were sent back to the West, as well as recruits who had not even traveled in the first place, al-Qaeda's short-term goal would be to let them perpetrate as many as possible of the aforementioned unsophisticated attacks. What kind of attacks al-Qaeda had in mind became apparent in another document found on Lodin's memory device. In contrast to "Future work," which was written in English, "Notizen" was written in German and contained short notes, not all of them entirely clear in their meaning, detailing possible modes of action. As it appears from this document, the two recruits had been tasked to take hostages, film them while killing them and to afterwards "send the video to al-Qaeda."[9]

The author of future works was well aware, however, that al-Qaeda's enemies might interpret simple attacks as indications of an organization's weakness, potentially suggesting that al-Qaeda was no longer able to meet its own high standards established on September 11, 2001. Therefore, al-Qaeda should not neglect the long-term planning of large operations. To this end, he argued, the network primarily needed an infrastructure taking care of recruitment, communication and financing. As a consequence, al-Qaeda would have to build cells which would exclusively support the groups actually perpetrating attacks – especially by providing them with explosives. This was new insofar as these recruiters and logisticians should be based in Western countries. And after their establishment, al-Qaeda could focus on new attack plans. Although the text is somewhat inconclusive in this regard, the author seems to have preferred seaborne attacks ("jihad in the seas"), but also discussed high-profile attacks on prisons in order to free jailed jihadists.

Although the identity of the author of this strategy paper has yet to be established, there are certain hints. In the text, the author claims to work for an institution named "Maktaba Dawa" (office of the propagation of faith). From the context, it is clear that this is the name of a unit close to al-Qaeda's leadership, even though it had remained unknown under this name until the emergence of this document. This was or is quite likely a part or the whole of a unit frequently called "external operations" by al-Qaeda and Western security authorities for lack of a better term and in which English was known to be spoken – as opposed to most other parts of the organization, where Arabic remains the lingua franca. As the text in question is authored in English, it is very likely that one of the few English-speaking operational chiefs of the organization penned it. In 2009, these were the Englishman of Pakistani origin Rashid Rauf, US-raised Saudi Adnan Shukrijumah and the Pakistani Ilyas Kashmiri. If Rauf was not the author, the latter seems to have stood in connection with the former, as he frequently refers to the documents "Report on operations" and "Lessons learnt from previous operations," of which at least the first one quite definitely had been written by Rauf.

As the text (together with the other documents) was found with a European recruit, it is very likely that the Austrian had travelled back to Europe with the task of building new, partly independent structures for al-Qaeda. Providing non-Arab recruits with only a short history of engagement in the jihadist movement with important documents like these was evidence that al-Qaeda in 2009 accepted the need for a greater decentralization if it wanted to survive – even at the expanse of operational security.

"Report on operations" and "Lessons learnt from previous operations"

The strategic and tactical recommendations in "Future work" were based on extensive assessments about former plots, both failed and successful, which are contained in two other documents found with Lodin, entitled "Report on operations" and "Lessons learnt from previous operations" respectively. The first deals with three al-Qaeda plots in Britain, namely:

- the London bombings of July 7, 2005, which targeted the public transport system and killed 52 persons and wounded several hundred;
- the attack on the London public transport system of July 21, 2005, which failed when the explosives did not detonate;
- the "transatlantic airliner plot," which was thwarted by the British authorities in August 2006 and in which nine suicide attackers planned to detonate liquid explosives hidden in bottles on planes starting from Heathrow airport.[10]

Perhaps most importantly, the document spectacularly proved that all these plots had been hatched by al-Qaeda Central in Pakistan – something which had been suspected for a while but which had not been public knowledge until 2011.[11] According to the text, a deceased al-Qaeda leader named "Hajji" had served as the emir of all three operations, most likely the Egyptian Abu Ubaida al-Masri who had died in 2007. His lieutenant in the actual organizing of the plot in Pakistan had been Rashid Rauf, who authored the document in slightly flawed English. This becomes clear in those parts where the author describes his own arrest by Pakistani authorities shortly before arriving in Bahawalpur in August 2006, by mentioning that Kashmiri was his mother tongue and that there was a British arrest warrant against him since 2002. The text provided a detailed narrative of the planning and organizing processes of the three plots, pointing out that the explosives instruction for the 7/7 plotters had been especially exhausting and methodical by qualified al-Qaeda trainers, how deep and friendly personal relations were between Rauf and the recruits partly due to their long stay in the tribal areas and frequent meetings, and how intensively Rauf had communicated with them about the targeting and problems with the production of the explosives after the cell's return to Britain.

In "Lessons learnt from previous operations," the author draws operative conclusions from the analysis of the planning, organization and execution of the three plots. He emphasizes the importance of attacks on US allies in order to weaken the West in its entirety and to force it to withdraw from the Muslim world:

> There are certain fronts we need to concentrate and improve in, to hasten the enemies defeat. One of these fronts is making operations in western countries and especially those who are in alliance with America in its war against the Muslims. This is one of the main strategies of the mujahideen to bring down the kufr system into chaos and force the retreat of the crusader alliance from the Muslim lands.[12]

While stressing the big successes of the years 2001–2005 like the 9/11 attacks, the Bali and the London bombings, he concludes that Western security authorities had greatly improved their countermeasures and worked more effectively in 2009 than before. This is one of the reasons why al-Qaeda had been weakened "on this front." Many recruits from Western countries had even lost hope that successful terrorist attacks in their home countries were possible at all.

The author then points to the great number of potential jihadists in Western countries and concludes that it would not be possible to keep all of them under surveillance for longer periods of time. He even mentions the case of two perpetrators of the London bombings who shortly appeared on the radars of the British authorities but were then let go because they were not considered to be dangerous.[13] In order to once again perpetrate successful attacks, so the argument goes, al-Qaeda would have to improve its tactics by drawing lessons from the failure of the last two of the three London plots. This held true, for example, for the inadequate explosives training of the 21/7 attackers, a lack of time for the preparation of the cell leader Muktar Said Ibrahim in Pakistan and flaws in the communication between al-Qaeda Central and the cell after Ibrahim's return to Britain.

The Muritani letter to bin Laden

After the documents were found with Maqsood Lodin, it became clear that the Austrian and his German comrade had been sent back to build new structures and perpetrate simple attacks in order to keep the security services busy. But still, "Future work" only contained rather general remarks about al-Qaeda's future strategy and it remained unclear to what extent the organization had decided to implement the deliberations into terrorist practice.

Some of these gaps were spectacularly closed during the trial of the Düsseldorf cell which started a few months after the Ocak-Lodin trial. First information about the return of Kebir from Pakistan had been passed on to the German authorities by their US partners, and in preparation for the trial the German prosecutor general asked the US attorney general for more information on the Moroccan. As a result, the American authorities provided the Germans with a previously unknown letter by Yunus al-Muritani, the mastermind of the Europlot, to Osama bin Laden, dated March 2010.[14] The content of the letter was clearly influenced by "Future work" and elaborated on some of the topics already raised there. Most strikingly, both texts contain a discussion of seaborne activities, which shows that the Muritani-letter can be seen as a more detailed and practice-oriented version of "Future work." All in all, the letter contains an ambitious and innovative, mostly pragmatic, but in parts also somewhat fantastic strategy for a new wave of attacks against important supply routes in the world economy.

The text must have been part of a much larger exchange of letters between Osama bin Laden, Muritani and Atiyatallah. Parts of this exchange have been published by the Counterterrorism Center at West Point.[15] In this letter, however, Muritani vividly describes the threat to al-Qaeda posed by the American drone campaign in the Pakistani tribal areas and uses it as an argument for a two-track strategy that had already been spelled out in "Future work." Some groups and cells would have to be tasked to keep Western security services busy with simple and small attacks in order to provide the al-Qaeda leadership in Pakistan with the opportunity to plan and organize new and bigger attacks. In a first step, al-Qaeda recruits should leave Pakistan for Iran and build new structures in Europe as well as Africa. The volunteers should establish new cells and train their members for sea- and airborne attacks in the future. For airborne attacks al-Qaeda members should be advised to undergo flight training and even work for commercial airlines as regular pilots, only to receive their orders to attack later – with the aim of destroying, for example, convention centers, chemical factories or government installations. Muritani specifically asked bin Laden to task al-Qaeda in Yemen's leader Nasir al-Wuhaishi (also known as Abu Basir) with finding recruits who were not known as al-Qaeda supporters and sending them to flight training in the Arab Gulf states. The aim could then be to eliminate the Saudi political leadership during the Janadiriya cultural festival or attack oil installations in the Eastern Province of the Kingdom or the interior ministry.

Apart from this quite obvious effort to address bin Laden's preoccupation with his home country Saudi Arabia, seaborne attacks seem to have been Muritani's passion. He especially aimed at damaging the international energy infrastructure by suggesting attacks against pipelines on land and under water or Qatari and Iranian gas tankers, but he also mentioned airports, tunnels, dams, bridges and internet cables. Besides the Persian Gulf he named Russia as a possible target region. He furthermore mentioned political, economic and intellectual leaders of the Western world as possible targets, thereby shedding some light as to who could have been among the victims of at least some of the returning Western recruits. Furthermore, Muritani stated that al-Qaeda Central would not be able to finance all these plans. Rather, it should bear the initial expenses and then ask al-Qaeda's regional organizations to cover the expenses. He explicitly mentioned al-Qaeda in the Islamic Maghreb and its group in the Sahel, allegedly the most wealthy part of the al-Qaeda network.

What does Muritani's letter reveal about al-Qaeda plotting?

Muritani's hitherto unknown letter was a spectacular revelation and remains, until today, the only document found in the Abbottabad raid to have surfaced, apart from the 17 documents officially published. It undoubtedly sheds much light on al-Qaeda's attempts at regaining striking capability at the time of its writing. But does it contain hints as to al-Qaeda's general ways and means of plotting and selecting targets?

The document, for example, oscillates between what can be described as fairly detailed sketches for plots and what could be called a vision. On the one hand there are passages outlining feasible and original ideas, like letting recruits train for jobs that may later be useful but remain sleepers for a long time, even years or possibly decades, before activating them; al-Muritani, for example, suggests that volunteers should seek training as drivers specialized in transporting dangerous goods, a capacity that could then be used in attacking targets like airports, "from within." In the same document, on the other hand, he speculates about how cadres of today will one day be military generals in the future worldwide Islamic state that he seeks to help establishing. Some passages seem to be a mixture of both: the detail and the vision. After describing some ideas for hitting Europe, al-Muritani for example suggests to Osama bin Laden that the al-Qaeda chief publish a speech after the first wave of attacks: "Please inform Europe!" Al-Muritani, in other words, was suggesting lines for a theoretical speech by Osama bin Laden, to follow on a theoretical first wave of attacks for which neither the appropriation of funds nor the recruitment nor the exact targeting had been completed at the time. Similarly, some other ideas he ventilates seem far-fetched and driven by imagination more than by pragmatism. Examples of that are his deliberations about planting submarine bombs in order to attack under water gas routes, using submarines in the process. He even goes as far as letting bin Laden know that these underwater pipelines have security valves built in every ten kilometers, a fact, he says, that needed to be taken into account. In passages like these it appears that the letter should, in parts, be seen as a brainstorming exercise protocol as well.

This in turn is interesting as it reveals the ways in which high ranking al-Qaeda members used to speak to one another. Bearing in mind that correspondence between al-Muritani and Osama bin Laden could not possibly have been uncomplicated but would have required the use of courier networks, it seems all the more telling that both felt this kind of visionary approach would help foster al-Qaeda's future. It appears as though the leadership of the organization was not only concentrating on 100 per cent feasible attack plans. They were also interested in exchanging very broad ideas, bordering on the fantastic.

This is not to diminish the meaning or substance of the letter in question. Al-Muritani uses the term "infiltration" quite a few times, it is his way of describing a very long-term sleeper strategy inside the West which can truly be called novel and original and potentially dangerous. Al-Muritani seemed to be concentrating on the combination of two ideas: taking the concept of sleeper cells to a new level while focusing on economic infrastructure as targets. It is just that for him, this part of his deliberations is part of a much larger vision which to a Western reader seems fantastic and megalomanic but need not be so for readers closer to a jihadist mindset. In any case, it seems as though al-Muritani failed on both accounts.

The Europlot and the turn towards individual jihad

In 2011, the ambitious scheme of the Europlot slowly unraveled. With the arrest of Ocak and Lodin and the members of El-Kebir's Düsseldorf cell, all German returnees seem to have been caught. Other returnees in Scandinavia and Britain were also apprehended, so that the scheme's only "success" were the attacks perpetrated by the Algerian-French Mohammed Merah, who killed three French Muslim soldiers and four Jewish civilians in Toulouse and Montauban in spring 2012. In parallel, most of the al-Qaeda leaders who were part of the planning process have either been killed – like Osama bin Laden in May and Atiyatallah al-Libi in August 2011 – or arrested, like Yunus al-Muritani in August 2011. Thus, al-Qaeda Central's quite possibly last effort to renew its ability to perpetrate large-scale attack worldwide seems to have failed for good.

As an alternative, the organization appears to have chosen a strategy that had originally been mentioned in "Future work" as an additional feature, namely that al-Qaeda should call upon its followers worldwide to execute operations by their own design, which would not even require them to travel to Pakistan for training. This concept became known as "individual jihad."

In June 2011, shortly after bin Laden's death in Abbottabad, al-Qaeda released a video in which its leading members presented this new strategy. Instead of going to Waziristan for training and formally joining the organization, potential recruits were called upon to remain in their home countries instead and perpetrate simple attacks there. They were told to provide themselves with the necessary information on possible and legitimate targets and possible modes of operations from pertinent publications on the internet.[16] The decision to support this new strategy was a major step for al-Qaeda, showing it was ready to part with its history of bottom-down and bureaucratic procedures and even willing to trade central command and control, which it had historically preferred, for effectiveness. It is of course difficult to say whether the parallel failure of the Europlot and death of bin Laden led to the release of the mentioned video. But it would seem that the organization had planned to call for individual jihad anyway and was caught by the arrest in Europe and the death of the leader during the production process.

Nevertheless, in summer 2011, lone wolf and small group attacks seemed to be al-Qaeda Central's last option to secure its survival. As of late 2013, it seemed as if the call did not go entirely unheeded. Lone wolf attacks in Boston, where two brothers of Chechen origin planted homemade bombs at the Marathon, and in London, where two jihadists of Nigerian origin killed a British soldier with knives, as well as other less spectacular plots, can be seen as evidence that the idea of individual jihad had gained traction. At the same time, however, the spectacular comeback of al-Qaeda affiliates in the Arab world following the Arab revolts of early 2011 provided the organization with a new lifeline, making sure that it will survive for at least another generation, even if it appears that its new focus is for the foreseeable future going to be within the Muslim world.

Notes

1 www.spiegel.de/politik/deutschland/terrorismus-al-qaida-droht-mit-anschlaegen-nach-der-bundestagswahl-a-649965.html.
2 Al-Sahab, *The Dark Tunnel*, Video, September 25, 2011.
3 Combating Terrorism Center at West Point, *Letters from Abbottabad. Bin Ladin sidelined?* May 3 2012, www.ctc.usma.edu/posts/letters-from-abbottabad-bin-ladin-sidelined.
4 www.spiegel.de/international/germany/german-jihadists-al-qaida-fighter-from-bonn-believed-dead-a-740326.html.
5 Not to be confused with German Jihadist Denis Cuspert who also calls himself Abu Talha al-Almani and is currently believed to be in Syria.
6 Yassin Musharbash, Holger Stark and Michael Bartsch, "Vorbild Mumbai," *Der Spiegel*, No. 47/2010, November 22, 2010, pp. 20–23.
7 This information stems from court documents, but has since been reported in German media.
8 Future_work.docx. n.p., n.d. [2009], unpublished document.
9 Notizen.txt, n.p., n.d. [2009], unpublished document.
10 Report_on_operations.docx, n.p., n.d. [2009], unpublished court document. It has long been assumed that the attackers planned to target planes to the US. In the text, however, the author only mentions flights from London to Pakistan.
11 Raffaello Pantucci was the first observer to comment on the documents. See his "Al-Qaeda's British Operative: A Biography of Rashid Rauf," in *CTC Sentinel*, Vol. 5, No. 7 (July 2012), pp. 12–16.
12 Lessons_learnt_from_previous_operations.docx, n.p., n.d. [2009] (errors in the original text).

13 Mohammed Siddique Khan and Shezad Tanweer had been in contact with other Pakistanis who were monitored during "Operation Crevice," which ended with the arrest of the group in March 2004.
14 Letter from Yunis al-Muritani to Zamara'i, March 28, 2010, n.p.
15 Combating Terrorism Center at West Point, *Letters from Abbottabad. Bin Ladin sidelined?* See especially the letters numbered 15 and 19.
16 As-Sahab Media, "You Are Held Responsible Only for Thyself", June 3, 2011.

12 Changing patterns in terror plots directed against Germany emanating from militant Uzbek networks in Waziristan

Yassin Musharbash and Guido Steinberg

Introduction

Of all Jihadist terror plots directed against Germany, the one that was closest to actual execution and would at the same time likely have been the most devastating was arguably the one hatched by the so-called Sauerland Cell, a group of ethnic Germans and Turks: the four young men who had sought and received training by the Uzbek "Islamic Jihad Union" (IJU) in Waziristan and were later dispatched by that group to return to Germany in 2007 in order to prepare terror attacks against German and US targets. In this chapter, we will compare this plot with the latest jihadist plot uncovered and foiled in Germany which was very different in character, although it was likely influenced, if not ordered, by another Uzbek organization in Waziristan, the "Islamic Movement of Uzbekistan" (IMU); in this case, the aim was primarily not an attack with explosive devices, but instead most likely a series of targeted murders of rightwing activists of an islamophobic political party. In contrast to the Sauerland group, the perpetrators were not trained in Pakistan and then sent back, but rather seem to have been "inspired" by a video produced by the IMU's media wing in which the German-Moroccan speaker called upon his followers in Germany to organize such murders. A comparison of the two plots shows very strikingly that German jihadists have changed their patterns of behavior considerably over the past years. Instead of joining terrorist organizations in Pakistan and Afghanistan, they "self-recruit" by following jihadist propaganda on the internet and the call from al-Qaeda and other organizations for individual jihad. Although it is too early to say whether this is a trend in German jihadist terrorism, "leaderless jihad" presents an alternative mode of action for would-be jihadists which influences the threat posed by these networks to the Western world.

The 2007 Sauerland plot

The Sauerland plot has been the most important and best-known jihadist terrorist plot targeting Germany to date.[1] It served as a wake-up call to many German politicians and officials who thought that Germany would be spared becoming a major target of al-Qaeda and its allies because the country had not taken part in the Iraq war in 2003 and had refused to send troops to stabilize that country as other European nations had done. For the jihadist scene in Germany, the Sauerland group paved the way for many more German recruits who now for the first time found in place a well-established route to the training camps of jihadist groups in Pakistan, something that had been missing until 2006 and which was now opened with the help of Turkish, Caucasian and Central Asian jihadist networks. Furthermore, the plot was a powerful hint at the radicalization of a substantial number of German jihadists from Turkish and Kurdish backgrounds as well as a clear indication of the consolidation of a German jihadist scene.[2]

The Sauerland group consisted of four German jihadists who travelled to North Waziristan together and there joined the Uzbek Islamic Jihad Union. The group comprised two ethnic Germans and two ethnic Turks, with the German convert to Islam Fritz Gelowicz (b. 1979) as the ringleader. The nucleus of the group seems to have formed in a Salafist cultural centre called Multicultural House (from the German *Multikulturhaus*) in Neu-Ulm in Bavaria, which had been an early centre of jihadist radicalization in Germany from around 2000. The young men's main point of reference at that time seems to have been the Egyptian preacher Yahia Yusuf, the early spiritual director of German jihadism.

Until now, it has not entirely been established how exactly the group ended up in Waziristan with the IJU. Although its four members confessed during their trial in Düsseldorf in 2009, they refused to talk much about their time in Germany and possible recruiters. Instead, they depicted their travels as a long line of accidents finally leading them to the Pakistani tribal areas. According to the group's number two, Adem Yilmaz (b. 1978), they had originally planned to go to Chechnya to join the fight against the Russians. Only when they failed to reach the North Caucasus via Turkey did they decide to go to Syria and study Arabic in Damascus instead. In the Syrian capital Yilmaz they came into contact with a group of Azerbaijani jihadists who promised to send the Germans to Chechnya, but only after they had gone through military training in Pakistan. The Germans changed their plans accordingly and travelled to Turkey, from where the Azerbaijanis organized their travel to Waziristan, where they would train with the IJU.

Only a few specialists and some Uzbeks had at that time even heard of this organization, which had splintered from the older and far stronger Islamic Movement of Uzbekistan. Under the leadership of the legendary Tahir Yoldashev (d. 2009), the IMU had been based in Afghanistan until the fall of the Taliban in 2001 and subsequently escaped to Pakistan. Shortly afterwards, the IJU leader Najmiddin Jalolov (d. 2009) and about a dozen followers left the mother organization in 2002 because of violent debates over ideology and strategy. While the IMU insisted on preparing for the coming struggle in Central Asia and did not take part in the insurgency in Afghanistan until 2007/2008, the IJU subscribed to jihadist internationalism and supported the Taliban and al-Qaeda in their struggle in Afghanistan. The group had become known to a wider public when it attacked security forces and the Israeli and American embassies in its native Uzbekistan in March and July, 2004, although without causing much damage. When the Sauerland group arrived at the IJU's headquarters in North Waziristan in 2006, the IJU was still small with between 100 and 200 fighters, but had built a network through long-standing contacts with like-minded jihadists in the Caucasus and Turkey. While still members of the IMU, Jalolov and some of his followers had trained in Chechnya in the late 1990s and had reached out to Turkish and Azerbaijani fighters – who later assisted Gelowicz and his friends to find their way to the IJU in Pakistan. The young Germans were thereby integrated into a Turkic network linking the Turkish jihadist scene to the Caucasus, Central Asia and the Afpak border region.

While the members of the Sauerland group primarily wanted to fight US troops in Afghanistan, the IJU leader soon asked them to perpetrate attacks in Germany. The reason might have been the poor military performance of at least three of the four recruits who all suffered from malaria, dysentery and hepatitis in the training camp. On a strategic level, the organization's goal was to force the German troops to withdraw from Afghanistan. Therefore, the attacks were planned to coincide with the debate over the extension of the Afghanistan mandates for the International Security Assistance Force (ISAF) and Operation Enduring Freedom (OEF) in the German Bundestag in October and November 2007. The IJU hoped that a successful attack in Germany would influence the debates and possibly the vote in parliament. Increased publicity that would translate into easier acquisition of funds may also have been an aim.

Although the young men were given a certain leeway in choosing the targets and organizing the attacks, Jalolov asked them to focus on American and Uzbek targets. When they were arrested, the Sauerland four had not ultimately decided where exactly they wanted to launch their attack, but the American airbase in Ramstein and off-base facilities frequented by Americans were certainly the focus of their plans, which were to be carried out by using three large car bombs detonated by remote control. The young men planned to flee to Turkey and on to Pakistan, where they wanted to rejoin the IJU after the attack. This plan was thwarted by the arrest of three members of the cell in September 2007 in a small town in the Sauerland region in North-Rhine Westphalia, giving the plot its name. The group had withdrawn to the countryside to prepare and test the explosives when German police arrested them after a joined police and intelligence operation which had lasted nearly a year.

Innovation and learning in the Sauerland plot

The Sauerland group was hardly a professionally trained terror cell. They had received basic training by the IJU in Waziristan, but it is known from their testimonies that training resources were mostly scarce, that there was a language barrier between them and their trainers that could not always be bridged and that they didn't see much logic or structure in what was offered to them as their training program. They received some explosives training, however, which was apparently solid enough to enable them to actually produce a theoretically sound batch of explosive material after they travelled back to Germany to pursue the plot they had been asked to perpetrate. Whenever questions arose, Fritz Gelowicz would check his notes against information available on the internet. But even so, the training quality was significantly lower than before 2001, when al-Qaeda trainers and their recruits had all the time and often ample resources to teach their students bomb-making skills.

Given this, the behavior of the cell members back in Germany is all the more interesting. It would seem fair to say that the ring leader Fritz Gelowicz especially professed a high level of awareness for operational security, even though he does not seem to have been trained in this field. For example, he did most of his research relating to bomb-making from a number of different internet cafes in different places, the only logical reason for which would have been to leave as few traces as possible should his research be traced and tracked by security institutions.

This was echoed in the ways in which the Sauerland group members acquired the chemical substances they needed for the bomb-making. They used fake identities and disguises when they picked up the chemicals that they could not order over the internet. While travelling, they even used fake and changing licence plates. When they finally decided to start producing what was likely to be a test batch of their explosives, they chose to rent a holiday home in a sparsely populated area far from their home towns, trying to avoid looking suspicious to people in their original environment.

Throughout the operation, Fritz Gelowicz displayed a high level of confidence in his group's operational security. There was one instance, roughly a six months prior to their arrest, when a press report alerted the group that security services were on the track of a jihadist cell with connections to the Islamic Jihad Union. Nevertheless, the group continued plotting confident that surveillance had already stopped and that police and intelligence services would not be able to find out about their plotting.

But even though some of Gelowicz's ideas and provisions would have to be described as sensible and intelligent, they did not work – for one simple reason: German police and domestic intelligence had been on his trail all the way, owing to a notification from US intelligence. An officially unnamed US agency, most likely the National Security Agency (NSA), had picked up parts of messages that amounted to a sketchy piece of information suggesting that two or more Islamists from Germany had been to Pakistan and had been sent back to Germany to perpetrate attacks. While Gelowicz and his friends had sometimes realized that they were followed, surveillance became more professional in the following months until the Sauerland group believed that they were not watched any more.

Besides the plot itself, the Sauerland four served as pioneers of the German jihadist scene. While preparing the attack in Germany from late 2007, the members recruited a small group of friends and acquaintances for the IJU. From 2007 on, dozens of volunteers traveled along the now well-established route via Istanbul, Teheran, Mashhad, Zahedan, Quetta and Bannu to Waziristan, where they joined the IJU and later other organizations. The most prominent of the young men was the German Turk Cüneyt Çiftçi (b. 1980), who became famous as the "first German suicide bomber" after he perpetrated a suicide attack on an American-Afghan military base in Khost province on March 3, 2008, killing two Americans and two Afghans. In the following years, the number of German recruits using the Turkic networks to bring them to Pakistan rose until they formed the largest national group among Westerners in the jihadist training camps there in 2009/2010. One of the groups benefitting from the influx of volunteers was the IMU, which was quite likely involved in a later plot in Germany which mirrored the development of the German jihadist scene in general and its consolidation in Germany itself especially.

The 2013 plot to assassinate pro NRW politicians

In March 2013, German police arrested four men aged between 23 and 43 years in the Western state of North Rhine-Westphalia. The arrests followed weeks of surveillance but they were prompted by a specific act: that particular night, three of the men had driven in a car to a little town on the outskirts of the city of Leverkusen, closing in on the home of Marcus Beisicht, the leader of "pro NRW" (NRW being the acronym for North Rhine-Westphalia, the biggest German state), a rightwing and explicitly anti-Islamic political party. While driving, the men had a conversation that alarmed those who listened in on it. "If we kill the leader, everyting will collapse," they heard one of the men say. That was when the police decided to intercept the car. In the subsequent searches of the homes of Marco G., Taifun S., Koray D. and Enea B. they discovered a pistol, several gas guns, 616 grams of a potentially explosive chemical substance – and a list of pro NRW party members. Nine of the names on it, including Beisicht's, were circled.

In May of the previous year, in the nearby cities of Solingen and Bonn, German Salafists had violently clashed with the police and pro NRW activists who had decided to publicly show cartoons of the prophet Mohammed. Was the trip to Beisicht's home an attempt to assassinate the party leader as an act of revenge or to prepare for such an act? As of the writing of this chapter, the trial of the four men has not yet begun. But it seems rather likely that they intended to commit one or more violent acts aimed at pro NRW.

There is, however, another angle to the alleged murder plot: what if the group had not acted entirely on their own initiative but instead had felt that they were taking on a mission that had been asked for rather then been born out of their own imagination?

There is ample reason to assume this, as in May 2012, in a direct reaction to the aforementioned riots in Bonn and Solingen, a video message from a German cadre of the Waziristan-based Islamic Movement of Uzbekistan had asked for just that sort of action, in a seven-minute German language message distributed over the internet. This message, delivered by Yassin Chouka, also known as Abu Ibraheem, had called for the assassination of pro NRW cadres in no uncertain terms: "We suggest that you … single out persons of pro NRW using the methods of secret services," the call went. "And then hit them!" The message went on to say that killing them would be advisable and that trying to accomplish this would best be served by doing it early in the morning or late at night. Chouka also stressed that an attempt should only be made after "good and sufficient research."[3]

Marco G., a German convert to Islam who is assumed to be the ring leader, is also believed to have been in touch with Yassin Chouka and his brother Mounir, who has also appeared in numerous German-language propaganda videos of the IMU.[4] Both brothers grew up in the city of Bonn, one of the hotspots of Salafist and jihadist activity in Germany, and both are reported to be in regular contact with followers there and in other places in Germany to this day.

It is not yet clear whether Marco G. or any other member of his cell had direct contact with the IMU in Waziristan. But be that as it may – it seems quite likely that the propaganda video calling for the assassination of pro NRW cadres did have some sort of effect on the cell.

This video was one of very few instances in which the German IMU cadres had actually called for violence in Germany. In most of their previous messages they had tried to explain why they had chosen the path of the warrior and had called on other German jihadists to also come to Waziristan. They had tried to paint an image of the IMU as an organization that offers a family friendly environment for Jihadists, including such institutions as clinics and kindergartens. They had glorified the fighters of the IMU and portrayed themselves as active fighters against the Pakistani army. They even had published traditional fighting songs in German ("Your Son is Fighting Jihad") and what amounts to an ode to the mothers of those fighting in jihad. Their messages did not go unheeded and it seems that after the German members of the IJU, the Sauerland Cell, had paved the way for German jihadists to get to Waziristan, the IMU profited from that influx and became attractive partly because of their emphasis on propaganda in the German language. German fighters in the vicinity of the IJU tried the same,[5] but IMU propaganda was always glossier, higher in production quality and more sophisticated. We now know from several trials of former German members of the IMU that a good part of what was shown in these videos was merely show and that actual living conditions were harsh and resources scarce. But, especially in the years 2009 and 2010, this was unknown to many members of the jihadist scene in Germany who willingly heeded the call to join the IMU.

But why did Yassin Chouka call for the murder of pro NRW politicians? It would seem that for him the display of the cartoons of the prophet Mohammed amounted to a "Danish situation" in Germany – an instance that needed to be reacted to. But rather than try to get the IMU leadership to dispatch fighters from Waziristan to Germany, he chose to try to instigate an attack in Germany "by remote control" – probably assuming that this would be the more effective method, given that travel to Germany from Waziristan had ended in Pakistani, Turkish or German jails in many other cases and also given that it is rather likely he was very aware of the militant potential of the jihadist scene in Germany.

This last point needs to be raised especially because Marco G., the alleged ring leader in the alleged assassination attempt, has in the process of the investigation also been named as the prime suspect in another plot. In December 2012, an explosive device in a blue sports bag was placed on the tracks at the main Bonn railway station. It failed to explode, and for months investigators were unsure whether this plot had been hatched by right-wing extremists, jihadists or other militants. But after Marco G. was arrested and his belongings were checked, police and prosecutors announced that they were now confident in naming him as a suspect in this case based on DNA traces. If this holds true and if Yassin Chouka was perhaps aware that Marco G. was willing to engage in acts of violence, he would probably have had all the more reason to hope that his call for murder would not go unheard.

From traditional schemes to individual jihad – echoes of the Al-Qaeda Euro plot?

To leave operational planning to amateurs who have not been trained properly is a risk for any traditional terror organization. It only makes sense if the obvious first option – dispatching cadres who are properly trained – is even riskier. This is certainly the case for al-Qaeda Central and explains that organization's Europlot campaign of 2009 and 2010. As we have seen, there is indication that this is also the case for groups like the IMU, except that in that group's case any successful attack in the West would be of added value, whereas in the case of al-Qaeda attacks in the West are part of the identity and purpose of the group and therefore may rank even higher on the agenda.

In any case, it seems that the idea of relinquishing traditional forms of terror plotting in exchange for calls for "individual" or "leaderless" jihad can be found outside al-Qaeda as well. Terror groups tend to be pragmatic, given that they as a rule have limited resources. This may explain this trend, especially in the light of a massive US drone campaign that targets other networks apart from al-Qaeda in similar intensity, illustrated by the fact that both the leaders of the IJU and the IMU, Jalolov and Yoldashev, were killed by US drones in 2009 (as have been a number of German recruits).

We know little about the relation between jihadist networks in Waziristan on the ground. Our information is sketchy and sometimes contradictory. The IMU for example used to have less than friendly relation with al-Qaeda, the reasons for which were disputes that go back to the late 1990s when Taliban leader Mullah Omar put the IMU leader Juma Namangani (d. 2001) in charge of all foreign fighters in Afghanistan to the disdain of al-Qaeda. During the Europlot of al-Qaeda, on the other hand, several foreign fighters left the IMU for al-Qaeda – apparently much to the chagrin of the IMU which threatened some of its former recruits with reprisals should they lose the protection of the Arab organization.

But in either case it would appear that at least Yassin Chouka's call for murders in Germany echoes al-Qaeda's new tactic of calling for "individual" jihad in the West. It seems noteworthy that the entire idea of planning attacks at home rather than travel to Waziristan for training was first diseminated by the jihadist online magazine *Inspire* – officially a product of the English-speaking recruits of al-Qaeda's branch in the Arab Peninsula, but from the beginning a publication that reached and in fact "inspired" jihadists around the world. Marco G. and his cell members also had copies of bomb-making instructions from *Inspire* at their homes. It seems only logical, then, to assume that this idea has taken root within the entire spectrum of the global jihadist movement rather than being a strategy exclusive to al-Qaeda.

What is to be expected in the future?

Of the two militant Uzbek terror networks in Waziristan, only the IMU is today important in a German context. But while the IMU had as many as a dozen or so German members in 2010 and 2011, only a handful of them remain alive in Waziristan. It is unlikely that they have any importance as fighters on the ground; but they continue to have an enormous propaganda value.

If we are to expect more threats from this source, therefore, it will likely come in the form of renewed calls for violence by German propagandists of the IMU. And it is a definite possibility that German followers will heed the call and perpetrate attacks in their home country. Germany today has a vibrant jihadist scene and the messages of IMU cadres like Yassin Chouka are being looked at and perhaps listened to by a significant faction of this scene.

In a broader sense, it would appear that the equation by which many terrorist networks who find themselves under pressure from drone attacks, army and police and secret service action, or other challenges, is that their best chances of instigating an attack in the West can best be realized by trying to motivate sympathizers already there to take action.

In this sense, it seems fair to assume that the comparison between the two plots described in this chapter, the Sauerland Plot of 2007 and the (alleged) IMU plot of 2013, reflects a more general learning curve true for the entire Waziristan-based jihadi movement rather than specific differences between two terror organizations. In other words, were the IJU still a real threat today with regard to Germany, with access to German-language propagandists, they would probably also call for attacks "by remote control" rather than try and repeat their own 2007 plot. One could call this trend innovation as a result of adaptation to circumstance.

Notes

1 See Chapter 11 on al-Qaeda's "Europlot" and its German dimension in this volume.
2 For a detailed discussion of the plot, see Guido Steinberg, *German Jihad: On the Internationalization of Islamist Terrorism*, New York: Columbia University Press, 2013, pp. 59–76.
3 "Die Islamische Bewegung Usbekistans: Tod der Pro-NRW," Video, May 2012.
4 "Bombenbau im Familienkreis," *Focus*, May 27, 2013.
5 This is particularly true for the short-lived "German Taliban Mujahidin," a small German contingent of IJU recruits who formed their own group that was effectively ended in April 2010 when three of the roughly one dozen members were killed in a firefight with Pakistani soldiers.

13 Conclusions

Magnus Ranstorp and Magnus Normark

The principal aim of this book was to examine the role of innovation and learning of al-Qaeda-inspired terrorist plots and attacks in Europe since the 9/11 attacks in 2001 in order to contribute with an enhanced insight into indicators of innovative behavior. An underlying basic assumption was that the evolution of al-Qaeda structures and emerging terrorist plots over time within the EU could provide us with enhanced insights into emerging indicators of terrorist plot innovation. On the surface it would appear there is significant learning from the trajectory of al-Qaeda-related terror plots in the EU. The terrorist attack in 2014 against the Jewish Museum in Brussels by Mehdi Nemmouche, a Frenchman who spent a year in Syria training with ISIS, seems to bear close resemblance to the 2012 Toulouse and Motauban shootings by Muhammed Merah, who spent some time in an al-Qaeda training facility in Waziristan, Pakistan. Both terrorists had a GoPro camera attached to their chests so that they could film their respective murders.[1] Whether Nemmouche had learnt this from studying the lessons of Merah or was advised by his ISIS trainers is still unknown.

On another level even terrorist *modus operandi* serve to directly inspire similar attack modes as starkly exemplified by the influence of 2008 Mumbai attacks by Lashkar-e-Taiba on the 2013 Westgate attack in Nairobi by the Somali al-Shabaab organization.[2] Since 2008 the contagion effect of so-called Mumbai-style terrorist attacks has been evident in the analytical thinking of many Western intelligence agencies, as evidenced by the so-called September 2010 plots to launch "commando-style" attacks on Britain, France and Germany. This terrorist alert stemmed from terrorist arrests with seized computers and the confession of a Pakistani Lashkar-e-Taiba member.[3] Mimicking attack modes was also evident in the grisly murder and beheading of British soldier Lee Rigby in Woolwich in May 2013 as the two perpetrators copied attack directives from the AQAP *Inspire* magazine. In it there were detailed instructions on how to run down targets with a vehicle before attacking them.

Terrorist plots have also intersected through key planners and operators. For example, David Headley, who provided advanced reconnaissance in Mumbai, scouting out the targets in advance of the Pakistani attack team, would later turn his focus towards planning a terrorist attack against *Jyllands-Posten* in Denmark in revenge for its publication of the Prophet Mohammed cartoons.[4] The terrorist plan involved storming the *Jyllands-Posten* offices, killing journalists and seizing responsible editors whose heads were to be chopped off in revenge. Headley was arrested by U.S. authorities in October 2009. Nevertheless, over a year later in December 2010, four Swedish terrorists were arrested in Copenhagen on their way to carry out a terrorist attack on *Jyllands-Posten* that involved storming their offices and taking editors and journalists hostage before beheading some of them. One of the Swedish terrorists had spent considerable time in Miranshah in Pakistan at a terrorist training camp. During the trial of the four Swedish terrorists it emerged that they had common connections to David Headley. Often extremist circles are small and overlap across conflict zones and milieus.

The list of terrorist operations that mimic terrorist tactics can be extended. The advance of technology and in particular information technology has revolutionized the speed at which terrorist groups observe and mimic successful attack modes elsewhere. In operational terms these attack modes simply work and terrorize effectively. However, the various contributions of this book illustrate that Brian Jenkins' statement in 1985 still holds true for today's terrorists as their "tactical repertoire has changed little over time. Terrorists appear to be more imitative than innovative."[5] As earlier observed by Bruce Hoffman terrorists are conservative in nature and continue to rely on two basic weapons systems: the gun and the bomb.[6] Terrorist groups learn from strategies and tactics of other organizations but they have to work hard to seek out a new tactic and operational capability that significantly departs from earlier patterns. This is particularly the case with al-Qaeda inspired terrorism where there is a growing trend towards individualized terrorist actions or smaller terrorist cells put together for shorter periods as attack teams. Simpler tactics with shock-and-awe effects can have massive societal impact, as exemplified by Bouyeri's knife attack on Theo van Gogh in 2004 or the grisly murder of Lee Rigby in Woolwich in 2013.

This conclusion should not to be interpreted as a reason for academics and government agencies to exclude the issue of indicators and trends in terrorist efforts to reinvigorate and develop their deadly tradecraft. There are clear and multiple reasons for vigilance within this field in the future.

The chapters in this volume provide some interesting emerging themes for further investigation as the al-Qaeda phenomenon increasingly become fragmented, affected by prolonged societal conflict in the greater Middle East and beyond, the spread of the ideology of al-Qaeda's brand of nihilistic extremism, and the regional wars in Syria and Iraq involving the Islamic State as well as the specter of foreign fighters. A more complex, chaotic and uncertain strategic landscape is unfolding which is greatly affected by the explosion of new technologies and social media. This was starkly illustrated by Lt. Gen. Michael Flynn, the U.S. Defence Intelligence Agency chief, who observed in the Aspen Security Forum 2014 that "in 2004, there were 21 total Islamic terrorist groups spread out in 18 countries. Today, there are 41 Islamic terrorist groups spread out in 24 countries."[7] Similarly, the UN Analytical Support and Sanctions Monitoring Team underscored that "multiple al-Qaeda affiliates are evolving, often autonomously, with generational, geographical, ethnic, structural and operational changes in evidence."[8] At the same time there is growing evidence that these terrorist groups have imitative and greater operational capacity as well as intelligence gathering capability.

Throughout this volume, authors have provided important insights into factors that drive or inhibit terrorist learning and innovation. A number of themes emerge out of these that are useful in guiding our thinking about the future trajectory of terrorist *modus operandi* and the possibility of learning and innovation over learning. This is particularly relevant as the unfolding wars in Syria and Iraq have significantly changed the dynamics of terrorist behavior. The contours of al-Qaeda's brand of extremism are intrinsically becoming more complex with major conflict and rivalry with the Islamic State and the establishment of a Caliphate. This rivalry matters, as conflict as well as shifting allegiances among regional al-Qaeda affiliates, who have given *bayat* (oath of loyalty), towards IS will invariably impact dynamics of future terrorist behavior. Unlike other conflict zones, the foreign fighter issue differs in the scale and scope of how many Westerners joined ISIS and Jabhat al-Nusra and other jihadist groups in Syria and Iraq from 2012 onwards. The scale and scope of over 3,500 Westerners joining ISIS and Jabhat al-Nusra significantly impacts the prospect and likelihood of foreign fighters returning to Europe to conduct either individualized acts of terrorism or more complex attack scenarios over time. The EU Counterterrorism Coordinator Gilles de Kerchove has repeatedly warned about the escalating threat of foreign fighters in Syria/Iraq and its impact on security within Europe. Similarly Europol's annual TE-SAT reports provide a useful barometer of the latest emerging trends relating to the foreign fighter issue.

Emerging themes

"From top-down to bottom-up," al-Qaeda propaganda and ideology seems to have become increasingly shaped by practices in the field, as elaborated by Sören Hove. Has AQ ideology become a channel for learning and emulation rather than strategic guidance towards new means and ends? Rather Hove points to the increasing absence of a single coherent AQ propaganda for guiding affiliates and followers in what to target and by what means. There seems to have been increasing ideological fragmentation in the doctrinal or operational frame of reference as inspiration for attack modes.[9] Al-Qaeda's strategic literature does offer important guidance to followers in determining courses of violent action. Al-Qaeda strategists with Abu Ubayd al-Qurashi, Abd al-Aziz al-Muqrin (*A Practical Course for Guerilla War*), Abu Bakr Naji (*The Management of Savagery*) and Abu Musab al-Suri (*The Call to Global Islamic Resistance*) are important contributors that have shaped al-Qaeda's operational strategy and tactical behavior. Studying both Naji and al-Suri offers important insight into efforts to conduct regional insurgency and how to conduct self-organizing strikes against the West.[10] At the same time, more easily digestible jihadi magazines proliferate on social media, such as *Inspire* and *Resurgence*, offering do-it-yourself advice on how to conduct terrorist attacks. Bridging the al-Qaeda strategic literature – a helicopter perspective – with "advice from below" will be increasingly important for predicting terrorist behavior. An illustration of the value of a strategic helicopter perspective can be found in the journalist Fouad Hussein's book published in 2005 about al-Qaeda's 2020 strategy, which provided a synthesis of the then current strategic and tactical thinking within al-Qaeda ranks. In a seven phase strategic plan, al-Qaeda sought to establish an Islamic caliphate and pinpoints that Arab regimes would collapse, and that this Islamic caliphate would be declared between 2013 and 2016.[11]

Another important trend, as illustrated by Nico Prucha's contribution, is the accelerated development of the militant jihadist use of internet and social media for a wide variety of purposes. A virtual explosion of militant jihadi websites and social media platforms exists that provide infinite constellations and opportunities to mobilize support, spread their ideological message and narrative, recruit new members, use as psychological weapon and as an intelligence-gathering tool for terrorist operations.[12] For militants and prospective recruits and sympathizers it provides an echo-chamber where like-minded extremists feed off each other while providing a sense of "imaginary" belonging to a greater cause. It can also provide a platform for collective brainstorming about targeting ideas and opportunities. The acceleration of foreign fighters going to Syria/Iraq has been coupled with a great and unprecedented openness on social media sites by jihadists where it is possible to gauge their direct experience from the battlefield, their thoughts and reactions to local and foreign developments and their intentions for the future. Monitoring this social media and internet presence provides important new insights into the process of radicalization into extremism and the worldview of extremists. The symbols and representations of online jihad can provide new ways to gauge depth of commitment and involvement with extremists groups if they are analyzed and understood properly.[13] The extremist online presence is probably the largest game changer for the future development of al-Qaeda inspired extremism. Perhaps this phenomenon is the current ultimate development of modern jihadi tradecraft that will cultivate a trend shift in the future frequency and shape of terrorist plots and activities in Europe and elsewhere. The skillful way in which ISIS have packaged and projected its online psychological warfare is impressive and a harbinger of what is to come. Its massive twitter storm #weareallisis showed that they had global presence and potential capability while its choreographed videos of beheadings of Westerners clad in orange "Guantanamo-style" jumpsuits indicate the operational value of shock-tactics combined with low-tech tactics. This escalating ladder of violence and low-tech tactic of dismemberment of targets is likely to continue for the future.

The terrorist environment is not always characterized by opportunities but also by constraints. As shown by Paul Cruickshank's contribution, in an era of brutal and targeted counter-terrorism measures against terrorist safe havens with training infrastructure and competences, such as the tribal areas in Pakistan and Yemen, the current opportunities for terrorist training designed for terror plots against Europe is at its lowest. Raffaello Pantucci provided an insightful testament to the effect of these degraded opportunities by highlighting the increased trend of terrorist plots in the UK by lone actors with limited capabilities implementing or emulating al-Qaeda propaganda and instructions. A possible shift from the current weakened terrorist training capabilities might be manifest through the increasingly established safe havens for Islamist rebel groups in northern and eastern Syria as well as within pockets in Iraq.

The occurrence of actors and cells without clear links to other terrorist entities constitutes a significant potential to increase. Here perpetrators with individually unique skill sets and ideas for terrorism plots could be inspired to action through ideological propaganda such as *Inspire* magazine, jihadi YouTube videos and the forthcoming al-Qaeda magazine *Resurgence* and other similar propaganda material. Understanding the underlying behavior and triggers for lone actors will become an important focus, together with the dynamics of online and offline behavior.

In the wider context of terrorism tactics and techniques related to production and use of explosives for terrorist plotting there exists a clear pattern of innovative elements generally among terrorist groups, as illustrated by the contribution by John Allison. However, any analogous development amongst radical al-Qaeda-related terrorist cells is difficult to discern since 2001. This is particularly true with regard to the complete lack of creative innovative means in order to increase the effect in terms of casualties of each attack by using alternative or new combinations of weapons or hazardous materials. It seems emulation rather than innovation is influencing the choice of means and tactics used by contemporary radical and violent Islamists in Europe and there are very few indications of dynamics in motion that may change this trend in the near future.

The *modus operandi* of so-called Mumbai-style attacks have been prevalent in the multiple attempts at *Jyllands-Posten* in Denmark but also witnessed in the Kenyan Westgate attack in 2013 and in the provision of operational directives to self-organizing militants such as the increased role of *Inspire* magazine.

The documents found in Usama bin Laden's hideout in Abbottabad in Pakistan underscores the difficulties in planning and setting in motion a strategic, large-scale terrorist attack in Europe, such as the so-called Europlot, while simultaneously being under pressure and with limited resources. The Abbottabad documents did provide a clear picture of the role played by bin Laden personally as a micro-manager where "he was down in the weeds as far as best operatives, best targets, best timing"[14] involving himself in almost every directed al-Qaeda attack, often selecting symbolic dates. The intelligence haul and others like it provide unique insights into the innovative nature of al-Qaeda. This should by no means lead law enforcement agencies to exclude any potential future terrorist plot with innovative weapons and practices, as the tendency towards more isolated plots amongst unknown cells and individuals is heavily dependent on individual skillsets and access to useful resources and targets which will diverge between each plot.

The crucial role of bomb-designers will continue to be a critical factor for innovation in targeting whether on the battlefield or in overcoming aviation security measures. The role of these have been important in the most serious plots against the West, such as Rashid Rauf and his role in the 7/7 and 21/7 London attacks in 2005, as well as the 2006 Liquid Bomb plot, or the continued impact of AQAP's Ibrahim al-Asiri, allegedly responsible for the Umar Farouk Abdulmutallab's underwear plot in December 2009, the printer cartridge plot and other technically more sophisticated concealed devices targeted at aviation. Here the role of cyberspace as a weapon and a target is a potential wildcard as illustrated by the Rajib Karim case where "he may have been planning to crash BA's sophisticated computer systems causing chaos and huge financial losses."[15]

Similarly the ability to stay concealed and underneath intelligence community and law enforcement radars to avoid detection is advancing with shared operational security protocols and use of new technologies in their tradecraft. Efforts to mask terrorist activities and identities have long been quite advanced. More work needs to be done on the operational tradecraft and counter-intelligence capabilities of these extremist networks.[16] Similarly, understanding the multiple dimensions of jihadi social media forums provides crucial insight into targeting preferences.[17]

Safe havens for terrorist cells and organizations seems to have been of the utmost importance in order to develop capabilities designed to overcome security arrangements and countermeasures with capabilities to cause mass casualty attacks on high priority targets in the West. It is unlikely that cells in Europe will enjoy such advantages, despite the unique example of Anders Behring Breivik's attacks in Norway. This underscores the increasing risk of Western individuals returning from conflict zones and areas under control of jihadist organizations and groups with an international agenda. The mode of jihadi warfare in conflict zones therefore increasingly constitutes an important indicator of potential terrorist practices in Europe in the future.

The effect of foreign fighter returnees

The outbreak of civil war in Syria combined with the ferocity of its escalation has had a significant impact on the evolving foreign fighter phenomenon, drawing in fighters from over 80 countries across the globe. While a majority of foreign fighters come from neighboring states to Syria there is a significant contingent of Western nationalities among the recruits to ISIS and Jabhat al-Nusra.[18] While the humanitarian dimensions are a strong motivation and even an obligation for getting involved there is also a major eschatological and apocalyptic dimension as to why the conflict is attracting so many jihadist foreign fighters from around the world. This territory is Bilad al-Sham, which is part of a doomsday prophecy that the Judgement Day is near with the coming final battles between Isa and Djallal (false Messiah) and the conquering of al-Qods, Jerusalem. In this way becoming involved in Syria with the Islamic State enables people to become role models in a historic area for Muslims and it allows them to pursue offensive jihad against a range of enemies while simultaneously establish an Islamic caliphate built around strict Sharia law under the authority of the self-declared caliph Abu Bakr al-Baghdadi.

Much debate exists trying to quantify the risks of returning foreign fighters. The most serious scholarly effort by Thomas Hegghammer, based on foreign fighters in previous al-Qaeda related conflict zones, estimate that one in nine returning foreign fighters pose a serious risk for conducting terrorist violence.[19] The risks of returning foreign fighters from conflict zones are considerable as highlighted by Paul Cruikshank in this volume. The combined effect of increased capability, changed intentions, lower violence threshold, PTSD traumatization from experiencing conflict and violence and their international jihadist contacts and their radicalization influence on sympathizers when they return make this a very dangerous and potentially explosive cocktail. The escalating trajectory of violence and brutality by the Islamic State against its enemies with beheadings, crucifixion and mass killings is a major cause for concern in this respect.[20]

There is a high degree of probability that the trend towards the occurrence of lone actors, planning attacks in European countries without links to terrorist entities, will change in the coming years due to the sharp rise of European foreign fighters joining radical jihadi entities in conflict zones as they return to Europe. How will the growing presence of returning European fighters affect potential innovation in terrorism planning and tactics against targets in Europe within the coming years? To what extent can this phenomenon affect and facilitate the capacity to learn from or synchronize with other jihadists in designing potential future terrorist plots?

The basic assumption is that returnees have a lowered threshold for committing violent acts and are most likely more skilled operatives (small arms) with a higher ability to plan and execute an armed attack, with an increased success rate in comparison to jihadists without this experience.

However, these returnees will probably also be indoctrinated by traditional jihadist warfare technology, tactics and techniques – potentially impeding the incentives for and ability to innovative plots and approaches. One inhibiting factor is that war fighting conditions are radically different for returning foreign fighters – how will the experience of targeting armed combatants in the war torn Aleppo, Idlib, Raqqa or Damascus districts be converted to identify and pursue targets in a peaceful, orderly and civilian European context? The exceptionally and radically different settings between these contexts will furthermore demand a drastic change of *modus operandi*. What type of operational security skills and experiences will the returnees from these conflict zones be equipped with? They are likely to be extremely limited, as the challenges to terrorism tradecraft in the Syrian arena are more or less absent in the environment in which they will be operating in Western countries.

There will likely be increased risk of interconnectivity between returning jihadists as well as between returnees and jihadist organizations in conflict zones. Interconnectivity provides the jihadists increased capacity in terms of advice and knowledge as well as synchronization for increased effect against targets, but demands some form of communication providing opportunities for detection and intervention from a counter-terrorism perspective. The impact of increased individuals with experiences from the jihadi environment will raise the probability for future terrorist plots in Europe but will not necessarily constitute a foundation for innovative terror plots.

A fundamental and critical indicator of innovative and committed jihadist returnees with high degree of potential for terrorist acts will most likely be individuals with experience from leading positions or with specific training activities administrated by groups based on al-Qaeda-inspired ideologies such as ISIS and Jabat al-Nusrah. Jihadist fascination with iconic targets and the effects of instant human casualties has led to successive target hardening measures in Western societies during the last decade. The most obvious example of this is the security arrangements surrounding civil aviation today. These target hardening measures provide a driving force for terrorist cells to innovate (box cutters, liquid explosives concealed through innovative measures such as in underpants, printer cartridges etc). However, the limited success of plots with innovative components within the timeframe of this book's focus is indicative of the seeming lack of ability to invigorate terrorist tradecraft in pace with the Western states' dynamic levels of vulnerabilities and critical dependencies. That is good news for now. However, the unfolding societal conflicts in the Middle East and Africa are likely to generate an increasingly complex and turbulent security landscape on the doorstep to Europe. Prediction of terrorist behavior is likely to become more complicated as solo-terrorists and returning foreign fighters are likely to elevate the terrorist threat in the EU and the West for the foreseeable future. As such more interdisciplinary research into terrorist innovation, which combines behavioral and hard sciences, ought to be prioritized for academics and by the policymaking community, especially in an age of an ever increasingly interlocking, chaotic and complex world with rapidly changing technologies. The signs already exist that the "blowback" effect of the terrorist threat from foreign fighters in Syria/Iraq will continue to envelop us for the foreseeable future.

Notes

1 Marc Weitzmann, "Who is Mehdi Nemmouche, and Why did He Want to Kill Jews?" *Tablet*, July 15, 2014.
2 David Kilcullen, "Westgate Mall Attacks: Urban Areas are the Battleground of the 21st century," *Guardian*, September 27, 2013.
3 Richard Norton-Taylor and Owen Bowcott, "Mumbai-Style Terror Attack on UK, France and Germany Foiled," *Guardian*, September 29, 2010.
4 Sebastian Rotella, "The American behind India's 9/11 – And How U.S. Botched Chances to Stop Him," *Propublica*, January 24, 2013.
5 Brian Jenkins, *International Terrorism: The Other World War* (Santa Monica: RAND, November 1985).
6 Bruce Hoffman, *Inside Terrorism* (New York: Columbia University Press, 2006).
7 James Kitfield, "Flynn's Last Interview: Iconoclast Departs DIA with a Warning," http://breakingdefense.com/2014/08/flynns-last-interview-intel-iconoclast-departs-dia-with-a-warning.
8 Letter dated January 22, 2014 from the Chair of the Security Council Committee pursuant to resolutions 1267 (1999) and 1989 (2011) concerning Al-Qaida and associated individuals and entities addressed to the President of the Security Council, United Nations Security Council.
9 See Yoram Schweitzer and Aviv Oreg, "Al-Qaeda's Odessey to the Global Jihad," Institute for National Security Studies, Tel Aviv University, Memorandum No. 134 (March 2014).
10 For an excellent overview, see Michael W.S. Ryan, *Decoding Al-Qaeda's Strategy* (New York: Columbia University Press, 2013).
11 Yassin Musharbash, "What al-Qaida Really Wants," *Der Spiegel*, December 8, 2005.
12 See Magnus Ranstorp, "The Virtual Sanctuary of al-Qaeda and Terrorism in an Age of Globalization," in Johan Eriksson and Giampiero Giacomello (eds), *International Relations and Security in the Digital Age* (Routledge, 2007).
13 Rudiger Lohlker (ed.), *Jihadism: Online Discourses and Representations* (Vienna: Vienna University Press, 2013).
14 *The Future of al-Qaeda: Results of a Foresight Project* (Ottawa: Canadian Security Intelligence Service, 2013).
15 Vikram Dodd, "British Airways Worker Rajib Karim Convicted of Terrorist Plot", *Guardian*, February 28, 2011.
16 Blake M. Mobley, *Terrorism and Counterintelligence* (New York: Columbia University Press, 2012).
17 Ezzeldeen Khalil, "Jihad 2.0 – Internet Forums Remain Key to Militant Islamists," *Jane's Intelligence Review* (September 2013), pp. 14–19.
18 Raffaello Pantucci, "Foreign Fighters: Battle-Hardened Europeans Return from Syria," *Jane's Intelligence Review* (February 2014), pp. 30–34.
19 Thomas Hegghammer, "Should I Stay or Should I Go? Explaining Variation in Western Jihadists' Choice between Domestic and Foreign Fighting," *American Political Science Review* (February 2013).
20 Richard Barrett, *Foreign Fighters in Syria* (The Soufan Group, June 2014).

Index

Page numbers in *italics* denote tables, those in **bold** denote figures.

Numbers including "n", for example 169n72 refer to notes. Names with the prefix al- or el- are alphabetically filed under the first letter of the second element, for example al-Qaeda is filed under Q.

For Product Safety Concerns and Information please contact our EU
representative GPSR@taylorandfrancis.com
Taylor & Francis Verlag GmbH, Kaufingerstraße 24, 80331 München, Germany

www.ingramcontent.com/pod-product-compliance
Ingram Content Group UK Ltd.
Pitfield, Milton Keynes, MK11 3LW, UK
UKHW020016200926
475702UK00014B/1915
* 9 7 8 0 4 1 5 7 8 7 1 8 5 *